Sulfur Metabolism in Plants

Proceedings of the International Plant Sulfur Workshop

Series Editors

Luit J. De Kok, Heinz Rennenberg, and Malcolm J. Hawkesford

For further volumes:
http://www.springer.com/series/8885

Luit J. De Kok • Michael Tausz
Malcolm J. Hawkesford • Rainer Hoefgen
Michael T. McManus • Robert M. Norton
Heinz Rennenberg • Kazuki Saito
Ewald Schnug • Linda Tabe
Editors

Sulfur Metabolism in Plants

Mechanisms and Applications to Food Security and Responses to Climate Change

Springer

Editors
Luit J. De Kok
Laboratory of Plant Physiology
University of Groningen
Groningen, Netherlands

Malcolm J. Hawkesford
Department of Plant Science
Rothamsted Research
Harpenden, Hertfordshire, UK

Michael T. McManus
Institute of Molecular BioSciences
Massey University
Palmerston North, New Zealand

Heinz Rennenberg
Institut für Forstbotanik, Baumphysiologie
Universitaet Freiburg
Freiburg, Germany

Ewald Schnug
Institute for Crop and Soil Science
Julius Kühn-Institute (JKI)
Braunschweig, Germany

Michael Tausz
Department of Forest
and Ecosystem Science
University of Melbourne
Creswick, VIC, Australia

Rainer Hoefgen
Department of Molecular Physiology
Max Planck Institute of Molecular Plant
Physiology
Potsdam - Golm, Germany

Robert M. Norton
International Plant Nutrition Institute
Horsham, VIC, Australia

Kazuki Saito
Yokohama Research Promotion Division
RIKEN Plant Science Center
Yokohama, Kanagawa, Japan

Linda Tabe
CSIRO Plant Industry
Canberra, ACT, Australia

ISBN 978-94-007-4449-3 ISBN 978-94-007-4450-9 (eBook)
DOI 10.1007/978-94-007-4450-9
Springer Dordrecht Heidelberg New York London

Library of Congress Control Number: 2012941158

Printed on acid-free paper

Springer is part of Springer Science+Business Media (www.springer.com)

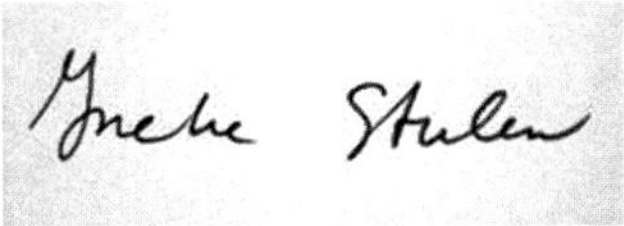

Preface

This proceedings volume contains the invited and a selection of the contributed papers of the 8th International Workshop on Sulfur Metabolism in Higher Plants, which was held at the Department of Forest and Ecosystem Science, University of Melbourne, Water Street, Creswick, Victoria 3363, Australia, from November 22 to 27, 2010. The meeting was co-organized by the University of Melbourne (Australia), the University of Groningen (The Netherlands), Massey University, Palmerston North (New Zealand), the International Plant Nutrition Institute, Horsham (Australia), and CSIRO Plant Industry, Canberra (Australia). The content of the volume shows that the understanding of sulfur metabolism in plants and the interaction of the environment are rapidly progressing. This volume covers various aspects of the regulation of sulfate uptake and assimilation in plants, from a cellular to a whole plant level, and additionally emphasizes interactions with other minerals. Moreover, the significance of sulfur metabolism in biotic and abiotic stress responses, in food security and quality, and in relation to interactions with global change factors is discussed in detail.

We are pleased to dedicate this book to A/Professor Ineke Stulen, Laboratory of Plant Physiology, University of Groningen, The Netherlands. She was one of the initiators of the Sulfur Workshop series in 1989 and was a member of the Organizing Committee of all succeeding workshops.

Luit J. De Kok
Michael Tausz
Malcolm J. Hawkesford
Rainer Hoefgen
Michael T. McManus
Robert M. Norton
Heinz Rennenberg
Kasuki Saito
Ewald Schnug
Linda Tabe

Contents

Sulfur Metabolism – Mineral Interactions

Sulfur Metabolism – Food Security and Environmental Interactions

Foreword: Exploring Interactions Between Sulfate and Nitrate Uptake at a Whole Plant Level

Ineke Stulen and Luit J. De Kok

Abstract Nitrogen and sulfur are essential for crop growth and quality, because both are needed for amino acid and protein synthesis. The organic N/S ratio on a molar basis is usually about 20. Plants, therefore, must have mechanisms to coordinate sulfur and nitrogen uptake and assimilation so that appropriate proportions of sulfur containing and other amino acids are available for protein synthesis. Experiments with vegetable crop plants grown at non-limiting nutrient supply showed that the uptake rates of nitrate and sulfate by the root are related to the growth rate of the plant. Reduced nitrogen and sulfur compounds as glutamine, glutathione and *O*-acetyl-L-serine, and/or nitrate and sulfate, might act signal molecules in regulation of the uptake of nitrate and sulfate. However, there is no evidence for a direct linkage between the uptake of nitrate and sulfate.

Introduction

In general most soils contain sufficient sulfur to cover the requirements of plants, whereas nitrogen is often limiting for plant growth. Nitrogen in soil is available in various forms, but in agriculture mostly nitrate and some ammonium are the main forms taken up by the root and used as source for growth (Miller and Chapman 2011). Sulfate taken up by the root appears to be the major sulfur source for growth (Hawkesford and De Kok 2006; Zhao et al. 2008; Haneklaus et al. 2007; De Kok et al. 2011). However, in industrialized areas atmospheric sulfur deposition may

I. Stulen • L.J. De Kok (✉)
Laboratory of Plant Physiology, University of Groningen,
P.O. Box 11103, 9700 CC Groningen, The Netherlands
e-mail: l.j.de.kok@rug.nl

L.J. De Kok et al. (eds.), *Sulfur Metabolism in Plants*, Proceedings of the International Plant Sulfur Workshop 1, DOI 10.1007/978-94-007-4450-9_1,

contribute to a significant extent to the sulfur fertilization of plants (De Kok et al. 2007, 2009, 2011), whereas nitrogen fertilization by atmospheric nitrogen deposition is limited (Wellburn 1990).

Nitrate and sulfate need to be reduced prior to their incorporation into various essential organic nitrogen and sulfur compounds. The uptake and assimilation of sulfur and nitrogen are strongly interrelated, since the major proportion of the reduced nitrogen and sulfur in plants is incorporated into amino acids and subsequently into proteins (Stulen and De Kok 1993; De Kok et al. 2011). The synthesis of cysteine from *O*-acetylserine and sulfide is a major reaction in the direct coupling between nitrogen and sulfur metabolism in the plant (Brunold 1993). Cysteine plays a key role in the synthesis of organic sulfur compounds; it is incorporated into proteins and the tripeptide glutathione and it is used as the sulfur compound for the synthesis of the essential amino acid methionine (Giovanelli 1990). Proteins contain both sulfur and non-sulfur amino acids and for this reason the availability of nitrogen and sulfur interacts with the utilization of nitrogen and sulfur for proteins and plant growth. Plants maintain their nitrogen and sulfur content in proteins within a certain range (Stulen and De Kok 1993) and the organic N/S ratio is generally around 20 on a molar basis (Stulen and De Kok 1993; Haneklaus et al. 2007). Plants, therefore, must have mechanisms to coordinate the uptake and reduction of sulfate and nitrate so that appropriate proportions of both sulfur containing and other amino acids are available for protein synthesis.

Nitrogen and sulfur research has mainly been focused on elucidating the pathways and characterizing the transporters and enzymes involved in their uptake and assimilation and their subsequent incorporation into organic compounds, from the molecular to the crop yield level. Although both laboratory and agronomic data indicate N/S interactions in metabolism, growth and plant composition, it is not known whether a direct mutual regulation as for instance proposed by Reuveny et al. (1980) based on experiments with isolated plant cell model systems growing under extreme nutrition conditions (nitrogen and sulfate deprivation), really occurs in whole plants in steady state grown under well-controlled nutrient conditions. Experiments on the effect of changes in nutrient supply and various environmental conditions have been performed from the molecular, biochemical and physiological to crop yield level. However, the question to what extent the measured changes in parameters as expression of the transporters and enzymes are of physiological significance in a whole plant context is not often addressed.

This foreword briefly evaluates the physiological mechanisms involved in the regulation of nitrogen and sulfur uptake, and the regulatory control of the coordination, on a whole plant level, based on results of experiments performed in our research group during our long cooperation.

Nitrate and Sulfate Uptake in Relation to Plant Growth Rate

The uptake of nutrients by the roots is generally adapted/in tune with the plant's need for growth (Hawkesford and De Kok 2006; Zhao et al. 2008; Haneklaus et al. 2007; De Kok et al. 2011). For plants in the vegetative phase, grown at non-limiting nutrients, the nutrient flux (N_{flux}) needed per gram plant biomass produced with time can be calculated as follows (Haneklaus et al. 2007; Zhao et al. 2008; De Kok et al. 2011):

$$N_{flux} = N_{content} \times \text{RGR}$$

where the N_{flux} is expressed as μmol g^{-1} plant day^{-1}, $N_{content}$ is the total nutrient content of the plant (μmol g^{-1} plant) and RGR is the relative growth rate of the plant during the growth period under investigation (g g^{-1} day^{-1}). RGR can be calculated by linear regression from the ln transformed weight data (Hunt 1982) or by an exponential fit of the weight data (Poorter 1989).

The uptake of nitrate and sulfate by the root are active processes and driven by a proton gradient maintained by a proton ATPase, mediated by transporter proteins. Distinct nitrate transporter groups have been characterized and plants contain inducible (iHATS) and constitutive (cHATS) high affinity nitrate transporters and constitutive low affinity nitrate transporters (LATS) (Touraine 2004; Miller and Chapman 2011). Likewise, different sulfate transporter proteins are involved in the uptake and distribution of sulfate in the plant, which may contain 12–14 different transporters classified in up to five different groups according to the possible functioning (Hawkesford and De Kok 2006; De Kok et al. 2011).

In experiments with *Spinacia oleracea* L. and *Plantago major* L. net nitrate uptake rate (NNUR) was measured in combination with RGR and plant nitrogen content. NNUR can be expressed on a plant weight basis (μmol g^{-1} plant day^{-1}) or a root weight basis (μmol g^{-1} root day^{-1}). Plants with a relatively low root weight ratio (RWR, root weight/plant weight) have a relatively high uptake rate on a root weight basis. These experiments showed that the measured NNUR was in accordance with the plant nitrogen flux, and therefore of physiological significance (Ter Steege et al. 1998, 1999; Fonseca et al. 1997). There was a linear relationship between RGR and measured NNUR, if plants were grown in non-limiting nutrient solution under the same environmental conditions and with similar plant nitrogen content and root weight ratio (Fig. 1). Part of the *Plantago* plants was grown at elevated CO_2, which resulted in an increase in RGR at the time of the measurement, Fig. 1 shows that NNUR in these species is closely linked to the RGR of the plant. Apparently, the NNUR is a well-regulated process, under the control of an internal regulating mechanism, which adjusts the NNUR to the nitrogen need of the plant, as determined by RGR and total plant nitrogen content (Touraine et al. 1994; Ter Steege et al. 1998, 1999).

There has been a long debate on the role of nitrate influx and efflux in the control of NNUR by roots under steady-state conditions. A double labeling design, with both ^{13}N- and ^{15}N-nitrate, made it possible to study the contribution of both fluxes to the regulation of NNUR rate in spinach (Ter Steege et al. 1998). These experiments showed that nitrate influx and efflux together regulate NNUR by roots, thereby providing a flexible and sensitive nitrate uptake system (Ter Steege et al. 1998, 1999; Miller and Chapman 2011; Fig. 2). It is unclear to what extent efflux of sulfate (*e.g.* via sulfate selective anion channels) has significance in the regulation of the net sulfate uptake by roots is still an open question (De Kok et al. 2011).

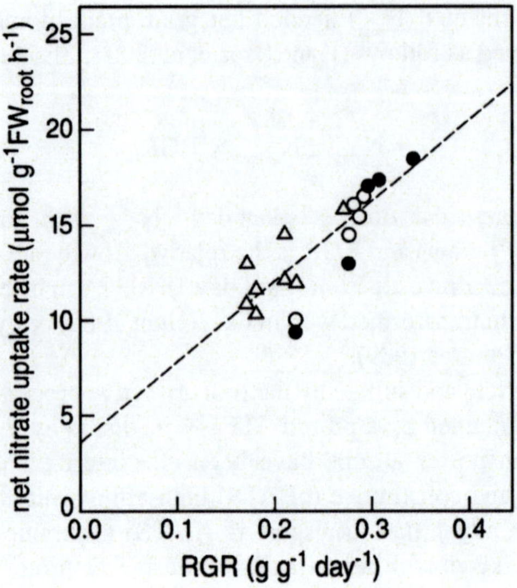

Fig. 1 Relationship between net nitrate uptake rate in *Spinacia oleracea* L. (Data from Ter Steege et al. 1998, 1999) and *Plantago major* L. (Fonseca et al. 1997), grown in nutrient solution with unlimited access to nitrate. *Spinacia* (Δ); *Plantago*, grown at an ambient (○) or elevated (•) atmospheric CO_2 concentration of 350 and 700 µl l^{-1}, respectively

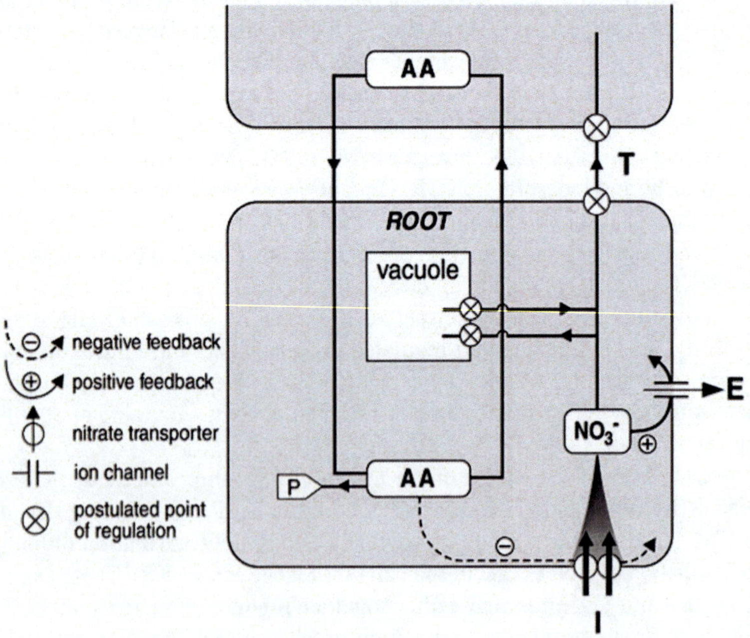

Fig. 2 Localization and regulation of processes of uptake, translocation and storage of nitrate in plant roots. *I* nitrate influx, *E* nitrate efflux, *T* nitrate translocation, *AA* amino acids, *P* protein (Adapted from Ter Steege 1996)

Regulation of Nitrate Uptake – Comparison with Sulfate Uptake

The regulation of activity of the nitrate and sulfate transporters may be controlled at a transcriptional, translational and/or post-translational level (*e.g.* activation/deactivation; Hawkesford and De Kok 2006; De Kok et al. 2011). However, the expression and activity of the nitrate and sulfate transporters are differently regulated. Sulfur-deprived plants are characterized by a high expression and uptake capacity of the sulfate transporters (Westerman et al. 2000; Buchner et al. 2004; Hawkesford and De Kok 2006; Koralewska et al. 2008, 2009), whereas nitrate-deprived plants, show a lag phase in nitrate uptake capacity, related to an induction phase of the nitrate transporter proteins (Clarkson 1986).

It has been postulated that a shoot-derived signals down-regulate nitrate and sulfate uptake by negative feedback control at the level of the nitrate transporter proteins (Touraine et al. 1994; Ter Steege 1996; Ter Steege et al. 1999) and sulfate transporter proteins (Hawkesford and De Kok 2006). At present glutamine seems to be the most likely signal molecule for the regulation of nitrate influx (Touraine et al. 1994). Nitrate efflux might offer a mechanism for rapid reactions to increased cytoplasmic root nitrate concentrations, and nitrate itself might act as signal molecule (Ter Steege 1996; Hawkesford 2011). The signal transduction pathway involved in the regulation of the uptake sulfate uptake it still largely unsolved; it might be signaled or mediated by sulfate itself or products of the assimilatory reduction pathway (*e.g.* H_2S, cysteine or glutathione; Hawkesford and De Kok 2006; De Kok et al. 2011). Moreover, the cysteine precursor *O*-acetyl-L-serine (OAS) is thought to play an important role in the induction of sulfate uptake (Clarkson et al. 1999). However, the majority of plant cells, including root cells, have the capacity to both reduce and assimilate nitrate and sulfate, presumably facilitating local signaling of nitrate and sulfate uptake at a cellular level, which makes it difficult to separate local signaling at a local cellular root level from signaling at an integrated tissue *viz.* shoot to root level (Hawkesford and De Kok 2006; De Kok et al. 2011). Besides, it remains obscure to what extent changes concentrations of potential signal compounds and expression of the nitrate and sulfate transporters, both determined at the whole organ level, provides sufficient insight into the actual regulatory control of the sulfate uptake at the root cellular level (De Kok et al. 2011).

There is apparently no direct linkage between the uptake of nitrate and sulfate in roots. Sulfate deprivation of *Brassica* resulted in a decreased growth and nitrate uptake rate, whereas the expression and activity if the sulfate transporters rapidly increased (Westerman et al. 2000; Buchner et al. 2004; Yang et al. 2006; Koralewska et al. 2008, 2009). When sulfate-sufficient *Brassica* plants were exposed to atmospheric H_2S, both nitrate uptake rate and RGR were unaffected, while the sulfate uptake was decreased (Westerman et al. 2000, 2001). Exposure of *Brassica* to atmospheric NH_3 resulted in a downregulation of the nitrate uptake, whereas the uptake of sulfate remained unaffected (Castro et al. 2006). If *Brassica* plants were exposed to elevated Cu^{2+} concentrations in the root environment, it resulted in both a decrease in plant growth and nitrate uptake, however, the sulfate uptake was increased (Shahbaz et al. 2010).

The latter was probably due to a direct interference of the Cu with the signal transduction pathway involved in the regulation of the expression and activity of the sulfate transporters (Shahbaz et al. 2010).

Conclusions

Plants maintain their nitrogen and sulfur content within a certain range, since both are essential for synthesis of proteins. Protein synthesis requires inorganic carbon, and reduced nitrogen and sulfur. Co-ordination of the assimilatory reduction pathways of nitrate and sulfate is therefore necessary, so that appropriate proportions of both sulfur containing and other amino acids are available for protein synthesis. This might implicate a mutual regulation of the nitrate and sulfate uptake by the root. However, from our studies it is evident that changes in nitrate uptake rate are related to changes in growth, and that there is no direct linkage between the uptake of nitrate and sulfate.

References

Brunold C (1993) Regulatory interactions between sulfate and nitrate assimilation. In: De Kok LJ, Stulen I, Rennenberg H, Brunold C, Rauser WE (eds) Sulfur nutrition and assimilation in higher plants: regulatory, agricultural and environmental aspects. SPB Academic Publishing bv, The Hague, pp 61–75

Buchner P, Stuiver CEE, Westerman S, Wirtz M, Hell R, Hawkesford MJ, De Kok LJ (2004) Regulation of sulfate uptake and expression of sulfate transporter genes in *Brassica oleracea* as affected by atmospheric H_2S and pedospheric sulfate nutrition. Plant Physiol 136:3396–3408

Castro A, Stulen I, Posthumus FS, De Kok LJ (2006) Changes in growth and nutrient uptake in *Brassica oleracea* L. exposed to atmospheric ammonia. Ann Bot 97:121–131

Clarkson DT (1986) Regulation of the absorption and release of nitrate by plant cells: a review of current ideas and methodology. In: Lambers H, Neeteson JJ, Stulen I (eds) Physiological, ecological and applied aspects of nitrogen metabolism in higher plants. Martinus Nijhoff Publishers, Dordrecht, pp 3–28

Clarkson DT, Diogo E, Amâncio S (1999) Uptake and assimilation of sulphate by sulphur deficient *Zea mays* cells: the role of *O*-acetyl-L-serine in the interaction between nitrogen and sulphur assimilatory pathways. Plant Physiol Biochem 37:283–290

De Kok LJ, Durenkamp M, Yang L, Stulen I (2007) Atmospheric sulfur. In: Hawkesford MJ, De Kok LJ (eds) Sulfur in plants – an ecological perspective. Springer, Dordrecht, pp 91–106

De Kok LJ, Yang L, Stuiver CEE, Stulen I (2009) Negative vs. positive functional plant responses to air pollution: a study establishing cause-effect relationships of SO_2 and H_2S. In: Legge AH (ed) Air quality and ecological impacts: relating sources to effects, 9th edn, Developments in environmental sciences. Elsevier B.V, Amsterdam/Boston/London, pp 121–134

De Kok LJ, Stulen I, Hawkesford MJ (2011) Sulfur nutrition in plants. In: Hawkesford MJ, Barraclough P (eds) Molecular and physiological basis of nutrient use efficiency in crops. Wiley-Blackwell, Chichester/Ames, pp 295–309

Fonseca FG, Bowsher CG, Stulen I (1997) Impact of elevated CO_2 on nitrate reductase transcription and activity in leaves and roots of *Plantago major*. Physiol Plant 100:940–948

Giovanelli J (1990) Regulatory aspects of cysteine and methionine synthesis. In: Rennenberg H, Brunold C, De Kok LJ, Stulen I (eds) Sulfur nutrition and sulfur assimilation in higher plants; fundamental, environmental and agricultural aspects. SPB Academic Publishing, The Hague, pp 33–48

Haneklaus S, Bloem E, Schnug E, De Kok LJ, Stulen I (2007) Sulfur. In: Barker AV, Pilbeam DJ (eds) Handbook of plant nutrition. CRC Press, Boca Raton, pp 183–238

Hawkesford MJ (2011) An overview of nutrient use efficiency and strategies for crop improvement. In: Hawkesford MJ, Barraclough P (eds) Molecular and physiological basis of nutrient use efficiency in crops. Wiley-Blackwell, Chichester/Ames, pp 5–19

Hawkesford MJ, De Kok LJ (2006) Managing sulfur metabolism in plants. Plant Cell Environ 29:382–395

Hunt R (1982) Plant growth curves. The functional approach to plant growth analysis. E. Arnold Publ, London

Koralewska A, Stuiver CEE, Posthumus FS, Kopriva S, Hawkesford MJ, De Kok LJ (2008) Regulation of sulfate uptake, expression of the sulfate transporters Sultr1;1 and Sultr1;2, and APS reductase in Chinese cabbage (*Brassica pekinensis*) as affected by atmospheric H_2S nutrition and sulfate deprivation. Funct Plant Biol 35:318–327

Koralewska A, Buchner P, Stuiver CEE, Posthumus FS, Kopriva S, Hawkesford MJ, De Kok LJ (2009) Expression and activity of sulfate transporters and APS reductase in curly kale in response to sulfate deprivation and re-supply. J Plant Physiol 166:168–179

Miller AJ, Chapman N (2011) Transporters involved in nitrogen uptake and movement. In: Hawkesford MJ, Barraclough P (eds) Molecular and physiological basis of nutrient use efficiency in crops. Wiley-Blackwell, Chichester/Ames, pp 193–210

Poorter H (1989) Plant growth analysis: towards a synthesis of the classical and the functional approach. Physiol Plant 75:237–244

Reuveny Z, Dougall DK, Trinity P (1980) Regulatory coupling of nitrate and sulfate assimilation pathways in cultured tobacco cells. Natl Acad Sci USA 77:6670–6672

Shahbaz M, Tseng MH, Stuiver CEE, Koralewska A, Posthumus FS, Venema JH, Parmar S, Schat H, Hawkesford MJ, De Kok LJ (2010) Copper exposure interferes with the regulation of the uptake, distribution and metabolism of sulfate in Chinese cabbage. J Plant Physiol 167:438–446

Stulen I, De Kok LJ (1993) Whole plant regulation of sulfur metabolism: a theoretical approach and comparison with current ideas on regulation of nitrogen metabolism. In: De Kok LJ, Stulen I, Rennenberg H, Brunold C, Rauser WE (eds) Sulfur nutrition and assimilation in higher plants: regulatory, agricultural and environmental aspects. SPB Academic Publishing bv, The Hague, pp 77–91

Ter Steege MW (1996) Regulation of nitrate uptake in a whole plant perspective. Thesis, University of Groningen, The Netherlands

Ter Steege MW, Stulen I, Wiersema PK, Paans AJM, Vaalburg W, Kuiper PJC, Clarkson DT (1998) Growth requirement for N as a criterion to assess the effects of physical manipulation on nitrate uptake fluxes in spinach. Physiol Plant 103:181–192

Ter Steege MW, Stulen I, Wiersema PK, Posthumus F, Vaalburg W (1999) Efficiency of nitrate uptake in spinach: impact of external nitrate concentration and relative growth rate on nitrate influx and efflux. Plant Soil 208:125–134

Touraine B (2004) Nitrate uptake by roots – transporters and root development. In: Amâncio S, Stulen I (eds) Nitrogen acquisition and assimilation in higher plants. Kluwer Academic, Dordrecht, pp 1–34

Touraine B, Clarkson DT, Muller B (1994) Regulation of nitrate uptake at the whole plant level. In: Roy J, Garnier E (eds) A whole plant perspective on carbon-nitrogen interactions. SPB Academic Publishing, The Hague, pp 11–30

Wellburn AR (1990) Why are atmospheric oxides of nitrogen usually phytotoxic and not alternative fertilizers. New Phytol 115:395–429

Westerman S, De Kok LJ, Stulen I (2000) Interaction between metabolism of atmospheric H_2S in the shoot and sulfate uptake by the roots of curly kale (*Brassica oleracea* L.). Physiol Plant 109:443–449

Westerman S, Stulen I, Suter M, Brunold C, De Kok LJ (2001) Atmospheric H_2S as sulfur source for *Brassica oleracea*: consequences for the activity of the enzymes of the assimilatory sulfate reduction pathway. Plant Physiol Biochem 39:425–432

Yang L, Stulen I, De Kok LJ (2006) Impact of sulfate nutrition on the utilization of atmospheric SO_2 as sulfur source for Chinese cabbage. J Plant Nutr Soil Sci 169:529–534

Zhao FJ, Tausz M, De Kok LJ (2008) Role of sulfur for plant production in agricultural and natural ecosystems. In: Hell R, Dahl C, Knaff DB, Leustek T (eds) Sulfur metabolism in phototropic organisms. Springer, New York, pp 425–443

Part I
Sulfur Metabolism – Mechanisms

Sulfate Uptake and Assimilation – Whole Plant Regulation

Malcolm J. Hawkesford

Abstract Sulfur remains an important issue on the agenda for crop plant nutrition. In addition to avoidance of sulfur deficiency, which will impact on yield and quality, there are requirements for adequate fertilization of crops for resistance to biotic and abiotic stresses. Equally importantly, there are clear consequences for efficient nitrogen utilization and there are interactions with micronutrient acquisition (selenium and molybdenum). Substantial advances have been made at the cellular level, dissecting the signal transduction pathways linking cellular nutritional status with expression of sulfur regulated genes and pathways. However cellular processes need to be placed in the context of whole plant regulation of sulfur uptake and assimilation, which encompasses developmental, spatial and environmental factors, and which facilitates optimum growth and fecundity (and yield in the case of crops) with available sulfur supply. During development, adequate sulfur must be acquired for optimum growth, ideally with any excess being sequestered into re-mobilizable temporary stores. As the plant develops, efficient utilization of sulfur will require organ to organ transfer, and additionally degradation pathways, metabolic inter-conversions and multiple trans-membrane and vascular tissue mediated transport steps for both inorganic and organic sulfur compounds. For crops, efficient transfer of sulfur to harvested sink tissues and its incorporation into protein are important agronomic traits. Insufficient sulfur to meet the demand for growth results in a number of plant responses, targeted at optimising uptake and use of available sulfur. Notable early and specific responses are the up-regulation of transporters and key steps of the assimilatory pathways in sulfur-deficient tissues, and the allocation of resources to stimulate growth of root tissues compared to the shoots.

M.J. Hawkesford (✉)
Plant Science Department, Rothamsted Research,
Harpenden, Hertfordshire AL5 2JQ, UK
e-mail: malcolm.hawkesford@rothamsted.ac.uk

L.J. De Kok et al. (eds.), *Sulfur Metabolism in Plants*, Proceedings of the International Plant Sulfur Workshop 1, DOI 10.1007/978-94-007-4450-9_2,

Sulfate Uptake and Assimilation – Whole Plant Regulation

Malcolm J. Hawkesford

Abstract Sulfur remains an important issue on the agenda for crop plant nutrition. In addition to avoidance of sulfur deficiency, which will impact on yield and quality, there are requirements for adequate fertilization of crops for resistance to biotic and abiotic stresses. Equally importantly, there are clear consequences for efficient nitrogen utilization and there are interactions with micronutrient acquisition (selenium and molybdenum). Substantial advances have been made at the cellular level, dissecting the signal transduction pathways linking cellular nutritional status with expression of sulfur regulated genes and pathways. However cellular processes need to be placed in the context of whole plant regulation of sulfur uptake and assimilation, which encompasses developmental, spatial and environmental factors, and which facilitates optimum growth and fecundity (and yield in the case of crops) with available sulfur supply. During development, adequate sulfur must be acquired for optimum growth, ideally with any excess being sequestered into re-mobilizable temporary stores. As the plant develops, efficient utilization of sulfur will require organ to organ transfer, and additionally degradation pathways, metabolic inter-conversions and multiple trans-membrane and vascular tissue mediated transport steps for both inorganic and organic sulfur compounds. For crops, efficient transfer of sulfur to harvested sink tissues and its incorporation into protein are important agronomic traits. Insufficient sulfur to meet the demand for growth results in a number of plant responses, targeted at optimising uptake and use of available sulfur. Notable early and specific responses are the up-regulation of transporters and key steps of the assimilatory pathways in sulfur-deficient tissues, and the allocation of resources to stimulate growth of root tissues compared to the shoots.

M.J. Hawkesford (✉)
Plant Science Department, Rothamsted Research,
Harpenden, Hertfordshire AL5 2JQ, UK
e-mail: malcolm.hawkesford@rothamsted.ac.uk

L.J. De Kok et al. (eds.), *Sulfur Metabolism in Plants*, Proceedings of the International Plant Sulfur Workshop 1, DOI 10.1007/978-94-007-4450-9_2,

These responses involving root proliferation and transporter functionality are adaptations to improve pedospheric sulfur acquisition. A long standing question has concerned the existence and nature of inter-organ signals of nutrient status. It is possible that local sulfur availability, coupled with intrinsic cell specific programmed function, is sufficient to mediate local gene and pathway expression, influence organ responses and effect whole plant sulfur management without inter organ signals. Developmental cues will influence organ specific pathways, most clearly demonstrated in processes of leaf senescence and associated nutrient remobilization. Conversely, the recognition of possibly mobile phloem located miRNAs may be indicative of long distance regulatory mechanisms. Similarly, root proliferation will almost certainly have a hormonal basis.

Introduction

With respect to whole plant regulation of sulfur uptake and assimilation, the key questions to consider include: what regulation occurs, when is this regulation important, what signals are involved to initiate/control the regulation, what transduction pathways are required, how is cellular regulation integrated at the whole plant level and ultimately at the ecosystem level, and finally what are the consequences for agriculture? The primary function of regulatory mechanisms is to manage fluxes of sulfur in response to developmental and environmental cues (Hawkesford and De Kok 2006). The goal for the plant is to optimise the use of available sulfur to match the demands for growth and development, and for resistance to stress. Under limiting conditions this focuses on survival and reproduction. For a crop, management of sulfur is essential for both yield and quality (Zhao et al. 1999).

There has been substantial progress in the elucidation of signal transduction pathways at the cellular and gene level (see the chapter "Molecular and Cellular Regulation of Sulfate Transport and Assimilation", this volume), although the key links between metabolism and the transduction pathways remain elusive. Regulation at the level of the whole plant is at least the sum of cellular regulation integrated throughout the whole plant, and comprises both spatial and temporal components, possible inter organ signaling, and a co-ordination with developmental processes.

Key Processes

Regulated processes extend from flux control in branches of the sulfur assimilatory pathways to whole plant distribution of sulfur pools. The uptake of sulfate into plant roots has long been known to be regulated (induced during sulfur-deficiency, repressed at adequate supply), and this has been shown to be due to the direct regulation (certainly transcriptional (Smith et al. 1995a, 1997) but also possibly post transcriptional control (Yoshimoto et al. 2007) and post translational control as evidenced by the presence of the STAS (sulfate transporter and anti-sigma antagonist) domain (Aravind and Koonin 2000; Shibagaki and Grossman 2004; Rouached et al. 2005) of sulfate transporters (STs) at the plasma membrane of root cells. The transport of sulfate is catalyzed by members of a transporter family (SulP), which for example comprises 14 genes in *Arabidopsis*: these may be subdivided into five groups (Hawkesford 2003), with respectively, high and low affinities, undetermined function, vacuolar efflux transport function and a specific involvement in Mo accumulation. Expression of the genes for many of these transporters (particularly Groups 1, 2 and 4) is influenced by sulfur-nutritional status and development and the differential expression and activity of the whole family will contribute significantly to distribution, storage and remobilization of sulfate and other oxyanion pools (see below).

In wheat or barley root tissues, the major Group 1 ST is up-regulated (transcript abundance and transporter activity as measured by root uptake capacity) under conditions of limiting sulfur and re-repressed upon re-supply (Smith et al. 1997; Buchner et al. 2010). In *Arabidopsis* and Brassica the situation is complicated by the occurrence of two related transporters having differential patterns of expression: both are more highly expressed under sulfur-limiting conditions but one (Sultr1;1) is highly induced from a situation of little or no expression (hence very many-fold), whilst the other (Sultr1;2) shows expression increased around a twofold, with a background level comparable to the induced state of Sultr1;1.

Substantial regulation of the assimilatory pathway is evident (see Chap. 3, Yoshimoto and Saito, this volume, and Fig. 1), facilitating coordination with C and N metabolism (and hence growth), and enabling partitioning between primary assimilation and sulfur-containing amino acid biosynthesis or to secondary metabolites such as glucosinolates. Substantial allosteric or gene expression regulation of ATP sulfurylase, APS reductase (APR) and APS kinase as well as the occurrence of compartment-specific isoforms of these enzymes facilitates biochemical partitioning of sulfur between different branches of the assimilatory pathway (*e.g.* (Vauclare et al. 2002; Mugford et al. 2009)).

An important coordination with C/N metabolism occurs at the level of cysteine biosynthesis, with the cysteine synthase complex (serine acetyltransferase (SAT) and *O*-acetylserine(thiol)lyase (OASTL)) acting as both a sensor and a regulator (Hell et al. 2002), mediated by a reversible association/dissociation of the complex. SAT is active when associated with OASTL, but inactive when dissociated. As the dissociation is promoted by excess OAS, the complex effectively senses both OAS and sulfur availability, and self regulates further OAS production accordingly. Thus OAS is a signal mediating between substrate availability and flux. OASTL, which is in excess, will always catalyze synthesis of cysteine given availability of OAS and sulfide.

The induction/repression of STs and initial steps of the assimilatory pathway was assumed initially to be a negative feedback controlling gene expression, mediated by downstream products of the assimilatory pathway (*e.g.* cysteine or glutathione, see Fig. 1). Subsequently a model was proposed in which OAS mediates the up-regulation of ST gene (and others) expression (also shown on Fig. 1). This was proposed by analogy with the prokaryotic model, for example the *cysB* mediated

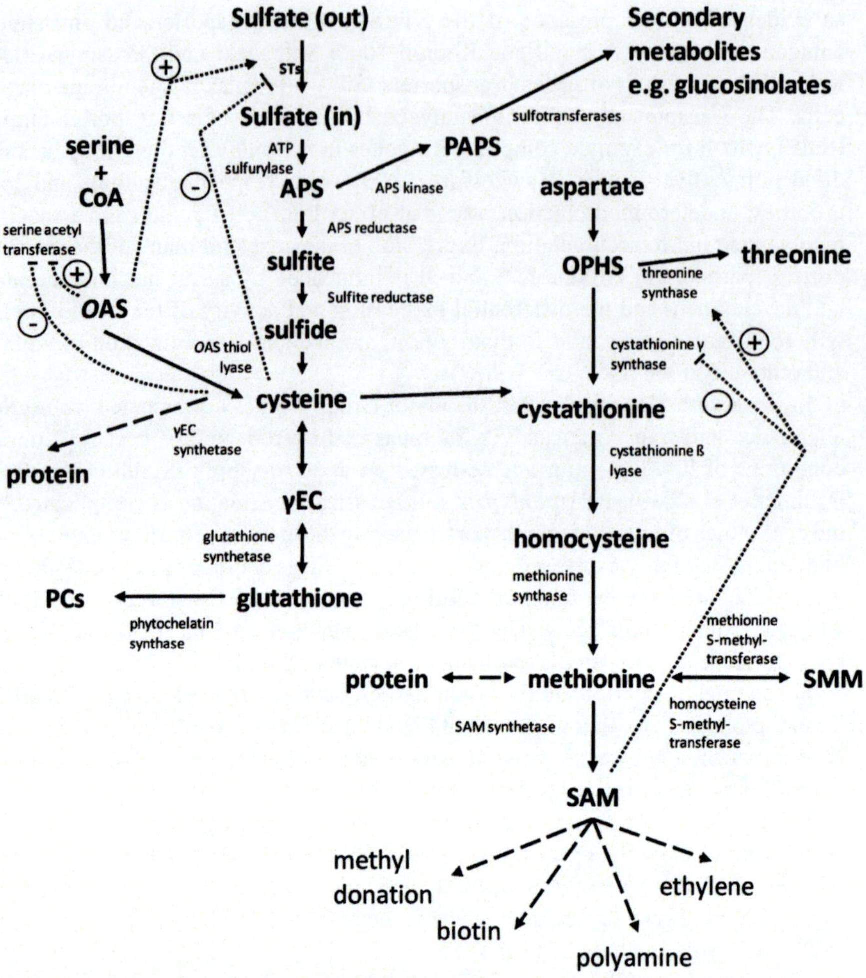

Fig. 1 Cellular regulation: S metabolism is part of a metabolic web (Modified from (Hawkesford et al. 2006)). *Solid lines* represent metabolite pathways (*dashed*=multiple steps), *dotted lines* represent possible feedback loops, which may be allosteric or may act via control of gene expression. *ST* sulfate transporter, *APR* APS reductase, *OAS O*-acetyl serine, *APS* adenosine 5′-phosphosulfate, *PAPS* 3′-phosphoadenosine5′-phosphosulfate, *OPHS O*-phosphohomo-serine, *SAM S*-adenosyl methionine, *SMM S*-methylmethionine, *PCs* phytochelatins

regulation in *Salmonella typhimurium* and *Escherichia coli* (Kredich 1993) in which OAS bound to cysB protein facilitates transcription, whilst sulfide is inhibitory. Evidence to support such a mechanism in plants was initially obtained by OAS feeding experiments to young seedlings grown in hydroponics, which showed enhanced ST expression in roots, with consequent high levels of reduced sulfur compounds (cysteine and glutathione) which might have been expected to mediate repression (Smith et al. 1997). A similar induction by OAS was seen in potato, and furthermore, transgenic over-expression of serine acetyltransferase (SAT), which modestly enhanced endogenous root OAS levels, also resulted in induction of a high affinity ST in roots (Hopkins et al. 2005). However a sulfur-deprivation treatment (up to 8 days) eventually leading to a huge OAS accumulation, only resulted in a small transitory increase in ST expression, soon after the removal of the external sulfur supply and before there was any OAS accumulation. In spite of these anomalies, regulation mediated by OAS or thiol compounds remains a viable model, but critically is not yet proven in plants. Alternatively sulfate may be a candidate as a signal as tissue sulfate contents show an inverse relationship with measured specific mRNA pools for STs (Buchner et al. 2010). Another candidate may be sulfide as copper exposure induces ST and APR expression in Chinese cabbage (*Brassica pekinensis*), with one explanation being an interference with the induction/repression signal transduction pathway, possibly by binding sulfide (Shahbaz et al. 2010).

Combining pedospheric sulfur deprivation experiments with atmospheric supply of hydrogen sulfide indicated, for *Brassica napus* and *Brassica pekinensis*, that foliar absorbed hydrogen sulfide is a viable source of sulfur to meet demands for growth, however little repression of induced root STs was observed. Shoot to root communication was concluded to be inefficient (Buchner et al. 2004; Koralewska et al. 2008).

Whereas in barley and wheat there appear to be good correlations between transport activity, gene expression and tissue sulfate concentrations (Smith et al. 1997; Buchner et al. 2010), this is not always the case. Large changes in mRNA abundance are often not matched by large changes in transporter capacity. Upon S resupply in *Brassica oleracea*, expression of BolSultr1;1 (as indicated by mRNA abundance) is transiently repressed but then remains high, decreasing only after a number of days; transport activity also remains high in spite of tissue sulfate and thiol levels being restored to control (sulfur-replete) levels (Koralewska et al. 2009). Although it appears that sometimes there is no correlation between activity, gene expression and potential regulatory metabolites, all of these studies are severely limited by the imprecise spatial resolution of the tissue sampling, usually separating little more than shoots and roots.

Translating cellular responses up to whole plant responses to sulfur limitation is dominated by the concept of demand driven regulation (Lappartient and Touraine 1996; Lappartient et al. 1999). Demand may be defined at the cellular or whole plant level, and will depend on growth rate and 'extra' demands imposed by stress resistance mechanisms involving sulfur-containing compounds; sulfur nutritional status will be the balance of supply and demand (Fig. 2). Demand at the cellular level modifies metabolite levels, which in turn triggers the changes in gene expression, and which leads to modified pathway or transporter activity. As internal pools are depleted in cells or organs, a 'signal' is effectively propagated, without the need for a long distance signaling molecule. Confounding this simplistic model is the recent description of a micro RNA (miR395) which has a likely role in ensuring turnover of mRNAs for key sulfur pathways genes, including a sulfate transporter (2:1) and ATPS; it is possible that this may be phloem-mobile and could facilitate a long-distance signal function (Kawashima et al. 2009; Liang et al. 2010).

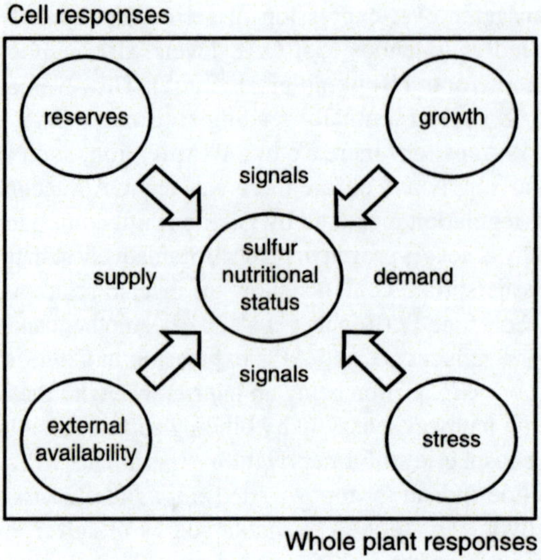

Fig. 2 Local or systemic regulation? The sulfur nutritional status will depend on both supply (external and internal pools) and demand (from growth and responses to stresses requiring S-containing compounds); cell responses involve changes in gene expression and modified fluxes through biochemical pathways which have a net impact on whole plant fluxes of sulfur, modified biomass partitioning and ultimately development. Whole plant regulation may be viewed as an extension of regulatory processes at the single cell or may involve signals between cells distributed throughout the plant

Coordination During Development

Sulfate transporter (ST) expression in relation to the management of plant sulfur reserves has been examined in detail in *Arabidopsis* (Yoshimoto et al. 2002; Kataoka et al. 2003, 2004a, b, c; Kataoka and Takahashi 2005), *Brassica* (Buchner et al. 2004; Parmar et al. 2007; Koralewska et al. 2008, 2009) and wheat (Buchner et al. 2010; Shinmachi et al. 2010). Cell specific expression data has been determined in *Arabidopsis* but is limited for the other species, and whilst many similarities exist between the different species (allowing straightforward assumptions of functional homologues), specific peculiarities are apparent, for example wheat has no close homologue of Atsulr1;2. In addition, the simplistic assumption of individual STs being expressed in a cell or organ specific manner is inadequate and almost all transporters are expressed in all tissues at some point in development, albeit at quite different levels and with differential responses to sulfur-nutrition (Buchner et al. 2010).

Sulfate is taken up to meet the needs for growth and metabolism throughout development, and that which is in excess of current demand is stored in vacuoles; generally it is assumed that cytoplasmic sulfate concentrations do not vary greatly. The vacuolar sulfate pools represent the tissue sulfate pools that respond to periods when demand outstrips supply, for example due to depletion in the pedosphere or unavailability caused by limiting water supply. However pre-programmed patterns of senescence also involve nutrient remobilization from senescing to growing tissues or storage tissues (*e.g.* seed). Remobilized sulfur may be in inorganic (sulfate) or organic (amino acids and their derivatives such as glutathione or *S*-methylmethionine) form, depending on whether pools are vacuolar or in protein. Movement of sulfate is catalysed by expression of members of the sulfate transporter gene family (Buchner et al. 2010), and their expression in individual tissues is also influenced by the S-nutritional status of that tissue. Overall levels of expression in individual tissues under all conditions examined show an exponentially decreasing relationship with tissue sulfate concentration. On the other hand, for wheat canopy tissue, protein degradation, whether as part of normal senescence or due to stresses, including sulfur-deficiency, is dominated by the breakdown of Rubisco and other components of the photosynthetic apparatus, with N and sulfur being transported to the grain (Gilbert et al. 1997) most likely as amino acids. Remobilization of both N and sulfur from the canopy to the grain occurs after anthesis, with grain sulfur accumulation being balanced with N accumulation. There is evidence for a small variable tolerance of excess of grain tissues with sulfur, as a flexibility exists in terms of N:sulfur ratio in grain protein (Godfrey et al. 2010). Within these constraints, N and sulfur export from the canopy are influenced also by plant N and sulfur nutritional status (Howarth et al. 2008).

In wheat grain, Group 1 STs are more highly expressed under low sulfur conditions reflecting the low tissue sulfate concentration, and the high expression will optimise sulfate acquisition by the grain. When plants are adequately supplied with sulfur, sulfur is remobilized in both organic and inorganic forms, and grain demand is determined by N supply to and N status of the grain, hence N nutrition may more directly control ST expression (see following discussion for Brassica). Supporting this mode of control, grain sulfate concentration (which is comparable to leaf concentrations) is only marginally reduced by restricted post anthesis sulfate supply, whilst Group 1 ST expression is substantially increased (Buchner et al. 2010).

In Brassica, sulfate accumulates in leaves (particularly in older leaves) of adequately fertilized plants, and is remobilized if supply is interrupted (slowest remobilization in older leaves); patterns of expression of sulfate transporters were related to these patterns of remobilization (Parmar et al. 2007). Specifically expression of Group 1 (high affinity, plasma membrane-type) and Group 4 (vacuolar efflux-type) STs are expressed in tissues with decreasing sulfate content (exporting sulfate); whether the increased expression is induced by the low tissue sulfate concentration, or is the cause of the low sulfate concentration is not clear. In addition, N-status impacts on sulfur remobilization patterns and differentially on the expression of the two Group 4 STs. Low sulfur and low N conditions result in enhanced senescence with the involvement of BnSultr 4;2, whilst low S conditions alone delayed senescence, and sulfate remobilization appeared more dependent upon BnSultr4;1 (Dubousset et al. 2009).

In summary, sulfur-remobilization from protein is driven by senescence processes and is coupled with N remobilization because of the fixed N:S ratio occurring within canopy protein. Superimposed upon this is the remobilization of inorganic sulfur (in the form of sulfate) which is strongly influenced by supply and demand.

Redistribution may occur to developing canopy tissues or to grain, and occurs in parallel with enhanced expression of vacuolar efflux transporters (Group 4) and Group 1 high affinity plasma membrane transporters (likely induced by the low sulfur content, but whose functionality is unclear unless expressed in a very cell specific manner to facilitate unloading from the leaf).

Are Response to Low Sulfur Availability Related to Sulfur Use Efficiency?

Sulfur use efficiency (SUE) can be defined similarly to nitrogen use efficiency (NUE), that is as crop harvestable yield per unit of available sulfur. Furthermore SUE, as with NUE, may be subdivided into components describing uptake efficiency (SUpE; the fraction of sulfur taken up as a fraction of the total available sulfur) and utilization efficiency (SUtE; the amount of harvestable biomass produced per unit of sulfur taken up). In addition to these measures, the sulfur harvest index (SHI; the fraction of sulfur taken off in the harvested part of crop divided by total sulfur taken up by the crop) is an important agronomic trait in that it is indicative of protein quality and nutritional value of the crop product (Zhao et al. 1999; Fig. 3).

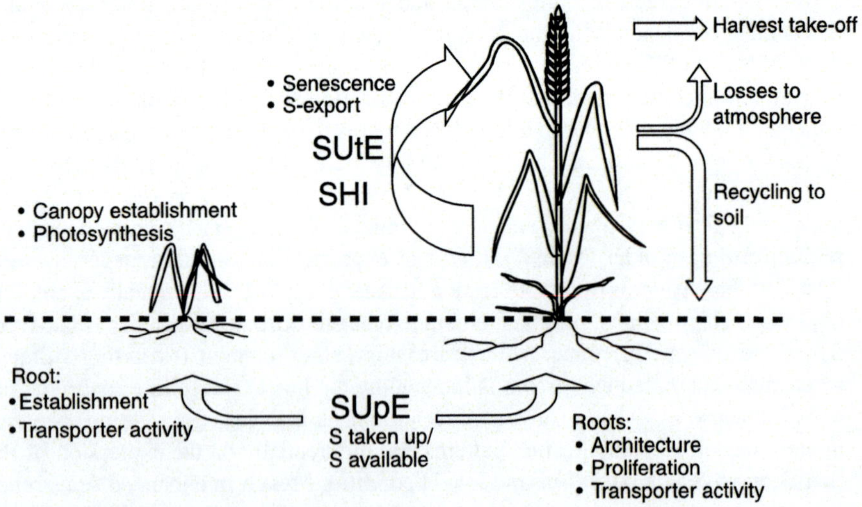

Fig. 3 Regulation at the plant and crop level: Sulfur Use Efficiency comprises both an element of the efficiency of uptake (SUpE), which is relevant throughout development and will contribute to successful canopy development, and sulfur utilization efficiency (SUtE), the amount of harvestable biomass produced per unit S taken up (most relevant at the generative stage of development). In addition to these measures, SHI, the sulfur harvest index, is a measure of how much sulfur is taken off in the harvested material as a ratio of the total taken up; ideally the S taken off is incorporated into beneficial compounds such as protein

Multi-level responses to sulfur deficiency are well documented (Hawkesford 2000; Nikiforova et al. 2005; Hawkesford and De Kok 2006); the question arises as to whether these responses are relevant to SUE in a crop and whether they represent legitimate targets for crop improvement. Increased ST expression and root proliferation will enhance capture of sulfate (thus enhanced SUpE), and will be effective from an early seedling stage. Tolerating a varying sulfur supply (for example, discrete applications of sulfur fertilizer and the subsequent depletion as it is used by the plant or possibly leached) requires uptake of excess, storage and effective remobilization, optimizing canopy development and generative tissue (seed) formation (SUtE). Remobilization efficiency is critical for delivery of sulfur to generative (and harvested) tissues such as seed (promoting SHI). SUpE, SUtE and SHI are critical components to ensure optimal fertilizer capture and conversion into valuable crop biomass, whilst minimizing losses to the environment (Fig. 3).

S/Se/Mo Imbalances

Sulfate transporters are not specific for sulfate; it has long been realized that selenate is transported by the same transporters, and suspected that molybdate might behave similarly. Selenate resistance was used to isolate sulfate uptake-deficient yeast mutants (Breton and Surdin-Kerjan 1977; Smith et al. 1995b) and in *Arabidopsis* (Shibagaki et al. 2002). Competition between sulfate and selenate uptake is well known at the whole plant level (Leggett and Epstein 1956). Recently a member of the SulP-sulfate transporter family, Atsultr5;2 (also known as mot1), was shown to have a specific role in molybdenum accumulation (Tomatsu et al. 2007; Baxter et al. 2008). A related gene, Atsultr5;1 (mot2), may also have a specific although less pronounced role. Both of these transporters occur on intracellular membranes and are therefore not responsible for primary uptake. It is probable that all members of the SulP family, which are able to transport sulfate, may also transport selenate and molybdate (Fitzpatrick et al. 2008) and will contribute to their uptake and distribution.

In a wheat field experiment (Broadbalk at Rothamsted Research, in the U.K.) in which control plots were compared to plots which had received no sulfur fertilizer over a period from 2001 to 2008, a consistent approximate 10% yield decrease was accompanied also by an approximate 10% decrease in grain sulfur, but several-fold increases in both grain Se and Mo concentrations (Stroud et al. 2010). A detailed analysis of vegetative tissues during development indicated very low sulfate pools in plants from the non-sulfur fertilized plots, and substantially increased Mo and Se concentrations in all tissues except roots (Shinmachi et al. 2010). Final partitioning to the grain indicated that Se was co-remobilized from the canopy with sulfur (resulting in a sevenfold increase in concentration in the grain relative to the control), whilst Mo was less efficiently remobilized, but still sufficiently to result in the 3.7-fold higher grain concentration relative to the control. Remarkably, although yield and grain sulfur were only slightly reduced in the non-sulfur fertilized plots, the substantial impact on vegetative tissue sulfate pools was a sufficient signal of low sulfur-nutritional status to result in an induction of gene expression of multiple members of the SulP family in many tissues. Critically, high expression of root STs in an environment with low sulfate, resulted in a disproportionate uptake of other oxyanions (Shinmachi et al. 2010).

Prospects Including Targets for Improvement

An important driver for plant science research is crop genetic improvement; ensuring a secure and sustainable food supply is the grand challenge for agriculture today. Whilst major issues are climate change and water availability, production in most agricultural systems is dependent upon adequate and balanced fertilizer application. Fertilizer production and use has significant costs to agricultural production and has a substantial environmental footprint. Optimizing inputs and fertilizer use are key targets for crop improvement, and both agronomic management and crop genetic improvement will have contributions to this trait. In this chapter, several components of whole plant sulfur regulation have been discussed and directly point to target areas for genetic improvement (Fig. 4). They are not mutually exclusive, and it is likely that stacking of traits will prove useful. In addition appropriate agronomy in terms of form, amount and timing of sulfur fertilizer application will be essential to maximize benefit. In no case can crops be grown without sulfur, but optimizing sulfur take-off from ecosystems to avoid wasteful losses due to, for example, leaching and run off are essential. In addition, appropriate partitioning of sulfur into harvested materials in a beneficial form is a legitimate target for improvement. As indicated, the accumulation of sulfate and its analogues is significantly affected by the relative abundances of the oxyanions; this is critical, as whilst a minimal level of Se is required for a healthy diet, excess intake is toxic (Hawkesford and Zhao 2007).

In the past, the fortuitous provision of sulfur from atmospheric pollutants has obviated the need to add sulfur fertilizers or seek sulfur-efficient genotypes. Currently the cost of sulfur fertilizers is not restrictive for commercial intensive farming, and generally the low applications required have a small environmental footprint and little potential to cause ecological damage. However as a general concept, there is a need to produce germplasm that will grow in low input conditions, and also that will be healthy and nutritious when grown in a range of agricultural environments. In truth, apart from when pedospheric sulfur is in high abundance, uptake is extremely efficient and plants are able to accumulate reserves in the form of either canopy protein or vacuolar sulfate. Furthermore, there is plenty of evidence for recycling of these reserves. Optimizing selectivity for anions will not be simple, but there are precedents for modified transporter substrate selectivity (Rogers et al. 2000). Attempts have been made to engineer high sulfur content of seed (Molvig et al. 1997; Tabe and Higgins 1998), but an overall increase in grain sulfur (and thus SHI) has been hindered by an apparent inability to deliver sufficient extra sulfur to the grain tissues. Such difficulties may require targeted expression of transporters

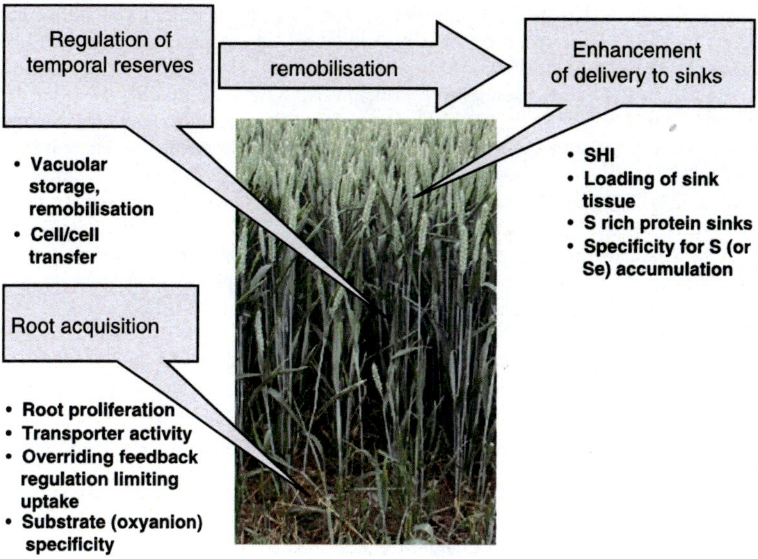

Fig. 4 Targets to improve sulfur use efficiency in crops (wheat as an example): germplasm improvement (by traditional breeding or transgenesis) may contribute (in addition to agronomic management) to efficient use of sulfur resources. As indicated this may include optimizing uptake, controlling the balance of oxyanion specificity, enhancing internal storage reserves and their mobilization, and ensuring optimal partitioning to critical (harvested) tissues

for sulfur-containing compounds such as *S*-methylmethionine (Tan et al. 2010). Opportunities exist to examine partitioning to seed tissues, interactions with nitrogen metabolism and the synthesis of specific storage proteins. Selection for high SHI should be straightforward but has never been pursued; this could be achieved in existing germplasm screening programmes.

Acknowledgments Rothamsted Research receives grant-aided support from the Biotechnology and Biological Sciences Research Council (BBSRC), U.K.

References

Aravind L, Koonin EV (2000) The STAS domain – a link between anion transporters and antisigma-factor antagonists. Curr Biol 10:R53–R55

Baxter I, Muthukumar B, Park HC, Buchner P, Lahner B, Danku J, Zhao K, Lee J, Hawkesford MJ, Guerinot ML, Salt DE (2008) Variation in molybdenum content across broadly distributed populations of *Arabidopsis thaliana* is controlled by a mitochondrial molybdenum transporter (MOT1). PLoS Genet 4:e1000004

Breton A, Surdin-Kerjan Y (1977) Sulfate uptake in *Saccharomyces cerevisiae*: biochemical and genetic study. J Bacteriol 132:224–232

Buchner P, Stuiver CEE, Westerman S, Wirtz M, Hell R, Hawkesford MJ, De Kok LJ (2004) Regulation of sulfate uptake and expression of sulfate transporter genes in *Brassica oleracea* as affected by atmospheric H_2S and pedospheric sulfate nutrition. Plant Physiol 136:3396–3408

Buchner P, Parmar S, Kriegel A, Carpentier M, Hawkesford MJ (2010) The sulfate transporter family in wheat: tissue-specific gene expression in relation to nutrition. Mol Plant 3:374–389

Dubousset L, Abdallah M, Desfeux AS, Etienne P, Meuriot F, Hawkesford MJ, Gombert J, Segura R, Bataille MP, Reze S, Bonnefoy J, Ameline AF, Ourry A, Le Dily F, Avice JC (2009) Remobilization of leaf S compounds and senescence in response to restricted sulphate supply during the vegetative stage of oilseed rape are affected by mineral N availability. J Exp Bot 60:3239–3253

Fitzpatrick KL, Tyerman SD, Kaiser BN (2008) Molybdate transport through the plant sulfate transporter SHST1. FEBS Lett 582:1508–1513

Gilbert SM, Clarkson DT, Cambridge M, Lambers H, Hawkesford MJ (1997) SO_4^{2-} deprivation has an early effect on the content of ribulose-1,5-bisphosphate carboxylase/oxygenase and photosynthesis in young leaves of wheat. Plant Physiol 115:1231–1239

Godfrey D, Hawkesford MJ, Powers SJ, Millar S, Shewry PR (2010) Effects of crop nutrition on wheat grain composition and end use quality. J Agric Food Chem 58:3012–3021

Hawkesford MJ (2000) Plant responses to sulphur deficiency and the genetic manipulation of sulphate transporters to improve S-utilization efficiency. J Exp Bot 51:131–138

Hawkesford MJ (2003) Transporter gene families in plants: the sulphate transporter gene family – redundancy or specialization? Physiol Plant 117:155–163

Hawkesford MJ, De Kok LJ (2006) Managing sulphur metabolism in plants. Plant Cell Environ 29:382–395

Hawkesford MJ, Zhao FJ (2007) Strategies for increasing the selenium content of wheat. J Cereal Sci 46:282–292

Hawkesford MJ, Howarth JR, Buchner P (2006) Control of sulfur uptake, assimilation and metabolism. In: Paxton W, McManus MT (eds) Control of primary metabolism in plants. Blackwell, Oxford, pp 348–372

Hell R, Jost R, Berkowitz O, Wirtz M (2002) Molecular and biochemical analysis of the enzymes of cysteine biosynthesis in the plant *Arabidopsis thaliana*. Amino Acids 22:245–257

Hopkins L, Parmar S, Baszczyk A, Hesse H, Hoefgen R, Hawkesford MJ (2005) O-acetylserine and the regulation of expression of genes encoding components for sulfate uptake and assimilation in potato. Plant Physiol 138:433–440

Howarth JR, Parmar S, Jones J, Shepherd CE, Corol DI, Galster AM, Hawkins ND, Miller SJ, Baker JM, Verrier PJ, Ward JL, Beale MH, Barraclough PB, Hawkesford MJ (2008) Co-ordinated expression of amino acid metabolism in response to N and S deficiency during wheat grain filling. J Exp Bot 59:3675–3689

Kataoka T, Takahashi H (2005) Sulfate transporters involved in redistribution and translocation of sulfate in root vasculature of *Arabidopsis thaliana*. In: Saito K, De Kok LJ, Stulen I, Hawkesford MJ, Schnug E, Sirko A, Rennenberg H (eds) Sulfur transport and assimilation in plants in the post genomic era. Backhuys Publishers, Leiden, pp 45–47

Kataoka T, Hayashi N, Takahashi-Watanabe A, Inoue E, Saito K, Yamaya T, Takahashi H (2003) Functional analysis of sulfate transporters expressed in pericycles of *Arabidopsis* root. Plant Cell Physiol 44:S173

Kataoka T, Hayashi N, Takahashi-Watanabe A, Inoue E, Yamaya T, Takahashi H (2004a) Functional analysis of sulfate transporters SULTR2;1 and SULTR3;5 in *Arabidopsis*. Plant Cell Physiol 45:S174

Kataoka T, Hayashi N, Yamaya T, Takahashi H (2004b) Root-to-shoot transport of sulfate in *Arabidopsis*. Evidence for the role of SULTR3;5 as a component of low-affinity sulfate transport system in the root vasculature. Plant Physiol 136:4198–4204

Kataoka T, Watanabe-Takahashi A, Hayashi N, Ohnishi M, Mimura T, Buchner P, Hawkesford MJ, Yamaya T, Takahashi H (2004c) Vacuolar sulfate transporters are essential determinants controlling internal distribution of sulfate in *Arabidopsis*. Plant Cell 16:2693–2704

Kawashima CG, Yoshimoto N, Maruyama-Nakashita A, Tsuchiya YN, Saito K, Takahashi H, Dalmay T (2009) Sulphur starvation induces the expression of microRNA-395 and one of its target genes but in different cell types. Plant J 57:313–321

Koralewska A, Stuiver CEE, Posthumus FS, Kopriva S, Hawkesford MJ, De Kok LJ (2008) Regulation of sulfate uptake, expression of the sulfate transporters Sultr1;1 and Sultr1;2, and APS reductase in Chinese cabbage (*Brassica pekinensis*) as affected by atmospheric H_2S nutrition and sulfate deprivation. Funct Plant Biol 35:318–327

Koralewska A, Buchner P, Stuiver CEE, Posthumus FS, Kopriva S, Hawkesford MJ, De Kok LJ (2009) Expression and activity of sulfate transporters and APS reductase in curly kale in response to sulfate deprivation and re-supply. J Plant Physiol 166:168–179

Kredich NM (1993) Gene-regulation of sulfur assimilation. In: De Kok LJ, Stulen I, Rennenberg H, Brunold C, Rauser WE (eds) Sulfur nutrition and assimilation in higher plants – regulatory agricultural and environmental aspects. SPB Academic Publishing, The Hague, pp 37–47

Lappartient AG, Touraine B (1996) Demand-driven control of root ATP sulfurylase activity and SO_4^{2-} uptake in intact canola. The role of phloem-translocated glutathione. Plant Physiol 111:147–157

Lappartient AG, Vidmar JJ, Leustek T, Glass ADM, Touraine B (1999) Inter-organ signaling in plants: regulation of ATP sulfurylase and sulfate transporter genes expression in roots mediated by phloem-translocated compound. Plant J 18:89–95

Leggett JE, Epstein E (1956) Kinetics of sulfate absorption by barley roots. Plant Physiol 31:222–226

Liang G, Yang FX, Yu DQ (2010) MicroRNA395 mediates regulation of sulfate accumulation and allocation in *Arabidopsis thaliana*. Plant J 62:1046–1057

Molvig L, Tabe LM, Eggum BO, Moore AE, Craig S, Spencer D, Higgins TJV (1997) Enhanced methionine levels and increased nutritive value of seeds of transgenic lupins (*Lupinus angustifolius* L) expressing a sunflower seed albumin gene. Proc Natl Acad Sci USA 94:8393–8398

Mugford SG, Yoshimoto N, Reichelt M, Wirtz M, Hill L, Mugford ST, Nakazato Y, Noji M, Takahashi H, Kramell R, Gigolashvili T, Flugge UI, Wasternack C, Gershenzon J, Hell R, Saito K, Kopriva S (2009) Disruption of adenosine-5′-phosphosulfate kinase in *Arabidopsis* reduces levels of sulfated secondary metabolites. Plant Cell 21:910–927

Nikiforova VJ, Kopka J, Tolstikov V, Fiehn O, Hopkins L, Hawkesford MJ, Hesse H, Hoefgen R (2005) Systems rebalancing of metabolism in response to sulfur deprivation, as revealed by metabolome analysis of *Arabidopsis* plants. Plant Physiol 138:304–318

Parmar S, Buchner P, Hawkesford MJ (2007) Leaf developmental stage affects sulfate depletion and specific sulfate transporter expression during sulfur deprivation in *Brassica napus* L. Plant Biol 9:647–653

Rogers EE, Eide DJ, Guerinot ML (2000) Altered selectivity in an *Arabidopsis* metal transporter. Proc Natl Acad Sci USA 97:12356–12360

Rouached H, Berthomieu P, El Kassis E, Cathala N, Catherinot V, Labesse G, Davidian J-C, Fourcroy P (2005) Structural and functional analysis of the C-terminal STAS (sulfate transporter and anti-sigma antagonist) domain of the *Arabidopsis thaliana* sulfate transporter SULTR1.2. J Biol Chem 280:15976–15983

Shahbaz M, Tseng MH, Stuiver CEE, Koralewska A, Posthumus FS, Venema JH, Parmar S, Schat H, Hawkesford MJ, De Kok LJ (2010) Copper exposure interferes with the regulation of the uptake, distribution and metabolism of sulfate in Chinese cabbage. J Plant Physiol 167:438–446

Shibagaki N, Grossman AR (2004) Probing the function of STAS domains of the *Arabidopsis* sulfate transporters. J Biol Chem 279:30791–30799

Shibagaki N, Rose A, McDermott JP, Fujiwara T, Hayashi H, Yoneyama T, Davies JP (2002) Selenate-resistant mutants of *Arabidopsis thaliana* identify Sultr1;2, a sulfate transporter required for efficient transport of sulfate into roots. Plant J 29:475–486

Shinmachi F, Buchner P, Stroud JL, Parmar S, Zhao FJ, McGrath SP, Hawkesford MJ (2010) Influence of sulfur deficiency on the expression of specific sulfate transporters and the distribution of sulfur, selenium, and molybdenum in wheat. Plant Physiol 153:327–336

Smith FW, Ealing PM, Hawkesford MJ, Clarkson DT (1995a) Plant members of a family of sulfate transporters reveal functional subtypes. Proc Natl Acad Sci USA 92:9373–9377

Smith FW, Hawkesford MJ, Prosser IM, Clarkson DT (1995b) Isolation of a cDNA from *Saccharomyces cerevisiae* that encodes a high-affinity sulfate transporter at the plasma-membrane. Mol Gen Genet 247:709–715

Smith FW, Hawkesford MJ, Ealing PM, Clarkson DT, Vanden Berg PJ, Belcher AR, Warrilow AGS (1997) Regulation of expression of a cDNA from barley roots encoding a high affinity sulphate transporter. Plant J 12:875–884

Stroud JL, Li HF, Lopez-Bellido FJ, Broadley MR, Foot I, Fairweather-Tait SJ, Hart DJ, Hurst R, Knott P, Mowat H, Norman K, Scott P, Tucker M, White PJ, McGrath SP, Zhao FJ (2010) Impact of sulphur fertilisation on crop response to selenium fertilisation. Plant Soil 332:31–40

Tabe L, Higgins TJV (1998) Engineering plant protein composition for improved nutrition. Trends Plant Sci 3:282–286

Tan QM, Zhang LZ, Grant J, Cooper P, Tegeder M (2010) Increased phloem transport of s-methylmethionine positively affects sulfur and nitrogen metabolism and seed development in pea plants. Plant Physiol 154:1886–1896

Tomatsu H, Takano J, Takahashi H, Watanabe-Takahashi A, Shibagaki N, Fujiwara T (2007) An *Arabidopsis thaliana* high-affinity molybdate transporter required for efficient uptake of molybdate from soil. Proc Natl Acad Sci USA 104:18807–18812

Vauclare P, Kopriva S, Fell D, Suter M, Sticher L, von Ballmoos P, Krähenbühl U, den Camp RO, Brunold C (2002) Flux control of sulphate assimilation in *Arabidopsis thaliana*: adenosine 5′-phosphosulphate reductase is more susceptible than ATP sulphurylase to negative control by thiols. Plant J 31:729–740

Yoshimoto N, Takahashi H, Smith FW, Yamaya T, Saito K (2002) Two distinct high-affinity sulfate transporters with different inducibilities mediate uptake of sulfate in *Arabidopsis* roots. Plant J 29:465–473

Yoshimoto N, Inoue E, Watanabe-Takahashi A, Saito K, Takahashi H (2007) Posttranscriptional regulation of high-affinity sulfate transporters in *Arabidopsis* by sulfur nutrition. Plant Physiol 145:378–388

Zhao FJ, Hawkesford MJ, McGrath SP (1999) Sulphur assimilation and effects on yield and quality of wheat. J Cereal Sci 30:1–17

Molecular and Cellular Regulation of Sulfate Transport and Assimilation

Naoko Yoshimoto and Kazuki Saito

Abstract Sulfur is an essential macronutrient for plants with important roles in biological structure and function. Although it has long been known that sulfate uptake, assimilation and metabolism are highly controlled by sulfur availability, the detailed mechanism of regulation has only recently begun to be elucidated. In this review, we highlight recent advances in our understanding of the molecular and cellular basis of plant response to sulfur limitation.

Introduction

Sulfur is a macronutrient essential for plant growth and survival. It is found in the amino acids cysteine and methionine, in sulfur-containing cofactors such as biotin, thiamine and coenzyme A, and in a variety of secondary metabolites such as glucosinolates in Cruciferae plants and *S*-alk(en)yl cysteine sulfoxides in *Allium* species. Although sulfur is found as both oxidized and reduced forms in nature, sulfate divalent anion, which is the most oxidized form of sulfur, is the major form

N. Yoshimoto
Graduate School of Pharmaceutical Sciences, Chiba University,
Chuo-ku, Chiba 260-8675, Japan
e-mail: naokoy@faculty.chiba-u.jp

K. Saito (✉)
Graduate School of Pharmaceutical Sciences, Chiba University,
Chuo-ku, Chiba 260-8675, Japan

RIKEN Plant Science Center, Tsurumi-ku, Yokohama 230-0045, Japan
e-mail: ksaito@faculty.chiba-u.jp

L.J. De Kok et al. (eds.), *Sulfur Metabolism in Plants*, Proceedings of the International Plant Sulfur Workshop 1, DOI 10.1007/978-94-007-4450-9_3,

of sulfur that plants utilize for the synthesis of various sulfur-containing organic compounds. It is taken up from the soil solution by the function of sulfate transporters bound in the plasma membrane of the root cells. Sulfate incorporated in the root cells is transported to the aerial parts of plants through vasculature, and enters the metabolic processes of assimilatory sulfate reduction followed by the synthesis of cysteine. It is well established that the expression of genes encoding sulfate transporters and enzymes for sulfur assimilation is coordinately controlled mainly at the transcriptional level in response to changes in the sulfur status (Leustek et al. 2000; Saito 2004). Throughout the last decade, great advances have been made in our understanding of how plants adapt to sulfur limitation at the molecular levels; multiple key regulatory components that contribute to metabolic regulations in response to sulfur limitation have been discovered, and the importance of protein-protein interaction among sulfur assimilatory proteins in controlling sulfur homeostasis has been demonstrated. Here we will summarize current progress toward understanding the molecular mechanisms of plant response to sulfur limitation, especially focusing on regulatory components involved in the control of sulfate uptake, assimilation and metabolism.

Transcription Factors

To date, several transcription factors have been reported as the regulator of sulfur assimilatory pathway. SLIM1 was identified as a central transcription factor for the regulation of sulfur limitation-responsive genes in *Arabidopsis* by the forward genetic approach (Maruyama-Nakashita et al. 2006). The transgenic *Arabidopsis* expressing GFP under the control of promoter of high-affinity sulfate transporter *SULTR1;2* significantly accumulated GFP during sulfur starvation. Using this transgenic plant as a parental line, ethyl methanesulfonate-mutagenized M2 seeds were generated, and the M2 plants showing reduced levels of GFP under low sulfur condition were screened. A family of allelic mutants was identified and named *slim1* after their sulfur limitation responseless phenotypes. *SLIM1* encoded the ETHYLENE-INSENSITIVE3-LIKE3 (EIL3), a putative EIL family transcription factor whose function in ethylene response has not been verified. Transcriptome analysis revealed that SLIM1 regulates the expression of a set of genes involved in primary and secondary sulfur metabolism. In response to sulfur limitation, SLIM1 activates sulfate assimilation and represses the synthesis of glucosinolates. These observations suggested that the physiological function of SLIM1 is to globally control the upstream signaling cascades of sulfur metabolism. Contrary to the function of SLIM1 for the response to sulfur limitation, the mRNA levels of *SLIM1* itself were not altered by sulfur deficit (Maruyama-Nakashita et al. 2006). It is suggested that a posttranscriptional mechanism is required for the control of function of SLIM1.

R2R3-MYB transcription factors, MYB28, MYB29, MYB34, MYB51, MYB76 and MYB122 are also involved in the regulation of sulfur metabolism in *Arabidopsis*.

They were discovered either by identification of the mutant showing altered levels of glucosinolates (Celenza et al. 2005; Gigolashvili et al. 2007a), by screening for the *trans*-activation potential toward glucosinolate biosynthetic genes (Gigolashvili et al. 2007b), or by omics-based approach (Hirai et al. 2007). While SLIM1 negatively regulates both aliphatic and indolic glucosinolate biosynthesis, each MYB positively controls only either aliphatic or indolic pathways. MYB28, MYB29 and MYB76 positively regulate aliphatic glucosinolate synthesis whereas MYB34, MYB51 and MYB122 exclusively activate the synthesis of indolic glucosinolates (Celenza et al. 2005; Hirai et al. 2007; Gigolashvili et al. 2007a, b, 2008). More recently, it has been found that these MYBs also activate the expression of ATP sulfurylase, APS kinase, APS reductase and sulfite reductase, enhancing the production of PAPS and reduced sulfur that are required for the synthesis of glucosinolate (Yatusevich et al. 2010). Intriguingly, the expression of MYB34 was negatively regulated by SLIM1 but the other MYBs were not (Maruyama-Nakashita et al. 2006), suggesting the presence of both SLIM1-dependent and independent mechanisms for regulating the synthesis of glucosinolate.

Cis-Acting Elements

A *cis*-acting element for the sulfur response was identified in the promoter of *SULTR1;1*, a gene encoding high-affinity sulfate transporter in *Arabidopsis*, by deletion and element-addition analyses (Maruyama-Nakashita et al. 2005). This 16-bp DNA sequence from –2,777 to –2,762 of *SULTR1;1* promoter was sufficient and necessary for the response to sulfur limitation, and was named SURE after sulfur-responsive element. The base substitution analysis indicated that the core sequence of SURE is GGAGACA. The SURE core sequences and its complementary sequences were found in the upstream regions of a number of sulfur limitation-responsive genes including sulfate transporter genes *SULTR2;1* and *SULTR4;2* and APS reductase gene *APR3*, suggesting the key role of this conserved *cis*-acting element in response to sulfur deficit. Although the transcription factor which can bind to the SURE sequence has not been identified yet, the fact that the SURE core sequence includes the auxin response factor (ARF) binding sequence (GAGACA) suggests the potential involvement of ARF-type transcription factor in the transcriptional activation of SURE-containing genes.

Another *cis*-acting element for the sulfur response was identified in the promoter region of *UP9C*, a tobacco gene, which is strongly induced by sulfur limitation (Wawrzyńska et al. 2010). A search for sequence motifs conserved in the promoter regions of *UP9C* and its *Arabidopsis* homologue genes identified the 20-nt palindromic DNA sequence (AGATACATTGAACCTGGACA). This motif, termed the UPE-box, was found also in the promoters of several genes up-regulated by sulfur deficiency, including APS reductase genes *APR1* and *APR3*. The presence of *tebs* element, a target sequence of NtEIL1/TEIL transcription factor, in the UPE-box strongly suggests the possibility that a member of EIL-family transcription factor binds to the UPE-box. Indeed, NtEIL2 and its *Arabidopsis* homologue SLIM1 were able to interact with UPE-box in the yeast one-hybrid system. Moreover, NtEIL2 and SLIM1 could *trans*-activate the *UP9C* promoter when ectopically expressed in *Nicotiana benthamiana*. It is notable that NtEIL2 could activate the *UP9C* promoter only under sulfur-deficient conditions, whereas SLIM1 was able to activate the *UP9C* promoter irrespective of sulfur conditions.

microRNA

MicroRNAs (miRNAs) are a class of small noncoding RNA molecules that post-transcriptionally regulate the stability of transcripts of their target genes by means of complementary base pair interactions. Among the plant miRNAs identified to date, miR395 in *Arabidopsis* was revealed to be involved in the regulation of sulfate transport and assimilation. Computational analysis identified the presence of miR395 which is complementary to mRNAs of three isoforms of ATP sulfurylase *ATPS1*, *ATPS3* and *ATPS4*, and low-affinity sulfate transporter *SULTR2;1* which facilitates inter-organ transport of sulfate (Jones-Rhoades and Bartel 2004; Bonnet et al. 2004). The miR395-guided cleavage of target genes was confirmed by rapid amplification of cDNA ends (Jones-Rhoades and Bartel 2004; Kawashima et al. 2009), and by the analysis of the levels of target transcripts in transgenic plants overexpressing miR395 (Liang et al. 2010). It is worth noting that *ATPS1*, *ATPS4* and *SULTR2;1* were targets of miR395 in both leaves and roots, whereas *ATPS3* was cleaved only in the leaves (Kawashima et al. 2009).

The expression of miR395 was strongly induced with low sulfur in a SLIM1-dependent manner (Kawashima et al. 2009). As expected from this, *ATPS4* transcript was significantly reduced by sulfur deficit. However, the mRNA levels of *ATPS1*, *ATPS3* and *SULTR2;1* were not regulated in the expected manners by sulfur starvation (Kawashima et al. 2009, 2011). During sulfur starvation, *ATPS1* transcript levels were not significantly affected, *ATPS3* mRNA levels were elevated, and *SULTR2;1* mRNA levels were reduced in leaves but were elevated in roots. These inconsistencies can be explained by the presence of miR395-independent regulatory pathway(s) for the control of the expression of miR395 target genes, and/or by the difference of the spatial expression pattern between miR395 and its target transcripts. In the case of *ATPS1*, its transcription was enhanced by SLIM1 during sulfur deficit, in parallel with the accumulation of miR395. Consequently, the steady-state levels of *ATPS1* mRNA and thus the total ATP sulfurylase activity in plants were not altered by sulfate limitation. The opposing actions of SLIM1 and miR395 on the expression of *ATPS1* may be necessary for optimizing the rate of sulfur assimilation in response to the fluctuation of sulfur conditions. This is supported by the recent analysis comparing sulfur-limitation response among wild-type plants, miR395-overexpressing plants, transgenic plants in which miR395 activity was reduced, and *slim1* mutant (Kawashima et al. 2011). In the case of *SULTR2;1*, its transcript levels were attenuated in leaves of sulfur-starved plants, which can be explained at least partially by miR395 accumulation. By contrast, the levels of *SULTR2;1* mRNA were markedly elevated in roots under sulfur-limited conditions. This upregulation was observed also for *slim1* mutant, indicating the presence of SLIM1-independent pathway for the regulation of *SULTR2;1* expression (Kawashima et al. 2009). Interestingly, in roots, miR395 predominantly expressed in the companion cells of the phloem whereas *SULTR2;1* expressed mainly in the xylem parenchyma cells (Kawashima et al. 2009; Takahashi et al. 2000). Most recently, it was revealed that miR395 overexpressing plants showed enhanced rate of sulfate translocation from roots to shoots under normal sulfate conditions, as in the case of sulfur-starved wild-type plants (Kawashima et al. 2011). It seems that, during sulfate starvation, miR395 functions to cleave *SULTR2;1* mRNA expressing at a low level in the phloem to strictly restrict *SULTR2;1* expression in the xylem parenchyma cells of roots, for the efficient transport of sulfate from roots to shoots through the xylem.

The fact that miR395 expresses in the phloem companion cells suggests that miR395 can act as potential signaling molecule that can move between organs through the phloem. In sulfur-starved *Brassica napus*, miR395 was more drastically accumulated in phloem sap than in leaf, stem and root (Buhtz et al. 2008), suggesting that miR395 synthesized in the phloem companion cells was moved into the adjacent sieve tube via plasmodesmata. However, the involvement of miR395 in long-distance systemic information transfer via sieve tube remains unknown and awaits further study.

Protein-Protein Interaction

The synthesis of cysteine is carried out by a two-step pathway, catalyzed by serine acetyltransferase (SAT) and *O*-acetylserine (thiol)lyase (OASTL). In the first step, SAT transfers an acetyl-moiety from acetyl coenzyme A to serine to form *O*-acetylserine (OAS). Subsequently, OASTL exchanges the activated acetyl group with sulfide by a β-replacement reaction to produce cysteine. It is well established that OASTL and SAT physically interact to form a bi-enzyme complex known as cysteine synthase complex (CSC). The physiological function of the CSC is not metabolic channeling, but to sense cellular sulfur status and to adjust the rate of cysteine synthesis. The stability of CSC is reciprocally regulated by sulfide and OAS. When intracellular sulfur levels are high, sulfide that is produced by assimilatory sulfate reduction stabilizes the CSC. In the CSC, SAT is highly active whereas OASTL is almost inactive. By contrast, OAS which is accumulated during sulfur starvation, promotes dissociation of CSC, resulting in the formation of less-active SAT and active OASTL. Since OASTL is present in much larger quantities than SAT in the plant cells, OAS synthesized by SAT in the CSC is released from the CSC and is subsequently converted to cysteine by the action of surrounding free OASTL (Hell and Hillebrand 2001; Wirtz and Hell 2006).

Recently, the structural basis of protein-protein interactions between SAT and OASTL were probed by multiple approaches. Crystal structure analysis revealed that the C-terminus of SAT binds to the active site of OASTL and blocks access to the catalytic center of the enzyme (Francois et al. 2006). This model elegantly explains the mechanism by which association with SAT downregulates OASTL activity, and the reason for the dissociating effect of OAS on the CSC. SAT is present as a dimer of two trimers (Wirtz et al. 2010), and thus contains six C-terminal tails that would be the potential binding sites for OASTL dimer. However, recent computational modeling and biophysical studies indicated that the CSC most likely consists of one SAT dimer of trimer and only two OASTL dimers (Feldman-Salit et al. 2009; Wirtz et al. 2010). In this model, two SAT trimers associate to form the core of the CSC, and each SAT trimer interacts with one OASTL dimer. It is also demonstrated that association of OASTL dimer downregulates feedback sensitivity of SAT to cysteine (Wirtz et al. 2010).

In addition to direct interaction with SAT, cytosolic form of OASTL has recently been found to bind at the C-terminal cytoplasmic domain named STAS (sulfate transporter and anti-sigma factor antagonist) of plasma membrane-bound high-affinity sulfate transporter SULTR1;2, by yeast two-hybrid screening (Shibagaki and Grossman 2010). *In vitro* binding assay showed that the interaction between SULTR1;2 STAS domain and OASTL was dependent on OAS. The cysteine synthase activity of OASTL was enhanced by co-incubation with STAS domain of SULTR1;2. By contrast, OASTL downregulated the sulfate uptake activity of SULTR1;2. This inhibition effect was less prominent under low-sulfate conditions than under high-sulfate conditions. It is suggested that, under low-sulfate conditions, the conformation of SULTR1;2 may change not to interact with OASTL despite high OAS concentration, allowing high sulfate transport activity. The physiological role of the complex consisting of SULTR1;2 and OASTL would be to optimize the rate of sulfate uptake to the ability of the root cells to reduce sulfate to sulfide, mainly under sulfate-replete conditions. The molecular mechanism by which the association between SULTR1;2 and OASTL modulates their activity awaits further investigation.

Conclusions and Future Perspectives

The past decade has brought major advances in our knowledge of the ways that plants respond and acclimate to sulfur starvation. The SLIM1 and MYB transcription factors and miR395 have been identified, and the importance of their interplay in the regulation of sulfate transport and assimilation has been revealed. Also, physical interaction of OASTL with SAT and SULTR1;2 have emerged as essential mechanism both for sensing cellular sulfur status and for optimizing the rate of sulfate assimilation. It seems that plants have evolved a highly complex network of structural and functional interactions among multiple regulatory components to maintain sulfur

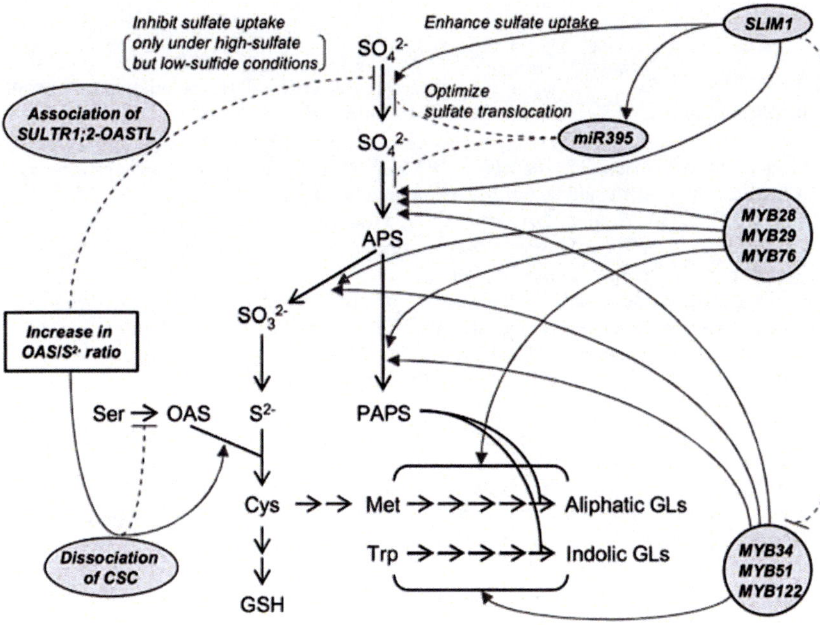

Fig. 1 Schematic representation for the regulatory pathways involved in sulfur deficiency response in plants

homeostasis. The current model for the molecular regulation of sulfate transport and assimilation in plants is depicted in Fig. 1.

Despite advances in the identification of key regulatory factors for the response to sulfur limitation, little is currently known about how plants perceive sulfur level and transmit signals both locally and systemically, and how this links to transcriptional machinery. Plasma membrane-bound sensory proteins and SNRK2 serine/threonine kinases for the signal transduction may be involved in the sensory and signaling systems in plants, as in the case of the unicellular green alga *Chlamydomonas reinhardtii* (Davies et al. 1996, 1999; Gonzalez-Ballester et al. 2008). Interestingly, *Arabidopsis* mutants lacking one of ten *SNRK2* genes was not able to induce sulfur-inducible *SULTR2;2* mRNA and showed increased accumulation of OAS during sulfur starvation (Kimura et al. 2006). In addition, although SLIM1 has been identified as a central transcription factor, the molecular mechanism of action of SLIM1 and the downstream components in the regulatory cascade of SLIM1 have not been fully elucidated yet. Discoveries on the targets of posttranslational modification and the regulatory mechanism controlling their modification process in response to sulfur limitation are the areas also needing further investigation. The recent advancement in multi-omics study and genome-wide association study will greatly accelerate the discovery of additional members involved in the regulatory networks.

References

Bonnet E, Wuyts J, Rouzé P, Van de Peer Y (2004) Detection of 91 potential conserved plant microRNAs in *Arabidopsis thaliana* and *Oryza sativa* identifies important target genes. Proc Natl Acad Sci USA 101:11511–11516

Buhtz A, Springer F, Chappell L, Baulcombe DC, Kehr J (2008) Identification and characterization of small RNAs from the phloem of *Brassica napus*. Plant J 53:739–749

Celenza JL, Quiel JA, Smolen GA, Merrikh H, Silvestro AR, Normanly J, Bender J (2005) The *Arabidopsis* ATR1 Myb transcription factor controls indolic glucosinolate homeostasis. Plant Physiol 137:253–262

Davies JP, Yildiz FH, Grossman A (1996) Sac1, a putative regulator that is critical for survival of *Chlamydomonas reinhardtii* during sulfur deprivation. EMBO J 15:2150–2159

Davies JP, Yildiz FH, Grossman AR (1999) Sac3, an Snf1-like serine/threonine kinase that positively and negatively regulates the responses of *Chlamydomonas* to sulfur limitation. Plant Cell 11:1179–1190

Feldman-Salit A, Wirtz M, Hell R, Wade RC (2009) A mechanistic model of the cysteine synthase complex. J Mol Biol 386:37–59

Francois JA, Kumaran S, Jez JM (2006) Structural basis for interaction of *O*-acetylserine sulfhydrylase and serine acetyltransferase in the *Arabidopsis* cysteine synthase complex. Plant Cell 18:3647–3655

Gigolashvili T, Berger B, Mock H-P, Müller C, Weisshaar B, Flügge U-I (2007a) The transcription factor HIG1/MYB51 regulates indolic glucosinolate biosynthesis in *Arabidopsis thaliana*. Plant J 50:886–901

Gigolashvili T, Yatusevich R, Berger B, Müller C, Flügge U-I (2007b) The R2R3-MYB transcription factor HAG1/MYB28 is a regulator of methionine-derived glucosinolate biosynthesis in *Arabidopsis thaliana*. Plant J 51:247–261

Gigolashvili T, Engqvist M, Yatusevich R, Müller C, Flügge U-I (2008) HAG2/MYB76 and HAG3/MYB29 exert a specific and coordinated control on the regulation of aliphatic glucosinolate biosynthesis in *Arabidopsis thaliana*. New Phytol 177:627–642

Gonzalez-Ballester D, Pollock SV, Pootakham W, Grossman AR (2008) The central role of a SNRK2 kinase in sulfur deprivation responses. Plant Physiol 147:216–227

Hell R, Hillebrand H (2001) Plant concepts for mineral acquisition and allocation. Curr Opin Biotechnol 12:161–168

Hirai MY, Sugiyama K, Sawada Y, Tohge T, Obayashi T, Suzuki A, Araki R, Sakurai N, Suzuki H, Aoki K, Goda H, Nishizawa OI, Shibata D, Saito K (2007) Omics-based identification of *Arabidopsis* Myb transcription factors regulating aliphatic glucosinolate biosynthesis. Proc Natl Acad Sci USA 104:6478–6483

Jones-Rhoades MW, Bartel DP (2004) Computational identification of plant microRNAs and their targets, including a stress-induced miRNA. Mol Cell 14:787–799

Kawashima CG, Yoshimoto N, Maruyama-Nakashita A, Tsuchiya YN, Saito K, Takahashi H, Dalmay T (2009) Sulphur starvation induces the expression of microRNA-395 and one of its target genes but in different cell types. Plant J 57:313–321

Kawashima CG, Matthewman CA, Huang S, Lee B-R, Yoshimoto N, Koprivova A, Rubio-Somoza I, Todesco M, Rathjen T, Saito K, Takahashi H, Dalmay T, Kopriva S (2011) Interplay of SLIM1 and miR395 in regulation of sulfate assimilation in *Arabidopsis*. Plant J 66(5):863–876

Kimura T, Shibagaki N, Ohkama-Ohtsu N, Hayashi H, Yoneyama T, Davies JP, Fujiwara T (2006) *Arabidopsis* SNRK2.3 protein kinase is involved in the regulation of sulfur-responsive gene expression and *O*-acetyl-L-serine accumulation under limited sulfur supply. Soil Sci Plant Nutr 52:211–220

Leustek T, Martin MN, Bick J, Davies JP (2000) Pathways and regulation of sulfur metabolism revealed through molecular genetic studies. Annu Rev Plant Physiol Plant Mol Biol 51:141–165

Liang G, Yang F, Yu D (2010) MicroRNA395 mediates regulation of sulfate accumulation and allocation in *Arabidopsis thaliana*. Plant J 62:1046–1057

Maruyama-Nakashita A, Nakamura Y, Watanabe-Takahashi A, Inoue E, Yamaya T, Takahashi H (2005) Identification of a novel *cis*-acting element conferring sulfur deficiency response in *Arabidopsis* roots. Plant J 42:305–314

Maruyama-Nakashita A, Nakamura Y, Tohge T, Saito K, Takahashi H (2006) *Arabidopsis* SLIM1 is a central transcriptional regulator of plant sulfur response and metabolism. Plant Cell 18:3235–3251

Saito K (2004) Sulfur assimilatory metabolism. The long and smelling road. Plant Physiol 136:2443–2450

Shibagaki N, Grossman AR (2010) Binding of cysteine synthase to the STAS domain of sulfate transporter and its regulatory consequences. J Biol Chem 285:25094–25102

Takahashi H, Watanabe-Takahashi A, Smith FW, Blake-Kalff M, Hawkesford MJ, Saito K (2000) The roles of three functional sulphate transporters involved in uptake and translocation of sulphate in *Arabidopsis thaliana*. Plant J 23:171–182

Wawrzyńska A, Lewandowska M, Sirko A (2010) *Nicotiana tabacum* EIL2 directly regulates expression of at least one tobacco gene induced by sulphur starvation. J Exp Bot 61:889–900

Wirtz M, Hell R (2006) Functional analysis of the cysteine synthase protein complex from plants: structural, biochemical and regulatory properties. J Plant Physiol 163:273–286

Wirtz M, Birke H, Heeg C, Müller C, Hosp F, Throm C, König S, Feldman-Salit A, Rippe K, Petersen G, Wade RC, Rybin V, Scheffzek K, Hell R (2010) Structure and function of the hetero-oligomeric cysteine synthase complex in plants. J Biol Chem 285:32810–32817

Yatusevich R, Mugford SG, Matthewman C, Gigolashvili T, Frerigmann H, Delaney S, Koprivova A, Flügge U-I, Kopriva S (2010) Genes of primary sulfate assimilation are part of the glucosinolate biosynthetic network in *Arabidopsis thaliana*. Plant J 62:1–11

Manipulating the Sulfur Composition of Seeds

Linda Tabe, Carlos Popelka, Pasquale Chiaiese, and T.J.V. Higgins

Abstract The amino acid composition of seeds is frequently sub-optimal for human and animal nutrition. We have attempted to modify the quantities of stored sulfur amino acids in seeds using three complementary strategies. Firstly, a transgene encoding a naturally sulfur amino acid-rich sunflower seed albumin (SSA), was expressed under control of a seed-specific promoter in narrow leaf lupin, chickpea, and field pea. In all cases, this strategy was successful in significantly increasing total sulfur amino acid concentrations of the mature seeds compared to non-transgenic controls. However, the abundance of some endogenous sulfur-rich seed proteins decreased in the transgenic seeds, indicating that the new storage sink competed to some extent with endogenous proteins for a limiting supply of sulfur amino acids in developing seeds. High levels of several enzymes of the sulfur amino acid biosynthetic pathway were detected in developing lupin and chickpea seeds, supporting the view that developing legume seeds are active in sulfur amino acid biosynthesis. A second strategy, of increasing sulfur supply to the plants, had limited success in further increasing seed sulfur amino acid content. A third strategy was used in an attempt to supplement the sulfur amino acid content of narrow leaf lupin seeds. A gene encoding a cysteine feedback-insensitive, plant serine acetyltransferase (SAT) was expressed under the control of a strong, seed-specific promoter in transgenic lupins. Transgenic developing seeds had greatly increased SAT activity, and some of the highest *in planta* concentrations of free cysteine yet reported. However, the free methionine concentrations of the developing seeds were unchanged by SAT over-expression, and the mature seeds contained no more total (protein-bound) sulfur

L. Tabe (✉) • C. Popelka • P. Chiaiese • T.J.V. Higgins
Plant Industry, CSIRO, P.O. Box 1600, Canberra, ACT 2601, Australia
e-mail: linda.tabe@csiro.au

L.J. De Kok et al. (eds.), *Sulfur Metabolism in Plants*, Proceedings of the International Plant Sulfur Workshop 1, DOI 10.1007/978-94-007-4450-9_4,

amino acids than non-transgenic controls. The results confirmed the activity of the cysteine biosynthetic pathway in developing lupin seeds, but indicated that biosynthesis of cysteine was not limiting for accumulation of free methionine in developing seeds, or for accumulation of sulfur-rich seed storage proteins in mature seeds.

Introduction

The pathways of sulfur assimilation and sulfur amino acid biosynthesis in plants are now well understood, although many questions remain regarding regulation of the pathway and its compartmentation, both within cells, and within the plant. A number of excellent reviews describe current knowledge of plant sulfur metabolism drawing from work in model systems, and increasingly, from crop plants (for example Saito 2004; Hawkesford and De Kok 2006 and references therein).

Seeds represent the terminal site of storage of nutrients assimilated by annual crop plants throughout their life cycle. Furthermore, seeds constitute an essential part of the diets of humans and animals, in many cases providing the bulk of daily nutritional intake. Nitrogen is stored in seeds primarily in the form of storage proteins, whose amino acid composition is frequently sub-optimal for human and animal nutrition (Shewry 2007). Two types of seeds are particularly important as food and feed, namely the cereals and grain legumes or pulses. The nutritionally limiting amino acid in cereal protein is generally lysine, while in grain legumes; it is the sulfur-containing amino acid, methionine (Tabe and Higgins 1998; Shewry 2007).

Here we review our work aimed at increasing storage of the sulfur amino acids, methionine and cysteine, in seeds of grain legumes. We manipulated plant sulfur nutrition, as well as transforming grain legumes with seed-expressed genes encoding either a sulfur-rich seed storage protein, or an enzyme of the cysteine biosynthetic pathway. Plants containing each transgene alone were crossed to combine the two modifications. Analysis of seeds of wild type and the transgenic grain legumes have contributed to our understanding of sulfur amino acid metabolism in these important crops.

Sulfur Metabolism in Developing Seeds of Grain Legumes

Sulfur is available to plants mainly as sulfate in the soil solution. It is taken up by roots through high affinity transporters whose expression is induced by sulfur deficiency (Buchner et al. 2004). Once in the plant, sulfate is moved between organs and between sub-cellular compartments largely via low affinity transporters, whose expression is generally less sensitive to sulfate concentration. Sulfate reduction occurs in the plastid; however, sulfur amino acid biosynthesis occurs in multiple sub-cellular compartments. It was recently demonstrated that exchange of pathway intermediates between sub-cellular compartments is involved in the efficient synthesis

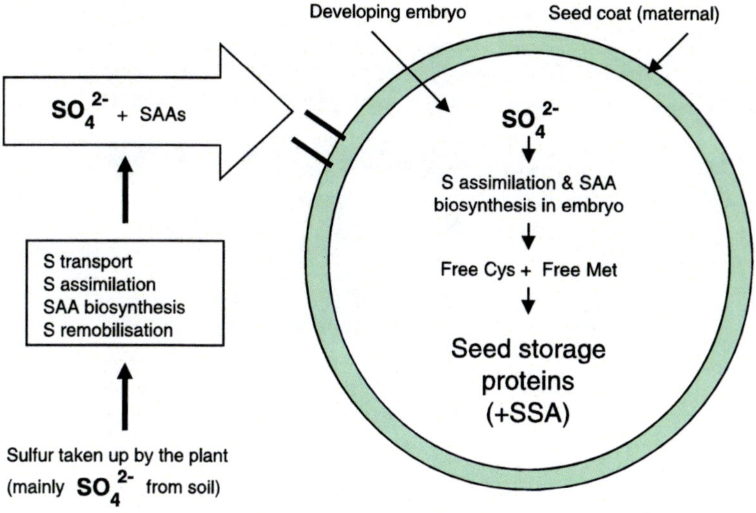

Fig. 1 Schematic diagram of sulfur pools in developing legume seeds. The boxed text represents processes in the vegetative parts of the plant. The *circle* represents a developing legume seed consisting of an embryo within a maternal seed coat (*shaded*), which is connected to the vasculature of the parent plant via the funiculus. The large, *horizontal arrow* represents the sulfur composition of the phloem feeding the developing seeds. The sizes of the different sulfur pools are indicated by the sizes of the fonts (see also text). *S* sulfur, SO_4^{2-} sulfate, *SAA* sulfur amino acids, *SSA* sunflower seed albumin

of cysteine, the first stable, reduced sulfur compound in the cell (Haas et al. 2008; Heeg et al. 2008; Krueger et al. 2009). Methionine is derived from cysteine and *O*-phosphohomoserine, a product of the aspartate amino acid pathway, via three further steps (Ravanel et al. 2004, 1998).

Developing seeds are supplied with water and nutrients via the vasculature of the parent plant. Water and nutrients delivered by the vasculature to the developing pod pass to the maternal seed coat of the seed, and from there, are delivered to the seed apoplastic fluid in which the embryo and endosperm develop. In grain legumes, the endosperm is a transient tissue, which disappears before the main phase of storage product accumulation, while the cotyledons of the embryo constitute the storage organs of the seeds. The embryo takes up nutrients such as sugars, amino acids, and inorganic ions via transporters in specialized transfer cells on its surface (Zhang et al. 2007).

Sulfur is reported to be transported in the plant vasculature in the reduced forms of glutathione (Herschbach and Rennenberg 2001; Kuzuhara et al. 2000) and *S*-methylmethionine (Bourgis et al. 1999). On the other hand, sulfate supplied exogenously to mature leaves of poplar was translocated to developing leaves in the form of sulfate (Hartmann et al. 2000). We determined that in the grain legume, narrow leaf lupin (*Lupinus angustifolius*), sulfur was delivered via the phloem to developing pods predominantly in the oxidized form of sulfate (Fig. 1; Tabe and Droux 2001).

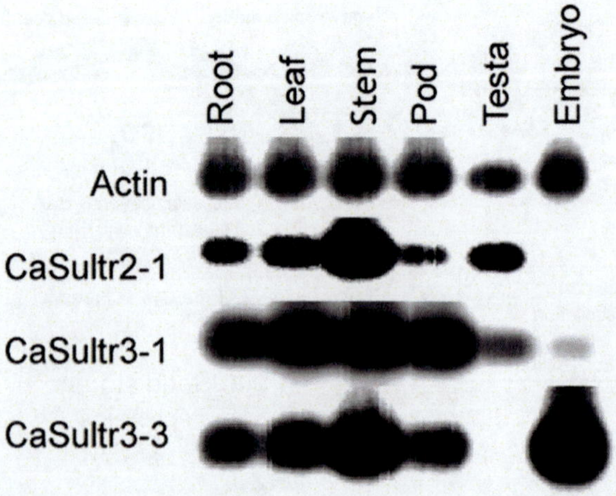

Fig. 2 Expression of putative sulfate transporters in organs of chickpea. Partial cDNAs encoding sequences with strong similarity (up to 84% identity at the amino acid level) to *Arabidopsis thaliana* sulfate transporter genes were isolated from chickpea using PCR with primers designed to conserved regions of plant sulfate transporters (Tabe et al. 2003). Primers specific to the chickpea sulfate transporter sequences were used in RT-PCR (30 cycles) to measure abundance of these transcripts in total RNA isolated from chickpea organs (listed along top of figure). Pod, testa and embryo were at mid-maturation, or approximately 25 days after anthesis. PCR products were resolved on agarose gels, transferred to nylon membranes and probed with ^{32}P-labelled probes consisting of the sequenced PCR products listed at *left* of the figure

Furthermore, isolated developing embryos of chickpea (*Cicer arietinum*) were able to take up sulfate *in vitro* in a saturable, pH-dependent manner (Tabe et al. 2003). Reverse transcriptase-PCR was used to demonstrate expression in developing chickpea embryos of transcripts with strong similarity to sulfate transporters from plant model systems (Fig. 2; Tabe et al. 2003). A sequence predicting a protein fragment of 220 amino acids with 84% amino acid identity to *Arabidopsis* AtSultr3-3 (At1g23090) was expressed at readily detectable levels in chickpea embryos (Fig. 2).

Large pools of oxidized sulfur were present in embryos of both lupin and chickpea during the process of seed maturation, when seed storage proteins accumulate (Figs. 1 and 3; Tabe and Droux 2001). At 20 days after anthesis (DAA), which corresponds to the beginning of the rapid phase of seed storage protein accumulation, approximately half the total sulfur pool present in chickpea developing embryos was in the oxidized form (Fig. 3). Most of the reduced sulfur would have been in the form of protein, therefore sulfate pools would have exceeded those of non-protein sulfur amino acids. For example, the concentrations of sulfate varied from 22.6 to 12 μmol g^{-1} dry weight (DW), while those of glutathione varied from 7.3 to 2.1 μmol g^{-1} DW in embryos at 20 and 40 DAA, respectively. Between 20 and 40 DAA, concentrations and total pool sizes of reduced sulfur increased with the accumulation of seed storage proteins.

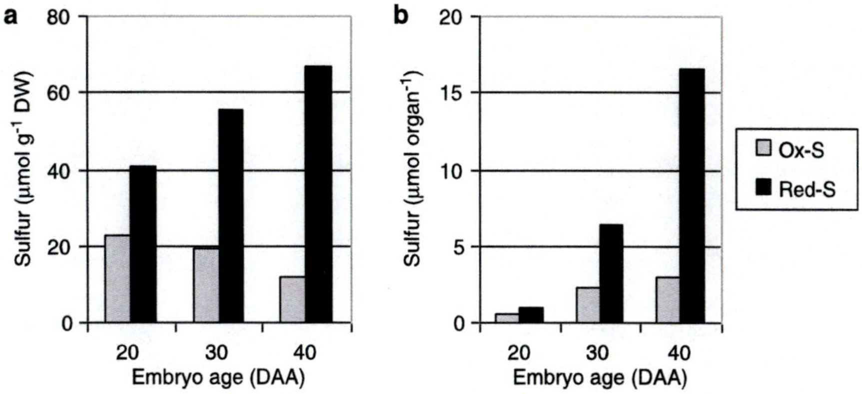

Fig. 3 Sulfur pools in developing embyros of chickpea. Oxidized and reduced sulfur levels were measured using X-ray fluorescence Spectrometry (XRFS, Pinkerton et al. 1989) in pools of approximately 30 powdered, freeze-dried chickpea embryos harvested at 20, 30 and 40 days after anthesis. (**a**): Sulfur concentration; (**b**): Sulfur amount per embryo. *DAA* days after anthesis

Table 1 Activities of enzymes of sulfur amino acid biosynthesis in chickpea organs. Serine acetyltransferase (SAT) and *O*-acetylserine (thiol) lyase (*O*AS-TL) activities were assayed in extracts from photosynthetic source leaves and developing embryos. Each value is the mean (+/− SD) of four assays on a total of two different extracts from the same pool of fresh tissue from five leaves or mid-maturation embryos

Organs	SAT (nmol cysteine min^{-1} mg^{-1} protein)	*O*AS-TL (nmol cysteine min^{-1} mg^{-1} protein)
Leaves	6.9 ± 0.5	298 ± 51
Developing embryos	22.7 ± 0.9	415 ± 52

Oxidized sulfur concentrations gradually declined in the developing embryos, but this decrease was offset by the increase in mass of the embryos, resulting in a slight accumulation of sulfate throughout maturation (Fig. 3).

The presence of large pools of oxidized sulfur in chickpea and lupin developing embryos was contemporaneous with high activities of the two enzymes of the cysteine synthase complex, serine acetyltransferase (SAT) and *O*-acetylserine (thiol) lyase (*O*AS-TL). SAT and *O*AS-TL activities were present in developing embryos in amounts comparable to those present in photosynthetic source leaves of the two grain legumes (Table 1; Tabe and Droux 2001). In addition, we demonstrated that isolated, developing chickpea and lupin embryos could incorporate into protein, labeled sulfur supplied as sulfate (Fig. 4; Tabe and Droux 2001). These complementary strands of evidence, along with published reports from soybean (Sexton and Shibles 1999; Chronis and Krishnan 2003; Phartiyal et al. 2008), support the suggestion that developing embryos of grain legumes are able to perform efficient reductive assimilation of sulfur, and synthesis of sulfur amino acids, for incorporation into seed storage proteins.

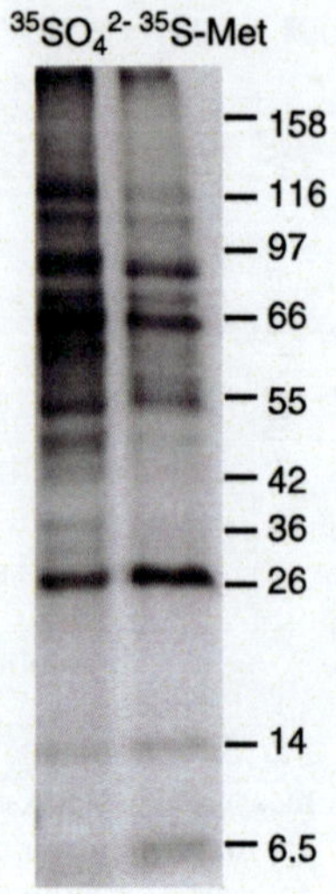

Fig. 4 Assimilation of ^{35}S-sulfate by developing chickpea embryos. Isolated mid-maturation chickpea embryos were incubated with [^{35}S] sulfate (track 1) or [^{35}S] methionine (track 2) for 4 h. Total proteins were extracted and fractionated on SDS-PAGE, and proteins containing ^{35}S were visualized by fluorography. Molecular marker sizes (kDa) are shown at *right*

Manipulating Storage of Sulfur Amino Acids by Transgenic Expression of a Sulfur-Rich Protein in Seeds

We have attempted to modify the quantities of stored sulfur amino acids in seeds using three complementary strategies. These were: (1) increasing the storage sink for sulfur amino acids in the seed, (2) increasing sulfur supply to plants, and (3) increasing the production of sulfur amino acids in the developing seed.

Initially, we created an increased storage sink for sulfur amino acids in seeds by expressing an additional sulfur-rich protein in developing embryos of grain legumes. A transgene encoding a naturally sulfur-amino-acid-rich sunflower seed albumin (SSA) was expressed under control of a seed-specific promoter in narrow leaf lupin and chickpea (Molvig et al. 1997; Chiaiese et al. 2004). In both cases, this strategy was successful in significantly increasing total sulfur amino acid concentrations of the mature seeds, compared to non-transgenic controls. In the case of lupins, transgenic events expressing SSA at levels estimated to represent around 4% of total seed protein had 100% more total seed methionine than non-transgenic controls. Similar results were achieved with transgenic chickpeas expressing SSA (Chiaiese et al. 2004). The transgenic SSA lupins were demonstrated to have increased nutritive value for chickens (Ravindran et al. 2002) and for ruminant (White et al. 2001), and non-ruminant animals (Molvig et al. 1997). However, further increases in total seed sulfur amino acids were needed in order to meet the full dietary requirements of animals for these essential nutrients.

Although the added SSA protein contained 8% cysteine residues as well as 16% methionine residues, total seed cysteine was not increased in either transgenic SSA lupins or chickpeas. In both cases, transgenic SSA seeds had reduced amounts of some endogenous cysteine-rich seed proteins, indicating that the new storage sink competed to some extent with endogenous proteins for an apparently limiting supply of sulfur amino acids. The concentrations of oxidized sulfur in the mature, transgenic SSA seeds were also less than in controls, suggesting that some inorganic seed sulfur was converted to additional sulfur amino acids in the SSA transgenic seeds.

In summary, comparison of wild type lupins and chickpeas with their transgenic counterparts expressing high levels of the introduced SSA protein suggested that the capacity of these developing grain legume seeds to supply sulfur amino acids for storage in protein could be increased by approximately 20–30%. This extra sulfur amino acid was stored exclusively in the form of methionine. Coupled with the findings detailed in the previous section, these results indicated that the transgenic, developing SSA seeds were able to synthesize more sulfur amino acids in response to the added demand, but that this capacity ultimately became limiting for sulfur-rich storage protein accumulation. In no case did expression of SSA increase the total amount of sulfur (reduced plus oxidized) that accumulated in the mature transgenic seeds.

Manipulating Sulfur Supply to Grain Legume Plants

High levels of sulfur nutrition were supplied to plants as a second strategy aimed at increasing storage of sulfur amino acids in seeds. Sulfur mineral nutrition of wild type and transgenic SSA lupins was varied in two different ways. Lupins were grown either in sand culture with controlled mineral nutrition, or in potting soil to which a supplement of solid calcium sulfate was added (Tabe and Droux 2002). In both cases, the sulfate supplement was added at the beginning of the reproductive growth phase. Supplementing lupin plants with sulfate increased the storage of total sulfur in mature seeds, but the extra stored sulfur was mostly in the oxidized form of sulfate. Lupins grown in the high sulfur nutrient (containing 3 mM $MgSO_4$) had higher concentrations of reduced sulfur in their mature seeds (14.6 μmol g^{-1} DW, or 25% more) than lupins grown with 0.3 mM sulfate. The increase in the oxidized sulfur fraction was 39.4 μmol g^{-1} DW, or 300% more in the high sulfur conditions. The plants grown in soil with a sulfate supplement produced mature seeds with concentrations of reduced sulfur (70.6 μmol g^{-1} DW) similar to those of the seeds grown in the high sulfur nutrient (73.4 μmol g^{-1} DW). Seeds grown in both these high sulfur conditions contained large concentrations of oxidized sulfur (48.8 μmol g^{-1} DW in soil-grown, 52.2 μmol g^{-1} DW in nutrient-grown). For comparison, mature seeds of lupins grown in normal field conditions contained 73.4 μmol g^{-1} DW reduced sulfur and 26.6 μmol g^{-1} DW oxidized sulfur. Supplementing transgenic lupins expressing SSA with high sulfate had similar results (Tabe and Droux 2002). Interestingly, similar effects were observed when lupin plants were supplemented with reduced sulfur in the form of 3 mM methionine during reproductive growth. Supplementation with methionine increased total seed sulfur, via a 53% increase in seed oxidized sulfur concentration, with only a 6% increase in seed reduced sulfur concentration (Popelka, unpublished). Further investigation will be required to determine whether the additional sulfur was transported to the developing lupin pods in reduced or oxidized form.

Manipulation of plant sulfur mineral nutrition also affected the sulfur content of chickpea seeds. Growth of the plants in nutrient with 2 mM sulfate produced mature seeds with 34% more total sulfur than chickpeas grown in nutrient with 0.2 mM sulfate (Chiaiese et al. 2004). This corresponded to an increase of 11% in reduced sulfur concentration, and an increase in oxidized sulfur concentration from undetectable levels in seeds grown in the lower sulfur conditions, to 22 μmol g^{-1} DW in those grown in high sulfate. The effect of sulfur mineral nutrition depended strongly on plant nitrogen nutrition. Mature seeds of plants grown with high sulfur and low nitrogen had very high concentrations of oxidized sulfur in their seeds (55 μmol g^{-1} DW). As was the case with lupins, transgenic chickpeas expressing the sulfur-rich SSA had more reduced sulfur but approximately the same total seed sulfur as non-transgenic chickpeas grown in the same nutritional conditions. Supplementing SSA transgenic chickpeas with sulfate had similar effects to those described above for the wild type chickpeas. Thus, total seed sulfur concentrations could be increased by growing the chickpea plants in nutrient with high sulfur, but this increase was almost all in the oxidized sulfur fraction (Chiaiese et al. 2004). In contrast to the situation with lupins, supplementing chickpea plants with 3 mM methionine during reproductive growth increased the reduced and oxidized sulfur fractions in the seeds by similar amounts (approximately 40–50%, Popelka, unpublished). This suggested that chickpeas had some capacity to deliver the extra methionine applied to the soil, from the roots to the developing seeds in the reduced form. We would expect that the additional reduced sulfur in the mature seeds was in the form of protein; however, further analysis will be required to determine this.

Manipulating Supply of Sulfur Amino Acids by Transgenic Expression of SAT in Seeds

Pools of free methionine, and in some cases, free cysteine and glutathione were lower in developing transgenic lupin seeds expressing SSA than in non-transgenic controls (Tabe and Droux 2002; Tabe et al. 2010). This apparent depletion of sulfur amino acids during seed development, by the added demand from SSA biosynthesis, was not ameliorated by increasing the supply of sulfate to plants (Tabe, unpublished). Since our previous studies indicated that developing lupin seeds were competent to synthesize sulfur amino acids, we developed a third strategy aimed at increasing the supply of sulfur amino acids available for storage protein synthesis during seed maturation.

SAT, a component of the cysteine synthase complex is reported to be a key, limiting enzyme of sulfur amino acid biosynthesis (Sirko et al. 2004). We targeted transgenic over-expression of SAT to developing lupin embryos by transfer of a gene encoding a cysteine feedback-insensitive, *Arabidopsis* SAT (AtSAT1 or AtSerat 2;1), expressed under the control of a strong, seed-specific promoter. Mid-maturation embryos of SAT transgenic lupins had greatly increased SAT activity, strongly increased concentrations of *O*-acetylserine, the immediate product of SAT, and some of the highest *in planta* concentrations of free cysteine yet reported (Tabe et al. 2010). Glutathione concentrations were also increased in the SAT transgenic mid-maturation embryos; however, their free methionine concentrations were unchanged relative to non-transgenic controls. Mature SAT transgenic seeds had approximately double the concentrations of glutathione, but no more protein-bound sulfur amino acids, than non-transgenic controls. Free cysteine was not increased in the mature seeds, indicating that the free cysteine that accumulated in developing embryos was metabolized by the end of seed maturation. Because the protein-bound fraction is by far the largest pool of sulfur amino acids in mature seeds, the net effect was no significant increase in total sulfur amino acids in the mature SAT transgenic seeds. The results confirmed the activity of the cysteine biosynthetic pathway in developing lupin embryos, and indicated that SAT activity, rather than sulfur reduction, was limiting for free cysteine accumulation in these organs. However, it was clear that while biosynthesis of cysteine limited glutathione accumulation, it was not limiting for accumulation of free methionine in the developing seeds, or for accumulation of sulfur-rich seed storage proteins in the mature seeds.

The two strategies of expressing SSA and SAT in developing seeds were combined by crossing transgenic lupins with high expression of each individual transgene. Analysis of segregating F_2 progeny seed, from F_1 plants confirmed by PCR to contain both transgenes, showed that over-expression of SAT was unable to prevent the depletion of free methionine seen in developing lupin embryos expressing SSA. Similarly, pooled mature F_2 seeds contained no more total sulfur amino acids than seed from a control cross between the SSA transgenic parent and a plant lacking SAT over-expression.

In summary, more sulfur amino acids could be "pulled" into seed protein by transgenic expression of SSA, but "pushing" through the pathway of sulfur amino acid biosynthesis by SAT over-expression in developing seeds could not add to the gain achieved by SSA alone. Our analysis of SAT- and SSA-expressing transgenic lupins showed that SAT over-expression in developing lupin embryos could indeed greatly increase the availability of free cysteine. However this extra cysteine was not incorporated into seed protein. Judging from the lowered concentrations of free methionine in developing lupin seeds expressing SSA, with or without SAT over-expression, free methionine was limiting for accumulation of the methionine- and cysteine-rich SSA in both cases. Free methionine is reported to be a regulator of the expression of sulfur-rich seed storage proteins in soybean (Holowach et al. 1986; Hirai et al. 2002). Our finding that levels of mRNA for the endogenous, cysteine-rich conglutin-delta were decreased in SSA transgenic seeds (Tabe and Droux 2002) indicates that free methionine may also regulate expression of lupin seed storage proteins that contain cysteine but no methionine. Future work aimed at increasing sulfur amino acid content of grain legume seeds will focus on modification of limiting enzymes in the pathway of methionine biosynthesis.

References

Bourgis F, Roje S, Nuccio ML, Fisher DB, Tarczynski MC, Li C, Herschbach C, Rennenberg H, Pimenta MJ, Shen TL, Gage DA, Hanson AD (1999) *S*-methylmethionine plays a major role in phloem sulfur transport and is synthesized by a novel type of methyltransferase. Plant Cell 11:1485–1498

Buchner P, Takahashi H, Hawkesford MJ (2004) Plant sulphate transporters: co-ordination of uptake, intracellular and long-distance transport. J Exp Bot 55:1765–1773

Chiaiese P, Ohkama-Ohtsu N, Molvig L, Godfree R, Dove H, Hocart C, Fujiwara T, Higgins TJV, Tabe LM (2004) Sulphur and nitrogen nutrition influence the response of chickpea seeds to an added, transgenic sink for organic sulphur. J Exp Bot 55:1889–1901

Chronis D, Krishnan HB (2003) Sulfur assimilation in soybean: molecular cloning and characterization of *O*-acetylserine (thiol) lyase (cysteine synthase). Crop Sci 43:1819–1827

Haas FH, Heeg C, Queiroz R, Bauer A, Wirtz M, Hell R (2008) Mitochondrial serine acetyltransferase functions as a pacemaker of cysteine synthesis in plant cells. Plant Physiol 148:1055–1067

Hartmann T, Mult S, Suter M, Rennenberg H, Herschbach C (2000) Leaf age-dependent differences in sulphur assimilation and allocation in poplar (*Populus tremula* x *P. alba*) leaves. J Exp Bot 51:1077–1088

Hawkesford MJ, De Kok LJ (2006) Managing sulphur metabolism in plants. Plant Cell Environ 29:382–395

Heeg C, Kruse C, Jost R, Gutensohn M, Ruppert T, Wirtz M, Hell R (2008) Analysis of the *Arabidopsis* *O*-acetylserine(thiol)lyase gene family demonstrates compartment-specific differences in the regulation of cysteine synthesis. Plant Cell 20:168–185

Herschbach C, Rennenberg H (2001) Significance of phloem-translocated organic sulfur compounds for the regulation of sulfur nutrition. Prog Bot 62:177–192

Hirai MY, Kim H, Hayashi H, Chino M, Naito S, Fujiwara T (2002) Independent roles of methionine and *O*-acetly-L-serine in the regulation of the β-subunit gene of β-conglycinin. Soil Sci Plant Nutr 48:87–94

Holowach LP, Madison JT, Thompson JF (1986) Studies on the mechanism of regulation of the mRNA level for a soybean storage protein subunit by exogenous L-methionine. Plant Physiol 80:561–567

Krueger S, Niehl A, Martin MC, Steinhauser D, Donath A, Hildebrandt T, Romero LC, Hoefgen R, Gotor C, Hesse H (2009) Analysis of cytosolic and plastidic serine acetyltransferase mutants and subcellular metabolite distributions suggests interplay of the cellular compartments for cysteine biosynthesis in *Arabidopsis*. Plant Cell Environ 32:349–367

Kuzuhara Y, Isobe A, Awazuhara M, Fujiwara T, Hayashi H (2000) Glutathione levels in phloem sap of rice plants under sulfur deficient conditions. Soil Sci Plant Nutr 46:265–270

Molvig L, Tabe LM, Eggum BO, Moore AE, Craig S, Spencer D, Higgins TJV (1997) Enhanced methionine levels and increased nutritive value of seeds of transgenic lupins (*Lupinus angustifolius* L.) expressing a sunflower seed albumin gene. Proc Natl Acad Sci USA 94:8393–8398

Phartiyal P, Kim WS, Cahoon RE, Jez JM, Krishnan HB (2008) The role of 5′-adenylylsulfate reductase in the sulfur assimilation pathway of soybean: molecular cloning, kinetic characterization, and gene expression. Phytochemistry 69:356–364

Pinkerton A, Randall PJ, Norrish K (1989) Estimation of sulfate and amino acid sulfur in plant material by X-ray spectrometry. Commun Soil Sci Plant Anal 20:1557–1574

Ravanel S, Gakiere B, Job D, Douce R (1998) The specific features of methionine biosynthesis and metabolism in plants. Proc Natl Acad Sci USA 95:7805–7812

Ravanel S, Block MA, Rippert P, Jabrin S, Curien G, Rebeille F, Douce R (2004) Methionine metabolism in plants – chloroplasts are autonomous for de novo methionine synthesis and can import *S*-adenosylmethionine from the cytosol. J Biol Chem 279:22548–22557

Ravindran V, Tabe LM, Molvig L, Higgins TJV, Bryden WL (2002) Nutritional evaluation of transgenic high-methionine lupins (*Lupinus angustifolius* L) with broiler chickens. J Sci Food Agric 82:280–285

Saito K (2004) Sulfur assimilatory metabolism. The long and smelling road. Plant Physiol 136:2443–2450

Sexton PJ, Shibles RM (1999) Activity of ATP sulfurylase in reproductive soybean. Crop Sci 39:131–135

Shewry PR (2007) Improving the protein content and composition of cereal grain. J Cereal Sci 46:239–250

Sirko A, Blaszczyk A, Liszewska F (2004) Overproduction of SAT and/or OASTL in transgenic plants: a survey of effects. J Exp Bot 55:1881–1888

Tabe LM, Droux M (2001) Sulfur assimilation in developing lupin cotyledons could contribute significantly to the accumulation of organic sulphur reserves in the seed. Plant Physiol 126:176–187

Tabe LM, Droux M (2002) Limits to sulfur accumulation in transgenic lupin seeds expressing a foreign sulphur-rich protein. Plant Physiol 128:1137–1148

Tabe L, Higgins TJV (1998) Engineering plant protein composition for improved nutrition. Trends Plant Sci 3:282–286

Tabe L, Venables I, Grootemaat A, Lewis D (2003) Sulfur transport and assimilation in developing embryos of chickpea (*Cicer arietinum*). In: Davidian J-C, Grill D, De Kok LJ, Stulen I, Hawkesford MJ, Schnug E, Rennenberg H (eds) Sulfur transport and assimilation in plants: regulation, interaction, signaling. Backhuys Publishers, Leiden, pp 335–337

Tabe L, Wirtz M, Molvig L, Droux M, Hell R (2010) Over-expression of serine acetlytransferase produced large increases in *O*-acetylserine and free cysteine in developing seeds of a grain legume. J Exp Bot 61:721–733

White CL, Tabe LM, Dove H, Hamblin J, Young P, Phillips N, Taylor R, Gulati S, Ashes J, Higgins TJV (2001) Increased efficiency of wool growth and live weight gain in Merino sheep fed transgenic lupin seed containing sunflower albumin. J Sci Food Agric 81:147–154

Zhang W-H, Zhou Y, Dibley KE, Tyerman SD, Furbank RT, Patrick JW (2007) Nutrient loading of developing seeds. Funct Plant Biol 34:314–331

Regulation of Expression of Sulfate Transporters and APS Reductase in Leaf Tissue of Chinese Cabbage *(Brassica pekinensis)*

Luit J. De Kok, C. Elisabeth E. Stuiver, Muhammad Shahbaz, and Aleksandra Koralewska

Abstract Leaf discs from sulfur-sufficient and sulfate-deprived Chinese cabbage plants were incubated at various levels of sulfate (ranging from 0 to 40 mM) for 24 h in the light. Sultr4;1 was the sole constitutively expressed sulfate transporter present in sulfur-sufficient leaf discs and its expression was enhanced upon incubation at ≤8 mM and decreased at higher sulfate concentrations. Leaf discs from sulfate-deprived plants were characterized by a high expression of Sultr1;2, Sultr4;1, Sultr4;2 and APS reductase. The high expression was down-regulated upon incubation of the leaf discs at >2 mM sulfate, whereas at 20–30 mM their expression was quite similar to that of sulfur-sufficient leaf discs. In both sulfur-sufficient and sulfur-deprived leaf discs there was an accumulation of sulfate upon incubation nearly linear with the sulfate concentration. The thiol content of leaf discs was slightly enhanced upon incubation at high sulfate concentrations. Evidently, the expression of the sulfate transporters and APS reductase were strictly controlled by the *in situ* sulfate concentration in the leaf tissue possibly via an interaction with the sulfate reduction in the chloroplasts. It is suggested that H_2S might function as endogenous gaseous transmitter in cross-talk between the sulfate reduction pathway in the chloroplasts and the transcription of sulfate transporters/sulfate reducing enzymes in the nucleus.

Distinct sulfate transporters are involved in the uptake and distribution of sulfate in plants, which expression and activity are regulated by the sulfur status of the plant (Hawkesford and De Kok 2006). For instance, the Group 1 transporters are high

L.J. De Kok (✉) • C.E.E. Stuiver • A. Koralewska
Laboratory of Plant Physiology, University of Groningen, P.O. Box 11103, 9700 CC Groningen, The Netherlands
e-mail: l.j.de.kok@rug.nl

M. Shahbaz
Institute of Agricultural Sciences, University of the Punjab, Lahore 54590, Pakistan

L.J. De Kok et al. (eds.), *Sulfur Metabolism in Plants*, Proceedings of the International Plant Sulfur Workshop 1, DOI 10.1007/978-94-007-4450-9_5,

affinity sulfate transporters responsible for primary sulfate uptake by the roots, but in *Brassica* they are also expressed in the leaf tissue upon prolonged sulfate deprivation (Buchner et al. 2004; Hawkesford and De Kok 2006; Koralewska et al. 2007, 2008, 2009; Stuiver et al. 2009). The Group 4 sulfate transporters are localized in the tonoplast and function in the vacuolar unloading of sulfate and their expression is highly up-regulated in both root and shoot upon sulfate deprivation (Buchner et al. 2004; Hawkesford and De Kok 2006; Koralewska et al. 2007, 2008, 2009; Stuiver et al. 2009). It has been assumed that the regulation of expression (transcriptional level) and/or activity (translational and/or post-translational level) of the sulfate transporters might be signaled or mediated by sulfate itself or by products of the assimilatory sulfate reduction pathway (*viz.* glutathione; Hawkesford and De Kok 2006). The expression and activity of APS reductase, the key regulating enzyme in the sulfate reduction pathway, is controlled by the plant sulfur status, with reduced sulfur compounds, *viz.* cysteine, glutathione as possible regulatory compounds (Westerman et al. 2001; Durenkamp et al. 2007; North and Kopriva 2007). However, in *Brassica* there was no clear relation between the overall tissue levels of sulfate and thiols and the activity of the sulfate transporters in the root (Buchner et al. 2004; Koralewska et al. 2007, 2008, 2009; Shahbaz et al. 2010). In the current study the regulation of expression of the Group 1 and 4 transporters and APS reductase was studied in leaf tissue of Chinese cabbage.

Chinese cabbage (*Brassica pekinensis* (Lour.) Rupr. cv. Kasumi F1 (Nickerson Zwaan, Made, The Netherlands)) was germinated in vermiculite and 10 day-old seedlings were grown on a 25% Hoagland nutrient solution (pH 5.9) containing 0.5 mM sulfate in 30 l containers (20 sets of plants per container, 2 plants per set) in a climate-controlled room for 11 days. Day and night temperatures were 21°C and 17°C, respectively, relative humidity was 60–70% and the photoperiod was 14 h at a photon fluence rate of 300 ± 30 μmol m^{-2} s^{-1} (within the 400–700 nm range) at plant height, supplied by PhilipsMaster 58 W/83 and 84 fluorescent tubes in a ratio of 1:1. Then plants were transferred to fresh 25% Hoagland solution at 0 (sulfur-deprived; all sulfate salts were replaced by chloride salts) or 0.5 mM sulfate (sulfur-sufficient) and grown for 4 days. Leaf discs (15 mm diameter) were punched out of the third and fourth leaf (between the largest veins) and incubated upside down with various levels of Na_2SO_4 (ranging from 0 to 40 mM; in tap water) in closed petri dishes (15 discs per 25 ml) in a climate-controlled room at 20°C and continuous light (light intensity 300 ± 20 μmol m^{-2} s^{-1}) for 24 h. For anion determination leaf discs were rinsed in ice-cold demi water (for 3 × 20 s) on a sieve, blotted dry between filter paper, weighed and frozen in liquid N_2. Frozen leaf discs were homogenized in demi water (5 leaf discs in 5 ml), the homogenate was filtered through one layer of Miracloth, incubated at 100°C in a boiling water bath for 10 min and the filtrate was centrifuged at 30,000 g for 15 min (0°C). Anions were separated by HPLC as described by Shahbaz et al. (2010). For analysis of the water-soluble non-protein thiols, fresh leaf discs (5 leaf discs in 5 ml) were homogenized in an extraction medium containing 80 mM sulfosalicylic acid, 1 mM EDTA and 0.15% (w/v) ascorbic acid, with an Ultra Turrax at 0°C. The homogenate was filtered through one layer of Miracloth and the filtrate was centrifuged at 30,000 g for

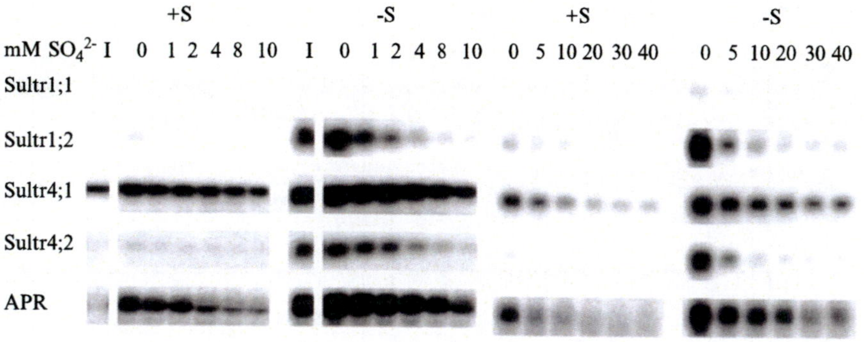

Fig. 1 Impact of Na_2SO_4 on the mRNA abundance of sulfate transporters (Sultr) and APS reductase (APR; Northern-blot analysis) in leaf discs from sulfur-sufficient (+S) and sulfate-deprived Chinese cabbage (–S). Leaf discs were incubated at various Na_2SO_4 concentrations for 24 h in light (I, the initial (t=0) mRNA abundance)

15 min (0°C). Total water-soluble non-protein thiol content was determined as described by Shahbaz et al. (2010). Total RNA isolation from leaf discs and Northern analysis of the sulfate transporters and APS reductase transcripts were performed as described by Shahbaz et al. (2010).

Sultr4;1 was the sole constitutively expressed sulfate transporter in leaf discs from sulfur-sufficient Chinese cabbage plants ("sulfur-sufficient leaf discs"), whereas APS reductase was hardly expressed (Fig. 1). A 24 h incubation of sulfur-sufficient leaf discs at ≤8 mM sulfate resulted in strongly enhanced expression of Sultr4;1 and APS reductase (Fig. 1). Moreover, there was a slight decrease in the sulfate content of the leaf discs as compared with the initial sulfate content (t=0; Fig. 2). Evidently, the regulation of expression of the sulfate transporters and that of APS reductase in leaf tissue was strictly triggered/controlled by the endogenous (likely cytoplasmic) sulfate concentration. Apparently, if the overall sulfate concentration in the leaf discs decreased to values lower than 8 mM, which was the initial sulfate concentration in the leaf tissue, it resulted in an initiation of the up-regulation of expression of the sulfate transporters and APS reductase. However, incubation of leaf discs at ≥10 mM sulfate resulted in a decreased expression of Sultr4;1, accompanied with an increase of the sulfate content, nearly linear with concentration (Fig. 2). At 40 mM sulfate there was a threefold increase in the sulfate content of the leaf discs. Incubation of the leaf discs at high sulfate concentrations also resulted in a slightly enhanced water-soluble non-protein thiol content (up to 1.5-fold; Fig. 2).

Leaf discs from sulfate-deprived Chinese cabbage ("sulfur-deprived leaf discs") were characterized by a strongly enhanced mRNA abundance of Sultr4;1 and APS reductase, a highly induced expression of both Sultr1;2 and Sultr4;2 (Fig. 1) and a low sulfate and thiol content (Fig. 2). The expression of Sultr1;2, Sultr4;1, Sultr4;2 and APS reductase were further up-regulated upon a 24 h incubation at 0 mM sulfate. The expression of the sulfate transporters and that of APS reductase were down-regulated upon incubation at >2 mM sulfate, whereas at 20–30 mM their expression

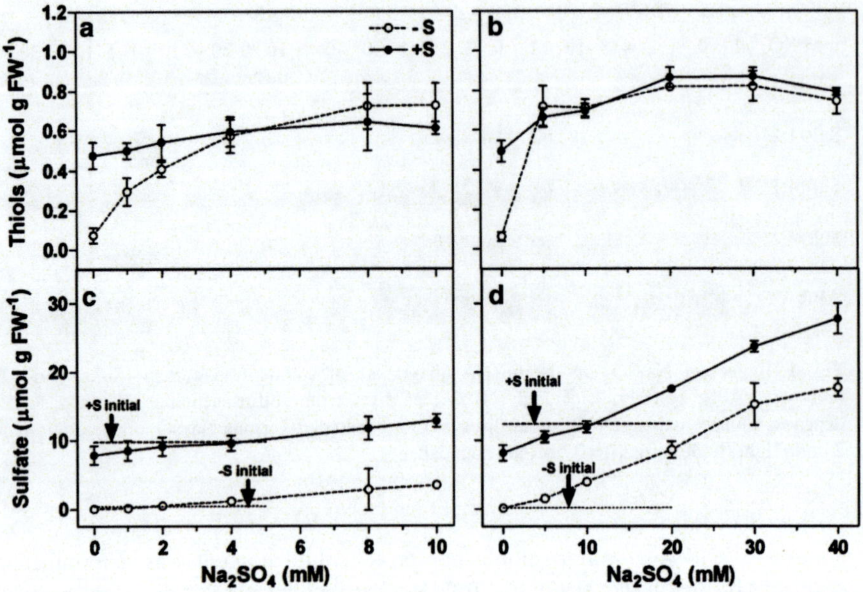

Fig. 2 Impact of Na_2SO_4 on the content of water-soluble non-protein thiols and sulfate in leaf discs from sulfur-sufficient (+S) and sulfate-deprived Chinese cabbage (–S). Leaf discs were incubated at various Na_2SO_4 concentrations for 24 h in light. Data on thiol content represent the mean of 2 (**a**) and 1 (**b**) experiments with 3 measurements on 15 leaf discs in each. Data on sulfate content (**c**, **d**) represent the mean of 3 experiments with 3 measurements on 15 leaf discs in each (± SD). Fresh weight of the leaf discs was 72±3 and 70±3 mg at t=0, and 89±2 and 81±2 at t=24 for sulfate-deprived and sulfate-sufficient leaf discs, respectively. The initial (t=0) sulfate levels of the leaf discs are shown by arrows

was quite similar to that of sulfur-sufficient leaf discs (Fig. 2). There was no direct relation between changes in expression of the sulfate transporters and APS reductase, and changes in thiol content of the leaf tissue upon sulfate incubation. The thiol content of the sulfur-deprived leaf discs was increased upon sulfate incubation and already at ≥5 mM its content was quite similar to that of sulfur-sufficient leaf discs (Fig. 2). There was an increase in sulfate content of the leaf discs upon sulfate incubation nearly linear with concentration. Despite the differences between sulfur-sufficient and sulfur-deprived leaf discs in the initial sulfate content, there was a rather similar accumulation of sulfate in the leaf discs upon incubation at various sulfate concentrations (Fig. 2). Noticeably, the observed down-regulation of expression of the sulfate transporters and APS reductase in sulfur-deprived leaf discs upon incubation at 20–30 mM sulfate coincidenced with an overall leaf disc sulfate concentration of approximately 8 mM, which appears to be quite characteristic for Chinese cabbage leaf tissue at this plant age (see chapter "Regulation of the Uptake of Sulfate by Chinese Cabbage (*Brassica pekinensis*) at Various Sulfate Concentrations in the Root Environment", this volume).

From the present study it was evident that in leaf tissue of Chinese cabbage the overall *in situ* leaf tissue sulfate rather than the thiol concentration was of great significance in the regulation of expression of the sulfate transporters and APS reductase. The expression of Sultr1;2, Sultr4;1, and APS reductase might be controlled by the *in situ* cytosolic sulfate concentration in direct interaction with the sulfate reduction in the chloroplasts. The affinity of ATP sulfurylase for sulfate appears to be rather low (K_m for sulfate ranges from 0.5 to 3 mM) and the *in situ* sulfate concentration in the chloroplast at the ATP sulfurylase site may play a significant role in the rate of sulfate reduction (Stulen and De Kok 1993).

From human and animal physiology it has become evident that in addition to nitric oxide (NO) and carbon monoxide (CO), H_2S might function as an endogenous gaseous transmitter where it may target K_{ATP} channel proteins and the cAMP-dependent protein kinase pathway (Wang 2002; Mancardi et al. 2009). In prokaryotes it has been shown that sulfide is involved in transcriptional regulation of the cys-operon, for genes involved in sulfur uptake and assimilation (Kredich 1993). Similarly, H_2S might function in plants as the first product of the sulfate reduction pathway, as an endogenous gaseous transmitter in the cross-talk between the sulfate reduction pathway in the chloroplasts/plastids and the transcription of sulfate transporters/sulfate reducing enzymes in the nucleus. Evidently, plants grown under normal conditions emit minute levels of H_2S, which emission has been presumed to be a regulatory step in the homeostasis of the sulfur pools in plants (Rennenberg 1984; Schröder 1993; Bloem et al. 2007). At the cellular pH, H_2S is largely undissociated and in this form it may easily pass through the membranes (De Kok et al. 2007). At a whole plant level, is has been shown that H_2S exposure may diminish the activity of sulfate reduction in the shoot by a down-regulation of the expression and activity of APS reductase, the key-regulating enzyme in the sulfate reduction pathway, and it may result in a down-regulation of the expression of sulfate transporters in the shoot (Westerman et al. 2000, 2001; Buchner et al. 2004; Durenkamp et al. 2007; Koralewska et al. 2008).

References

Bloem E, Haneklaus S, Salac I, Wickenhäuser P, Schnug S (2007) Facts and fiction about sulfur metabolism in relation to plant-pathogen interactions. Plant Biol 9:596–607

Buchner P, Stuiver CEE, Westerman S, Wirtz M, Hell R, Hawkesford MJ, De Kok LJ (2004) Regulation of sulfate uptake and expression of sulfate transporter genes in *Brassica oleracea* as affected by atmospheric H_2S and pedospheric sulfate nutrition. Plant Physiol 136:3396–3408

De Kok LJ, Durenkamp M, Yang L, Stulen I (2007) Atmospheric sulfur. In: Hawkesford MJ, De Kok LJ (eds) Sulfur in plants – an ecological perspective. Springer, Dordrecht, pp 91–106

Durenkamp M, De Kok LJ, Kopriva S (2007) Adenosine 5′ phosphosulphate reductase is regulated differently in *Allium cepa* and *Brassica oleracea* upon exposure to H_2S. J Exp Bot 58:1571–1579

Hawkesford MJ, De Kok LJ (2006) Managing sulphur metabolism in plants. Plant Cell Environ 29:382–395

Koralewska A, Posthumus FS, Stuiver CEE, Buchner P, Hawkesford MJ, De Kok LJ (2007) The characteristic high sulfate content in *Brassica oleracea* is controlled by the expression and activity of sulfate transporters. Plant Biol 9:654–661

Koralewska A, Stuiver CEE, Posthumus FS, Kopriva S, Hawkesford MJ, De Kok LJ (2008) Regulation of sulfate uptake, expression of the sulfate transporters Sultr1;1 and Sultr1;2, and APS reductase in Chinese cabbage (*Brassica pekinensis*) as affected by atmospheric H_2S nutrition and sulfate deprivation. Funct Plant Biol 35:318–327

Koralewska A, Buchner P, Stuiver CEE, Posthumus FS, Kopriva S, Hawkesford MJ, De Kok LJ (2009) Expression and activity of sulfate transporters and APS reductase in curly kale in response to sulfate deprivation and re-supply. J Plant Physiol 166:168–179

Kredich NM (1993) Gene regulation of sulfur assimilation. In: De Kok LJ, Stulen I, Rennenberg H, Rauser WE (eds) Sulfur nutrition and assimilation in higher plants: regulatory, agricultural and environmental aspects. SPB Academic Publishing, The Hague, pp 37–47

Mancardi D, Penna C, Merlino A, De Soldato P (2009) Physiological and pharmocological features of the novel gasotransmitter: hydrogen sulfide. Biochim Biophys Acta 1787:864–872

North KA, Kopriva S (2007) Sulfur in resistance to environmental stresses. In: Hawkesford MJ, De Kok LJ (eds) Sulfur in plants – an ecological perspective. Springer, Dordrecht, pp 143–168

Rennenberg H (1984) The fate of excess sulfur in higher plants. Annu Rev Plant Physiol 35:121–153

Schröder P (1993) Plants as source of atmospheric sulfur. In: De Kok LJ, Stulen I, Rennenberg H, Rauser WE (eds) Sulfur nutrition and assimilation in higher plants: regulatory, agricultural and environmental aspects. SPB Academic Publishing, The Hague, pp 253–270

Shahbaz M, Tseng MH, Stuiver CEE, Koralewska A, Posthumus FS, Venema JH, Parmar S, Schat H, Hawkesford MJ, De Kok LJ (2010) Copper exposure interferes with the regulation of the uptake, distribution and metabolism of sulfate in Chinese cabbage. J Plant Physiol 167:438–446

Stuiver CEE, Koralewska A, Posthumus FS, De Kok LJ (2009) The impact of sulfur deprivation on root formation, and activity and expression of sulfate transporters in Chinese cabbage. In: Sirko A, De Kok LJ, Haneklaus S, Hawkesford MJ, Rennenberg H, Saito K, Schnug E, Stulen I (eds) Sulfur metabolism in plants: regulatory aspects, significance of sulfur in the food chain, agriculture and the environment. Margraf Publishers, Weikersheim, pp 61–68

Stulen I, De Kok LJ (1993) Whole plant regulation of sulfate uptake and metabolism – a theoretical approach and comparison with current ideas on regulation of nitrogen metabolism. In: De Kok LJ, Stulen I, Rennenberg H, Rauser WE (eds) Sulfur nutrition and assimilation in higher plants: regulatory, agricultural and environmental aspects. SPB Academic Publishing, The Hague, pp 77–91

Wang R (2002) Two's company, three's a crowd: can H_2S be the third endogenous gaseous transmitter. FASEB J 16:1792–1798

Westerman S, De Kok LJ, Stulen I (2000) Interaction between metabolism of atmospheric H_2S in the shoot and sulfate uptake by the roots of curly kale (*Brassica oleracea*). Physiol Plant 109:443–449

Westerman S, Stulen I, Suter M, Brunold C, De Kok LJ (2001) Atmospheric H_2S as sulfur source for *Brassica oleracea*: consequences for the activity of the enzymes of the assimilatory sulfate reduction pathway. Plant Physiol Biochem 39:425–432

Response of Amino Acid Metabolism to Sulfur Starvation in *Synechocystis* sp. PCC 6803

Hiroshi Kiyota, Masahiko Ikeuchi, and Masami Yokota Hirai

Abstract Cyanobacteria are representative photoautotrophic organisms that are capable of regulating metabolic pathways under varying light and nutritional conditions. We investigated this metabolic regulation under sulfur starvation with a focus on amino acid metabolism. Gas chromatography–mass spectrometry (GC-MS) was used to quantify amino acid contents. The results revealed sulfur deprivation-induced temporal changes in the amount of free amino acids in *Synechocystis* sp. PCC 6803. This is the first study to report altered metabolite levels in response to sulfur starvation in cyanobacteria.

Sulfur is a macronutrient that is incorporated into proteins and other vital compounds such as vitamins and co-factors. In the unicellular cyanobacterium *Synechocystis* sp. PCC 6803, sulfur metabolism is associated with photosynthesis, respiration, and the redox-sensing systems. Despite the importance of sulfur

H. Kiyota
Depertment of Science, Tokyo University,
7-3-1 Hongo, Bunkyo-ku, Tokyo 113-0033, Japan

Metabolic Systems Research Team, RIKEN Plant Science Center,
1-7-22 Suehiro-cho, Tsurumi-ku,
Yokohama, Kanagawa 230-0045, Japan

M. Ikeuchi
Depertment of Science, Tokyo University, 7-3-1 Hongo,
Bunkyo-ku, Tokyo 113-0033, Japan

M.Y. Hirai (✉)
Metabolic Systems Research Team, RIKEN Plant Science Center,
1-7-22 Suehiro-cho, Tsurumi-ku,
Yokohama, Kanagawa 230-0045, Japan

Japan Science and Technology Agency, CREST, 4-1-8 Hon-chou,
Kawaguchi, Saitama 332-0012, Japan
e-mail: myhirai@psc.riken.jp

L.J. De Kok et al. (eds.), *Sulfur Metabolism in Plants*, Proceedings of the International Plant Sulfur Workshop 1, DOI 10.1007/978-94-007-4450-9_6,

metabolism, its regulation in *Synechocystis* has not yet been studied. It has been reported that the expression levels of various genes change under sulfur starvation conditions in *Synechocystis* (Zhang et al. 2008). However, it is currently unknown whether sulfur starvation also induces changes in metabolite levels. The majority of the absorbed sulfate is incorporated into sulfur-containing amino acids such as cysteine and methionine. Intracellular amino acid levels are controlled by multiple mechanisms, including allosteric effects of amino acids on enzyme activity and/or metabolite-mediated repression of transcription. In this study, we determined the intracellular amounts of amino acids to estimate the extent to which sulfur deprivation affects amino acid metabolism.

Synechocystis was grown under normal growth conditions at 30°C in BG11 medium (0.3 mM $MgSO_4$, supplemented with 20 mM HEPES-KOH, pH 7.8 (Rippka 1988) with bubbling CO_2 (1% v/v)) under continuous illumination by white fluorescent lamps (30–40 μmol m^{-2} s^{-1}). For medium change, cells at an OD_{730} of 1.0–1.5 grown in BG11 medium were harvested by filtration and resuspended in fresh BG11 medium or in sulfur-free BG11-S medium, in which $MgSO_4$ is replaced by 0.3 mM $MgCl_2$. Cells were grown under the same growth conditions following the medium change until collection of cells at various time points (1, 3, 6, 12, 24, 48, and 72 h) by filtration and freezing in liquid nitrogen. The 0 h samples were collected before the medium change. The frozen pellet was preserved at −80°C until extraction. One milliliter of 60% methanol/H_2O (v/v) and 1 g of zirconia beads were added to the frozen pellet at 4°C, followed by vortexing twice for 30 s with a 1 min interval. This mixture was centrifuged at 10,000 rpm for 5 min, and the extract was collected as the supernatant. The beads were washed with 500 μl of 60% (v/v) methanol/H_2O, and the washed extract was collected by centrifugation at 15,000 rpm for 5 min. The supernatant was added to the first extract, and the combined extracts were dried and stored until further use. Dried extracts were dissolved in 60% (v/v) methanol/H_2O containing 2 nmol of norvaline as an internal standard and derivatised using the EZ:faast kit (Phenomenex, California). Gas chromatography–mass spectrometry (GC-MS) was performed using the QP2010 Plus GC-MS (Shimadzu, Kyoto) equipped with a ZB-AAA column (Phenomenex, California). The analytical conditions were as follows: carrier gas, He (1.17 ml/min); ionization voltage: 70 kV; injector temperature: 280°C; and oven program, 110°C (1 min hold), increased at 20°C min^{-1} to 300°C (2 min hold). Analyses were carried out in the Selected Ion Monitoring mode. Each peak was identified as the retention time in the selected ion chromatogram and quantified. Quantified data were normalized to the fresh weight of the frozen pellet.

Figure 1 shows the fluctuation of 17 proteinogenic amino acids as well as the non-proteinogenic amino acid ornithine and the tripeptide glutathione (Fig. 1a and b) and cluster analysis of these temporal changes (Fig. 1c) under sulfur starvation conditions. The cysteine content was below the detection limit in this study. The amino acid pool in *Synechocystis* was dramatically altered by the change to sulfur-free medium. Surprisingly, the levels of 10 of the 19 amino acids analysed (alanine, glycine, valine, leucine, isoleucine, aspartate, threonine, serine, lysine, and tyrosine) increased during the 72 h following the initiation of sulfur deprivation (Fig. 1a). The levels of these amino acids, with the exception of isoleucine, lysine, and tyrosine, were at a maximum at 48 h after the medium change, whereas the isoleucine level

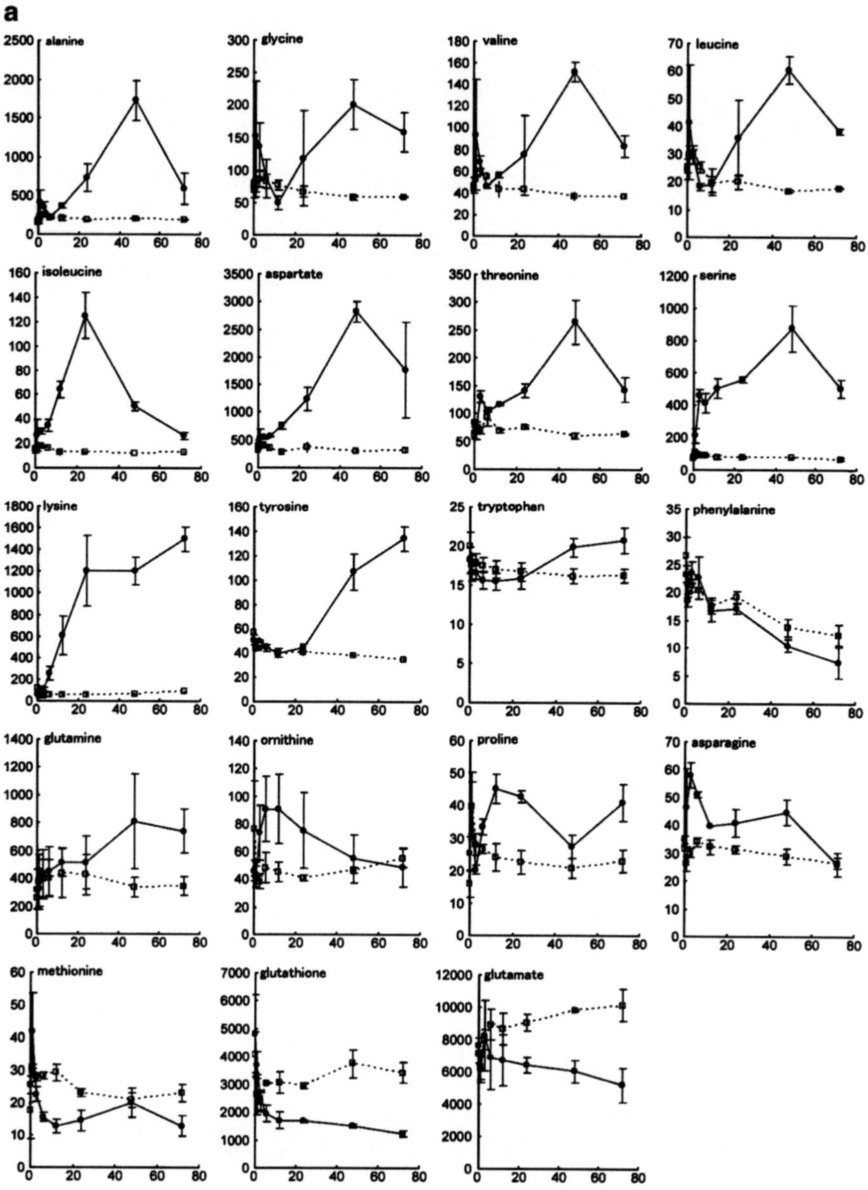

Fig. 1 (**a** and **b**)Temporal changes in amino acid contents during S deprivation. *Open square*: changed to BG11 medium; *filled circle*: changed to BG11-S medium. *Vertical axis*: amount of amino acid (pmol mg^{-1} FW); *horizontal axis*: time after medium change (h). The average and standard error (n=3) are shown. (**a**) 0–72 h, (**b**) 0–12 h. (**c**) Cluster analysis of temporal changes in amino acid contents during S deprivation. Hierarchical clustering was performed by complete linkage algorithm

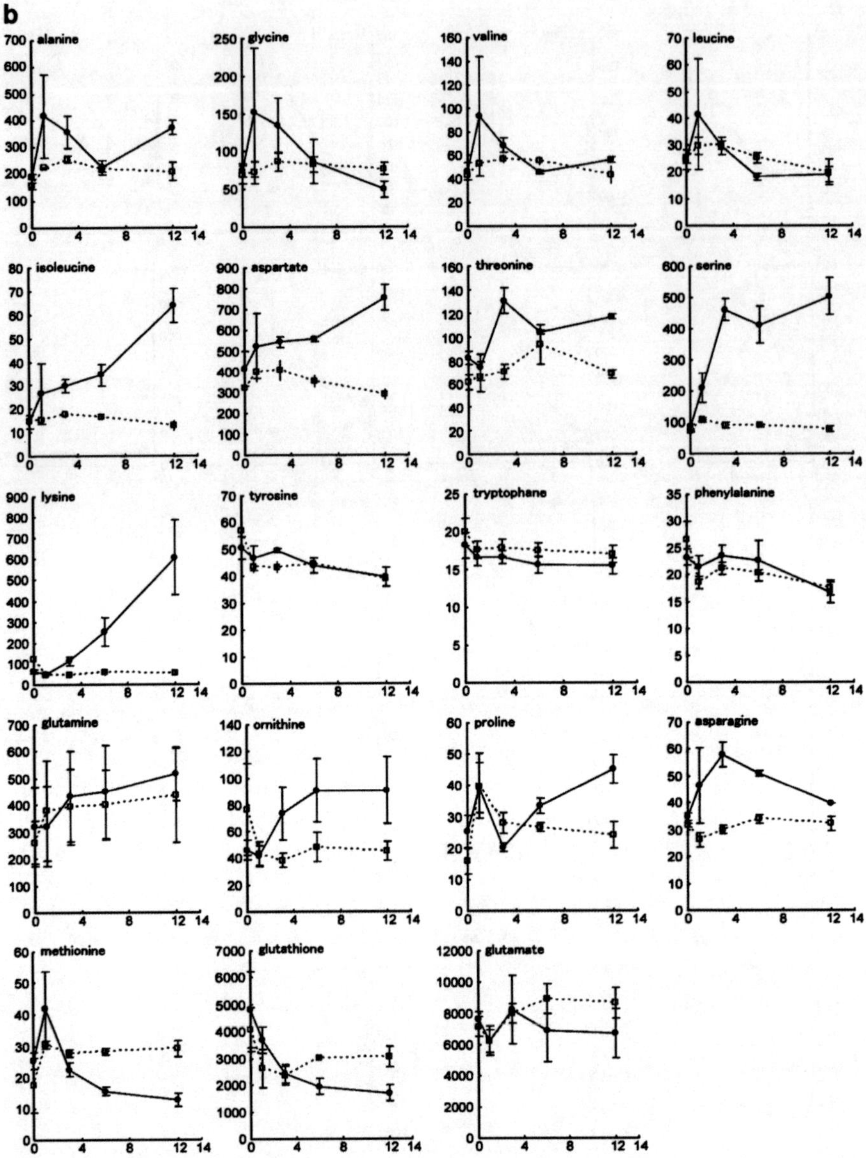

Fig. 1 (continued)

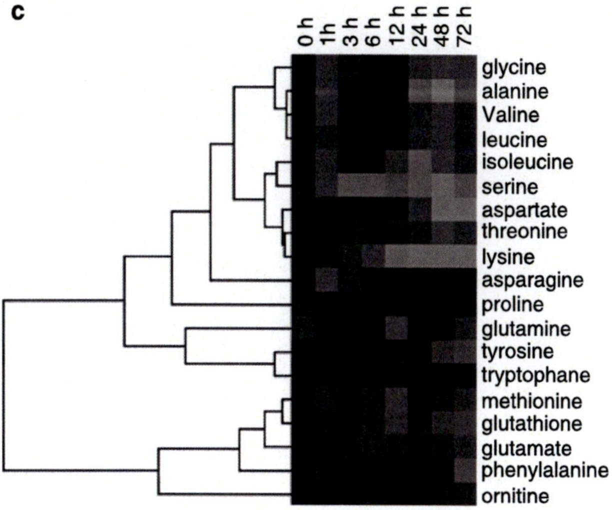

Fig. 1 (continued)

was at a maximum after 24 h. Lysine and tyrosine levels increased almost continuously during the 72 h following the initiation of sulfur deprivation, although the early responses differed; the tyrosine level remained unaltered during the first 24 h, whereas lysine increased linearly during this period. It is notable that serine levels showed an immediate increase within 3 h (Fig. 1b). The overall increase in the levels of these 10 amino acids presumably resulted from upregulation of *de novo* biosynthesis and/or protein degradation as well as downregulation of consumption for synthesis of other biomolecules including proteins. Especially serine and aspartate seemed to be accumulated for decreased use for cysteine, methionine and GSH. However, the genes encoding the biosynthetic enzymes of these amino acids were not significantly upregulated in a similar experiment (Zhang et al. 2008; see discussion in the next paragraph). On the other hand, the amounts of the sulfur-containing amino acid and tripeptide, methionine and glutathione, decreased. Interestingly, glutamate level also declined, although this amino acid contains no sulfur. Because the biosynthesis of glutamate requires nitrate and reducing power, the availability of reducing power may regulate nitrogen assimilation. The microarray data in a previous study (Zhang et al. 2008) suggests that genes related to the photosystems, which supply reducing power, and the gene encoding glutamate ammonia ligase, an enzyme that assimilates nitrogen, were downregulated under sulfur starvation conditions. Downregulation of these genes might result in a loss of both the reducing power and the enzymatic activity to synthesize glutamate in our study. The amounts of the amino acids tryptophan, phenylalanine, glutamine, ornithine, proline, and asparagine did not change significantly.

To estimate protein biosynthesis and degradation, we quantified total protein of *Synechocystis* with Bradford assay (Fig. 2). Protein content of *Synechocystis* under sulfur starvation decreased about 30 μg mg^{-1} FW at 72 h after the medium change.

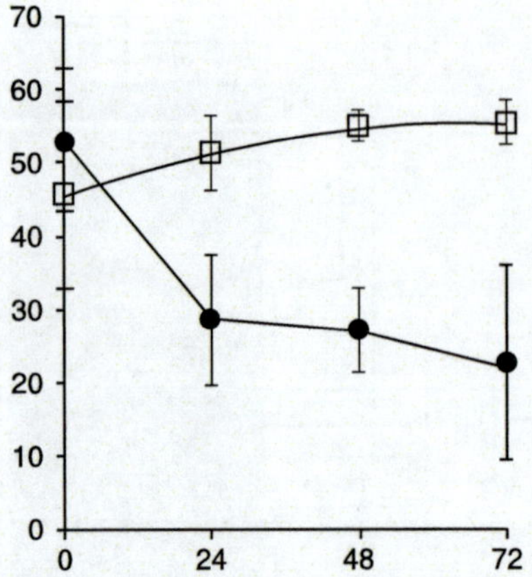

Fig. 2 Protein degradation under S deprivation. Amount of total protein was quantified with Bradford assay. *Open square*: changed to BG11 medium; *filled circle*: changed to BG11-S medium. *Vertical axis*: amount of total protein (μg mg^{-1} FW); *horizontal axis*: time after medium change (h). The average and standard error (n = 3) are shown

It suggests that protein biosynthesis was downregulated and/or protein degradation was upregulated. We suppose that 30 μg mg^{-1} FW of protein is sufficient to provide several pmols mg^{-1} FW of amino acids. So, to some extent, increase of certain amino acids were explained with protein degradation.

As mentioned above, the increase in amino acid contents under sulfur starvation could not be explained simply by changes in the transcript levels of genes involved in amino acid biosynthesis. *Synechocystis* might be unable to synthesize cysteine under conditions of sulfur deprivation owing to the shortage of sulfate supply. This might result in an immediate increase in serine, a precursor of cysteine, and in a decrease in methionine and glutathione, which are synthesized from cysteine. It might also result in decrease in *S*-adenosylmethionine, which is synthesized from methionine. The decrease in *S*-adenosylmethionine levels presumably enhanced aspartate kinase activity, which mediates the first step in lysine, threonine, and isoleucine *de novo* biosynthesis, via feedback inhibition by *S*-adenosylmethionine (Rognes et al. 1980). Consequently, the synthesis of lysine and isoleucine might be enhanced, leading to an accumulation of these amino acids. Sulfur deprivation might therefore result in an increased biosynthesis of amino acids in the aspartate family mediated by a regulation at the enzymatic level. This discussion is supported by clustering analysis. Serine, isoleucine, threonine and lysine were nearly clustered (Fig. 1c).

In addition to the metabolic regulation of *de novo* amino acid biosynthesis, protein synthesis and degradation might also affect amino acid pools. Actually, our result showed that certain amino acids increased and the total protein in *Synechocystis* decreased (Figs. 1 and 2). Under nutrient starvation conditions, cyanobacteria degrade phycobiliproteins, which are light-harvesting proteins involved in photosynthesis (Lau et al. 1977; Collier and Grossman 1992). We found that the immediate increase of certain amino acids under nitrogen starvation was abolished in a loss-of-function mutant of the *nblA* genes that trigger degradation of phycobiliprotein (our unpublished data). Furthermore, *Synechocystis* upregulates the *nblA* genes under sulfur starvation conditions (Zhang et al. 2008). Then we hypothesize that the increase in the amino acid contents under sulfur starvation conditions was caused by both a regulation of the metabolic flow and *nblA*-dependent phycobiliprotein degradation.

In this study, we report changes in the amino acid contents in response to sulfur deprivation in *Synechocystis* sp. PCC 6803 using GC-MS. Sulfur starvation transiently increased the levels of most amino acids, even that of the sulfur-containing amino acid methionine (Fig. 1b), and the levels of some amino acids increased continuously for up to 72 h after the initiation of sulfur deprivation. Because these responses resulted from a combination of complex factors such as gene expression and enzyme activity levels, the specific responses have to be analysed using gene-deficient mutants and/or metabolic inhibitors to understand the molecular mechanisms underlying the metabolic responses to nutritional stress.

References

Collier JL, Grossman AR (1992) Chlorosis induced by nutrient deprivation in *Synechococcus* sp. strain PCC 7942: not all bleaching is the same. J Bacteriol 174:4718–4726

Lau RH, Mackenzie MM, Doolittle WF (1977) Phycocyanin synthesis and degradation in the blue-green bacterium *Anacystis nidulans*. J Bacteriol 132:771–778

Rippka R (1988) Isolation and purification of cyanobacteria. Methods Enzymol 167:3–27

Rognes SE, Lea PJ, Miflin BJ (1980) *S*-adenosylmethionine – a novel regulator of aspartate kinase. Nature 287:357–359

Zhang Z, Pendse ND, Phillip KN, Cotner JB, Khodursky A (2008) Gene expression patterns of sulfur starvation in *Synechocystis* sp. PCC6803. BMC Genomics 9:344–357

Regulation of the Uptake of Sulfate by Chinese Cabbage *(Brassica pekinensis)* at Various Sulfate Concentrations in the Root Environment

Aleksandra Koralewska, Freek S. Posthumus, C. Elisabeth E. Stuiver, Muhammad Shahbaz, Ineke Stulen, and Luit J. De Kok

Abstract If Chinese cabbage was grown hydroponically at various sulfate concentrations ranging from 5 to 100 μM in the root environment, it was able to maintain its growth even at 5 and 10 μM, concentrations close to the Km of the Group 1 high affinity sulfate transporters. The sulfate uptake capacity of the roots was up-regulated at ≤25 μM sulfate, however, the expression of Sultr1;2, the primary sulfate transporter abundant in the root of Chinese cabbage, was hardly affected. There was no clear relation between the overall tissue levels of sulfate and thiols and the expression and activity of the sulfate transporters in root tissue upon variation in sulfate supply.

Brassica species are characterized by their high growth rate and sulfur demand and their seedlings may require an overall sulfate uptake of up to 40 μmol g^{-1} fresh weight root day^{-1} in order to maintain their growth (Koralewska et al. 2007, 2008, 2009a, b; Shahbaz et al. 2010). The uptake, transport and distribution of sulfate in the plant are mediated by different sulfate transporter transmembrane proteins. *Brassica* species contain 12–14 different sulfate transporter genes and the sulfate transporter family has been classified in up to five different groups according to their cellular and subcellular expression and possible functioning (Hawkesford 2003, 2007, 2008; Buchner et al. 2004; Parmar et al. 2007; Hawkesford and De Kok 2006). In Brassica, the Group 1 sulfate transporters Sultr1;2 appeared to be responsible for the primary uptake of sulfate by the root, whereas Sultr1,1 was only expressed upon sulfate deprivation (Koralewska et al. 2007, 2008, 2009a, b; Shahbaz

A. Koralewska • F.S. Posthumus • C.E.E. Stuiver • I. Stulen • L.J. De Kok (✉)
Laboratory of Plant Physiology, University of Groningen, P.O. Box 11103, 9700 CC Groningen, The Netherlands
e-mail: l.j.de.kok@rug.nl

M. Shahbaz
Institute of Agricultural Sciences, University of the Punjab, Lahore 54590, Pakistan

L.J. De Kok et al. (eds.), *Sulfur Metabolism in Plants*, Proceedings of the International Plant Sulfur Workshop 1, DOI 10.1007/978-94-007-4450-9_7,

Table 1 Impact of external sulfate concentration (5, 10, 25 and 100 μM) on biomass production, sulfate and thiol content, and sulfate uptake capacity in Chinese cabbage. Seedlings were grown on a 25% Hoagland nutrient solution at various sulfate concentrations for 4 days. Data on biomass production (g FW) and shoot to root ratio (S/R) represent the mean of 2 experiments with 12 measurements with 3 plants in each (±SD). On average plant growth rate was 0.34 $g\,g^{-1}$ FW day^{-1}. Sulfate content (μmol g^{-1} FW) represents the mean of two experiments with two to three measurements with six plants in each (±SD). Thiol content (μmol g^{-1} FW) represents the mean of three measurements with six plants in each (±SD). Sulfate uptake capacity (μmol g^{-1} FW root h^{-1}) was measured at 25 and 500 μM sulfate and represents the mean of three measurements with three plants in each (±SD). Different letters indicate significant differences between treatments ($p \leq 0.01$, Student's *t*-test)

Sulfate (μM)	5	10	25	100
Biomass production				
Shoot	0.150±0.018a	0.154±0.022a	0.159±0.023a	0.161±0.017a
Root	0.030±0.005a	0.029±0.004a	0.027±0.004a	0.029±0.004a
S/R	5.0±0.6a	5.5±0.7a	5.9±0.6a	5.6±0.6a
Sulfate content				
Shoot	13.8±2.9a	13.3±3.3a	14.3±3.3a	13.0±3.1a
Root	6.4±0.5a	6.8±0.4a	8.2±0.6b	8.2±0.2b
Thiol content				
Shoot	0.49±0.03a	0.50±0.04a	0.50±0.03a	0.49±0.03a
Root	0.37±0.02a	0.43±0.04ab	0.43±0.02b	0.39±0.09ab
Sulfate uptake capacity	3.96±0.41c	3.34±0.33c	2.14±0.30b	1.59±0.13a

et al. 2010). In the current paper the impact of concentration of sulfate in the root environment on plant growth and sulfate uptake is further evaluated in a case study with Chinese cabbage.

Ten day-old seedlings of Chinese cabbage (*Brassica pekinensis* (Lour.) Rupr. cv. Kasumi F1 (Nickerson Zwaan, Made, The Netherlands)) were grown on a 25% Hoagland solution at various sulfate concentrations, varying from 5 to 100 μM for up to 13 days in a climate-controlled greenhouse (see for experimental details Koralewska et al. 2007). Measurement of biomass production, sulfate, nitrate and thiol content, sulfate uptake capacity, and extraction of mRNA and determination of the expression of sulfate transporters were carried out as described previously (Koralewska et al. 2007).

Similar to previous observations with curly kale (*Brassica oleracea*; Koralewska et al. 2007), the uptake of sulfate by Chinese cabbage was controlled by the sulfur demand for growth and modulated by the external sulfate concentration. If Chinese cabbage was grown at various sulfate concentrations ranging from 5 to 100 μM, plants were able to maintain their biomass production and growth rate (on average 0.34 $g\,g^{-1}$ FW plant day^{-1}) even at 5 or 10 μM (Tables 1 and 2). The latter sulfate concentrations are close to the Km value of the Group 1 high affinity sulfate transporters (Hawkesford and De Kok 2006; Koralewska et al. 2007; Hawkesford 2008). The sulfate uptake capacity of the root was increased at ≤25 μM up to 2.5-fold at 5 μM sulfate (Table 1). The mRNA abundance of Sultr1;2, the constitutively

Table 2 Impact of external sulfate concentration (10 and 100 μM) on shoot and root biomass production and sulfate uptake capacity during growth of Chinese cabbage. Seedlings were grown on a 25% Hoagland nutrient solution at various sulfate concentrations for 7, 10 and 13 days. Data on biomass production (g FW), shoot to root ratio (S/R) and plant growth rate (g g^{-1} FW day^{-1}) represent the mean of 7–11 measurements with three plants in each (±SD). On average plant growth rate was 0.35 g g^{-1} FW day^{-1}. Sulfate uptake capacity (μmol g^{-1} FW root h^{-1}) was measured at 500 μM sulfate and represents the mean of three measurements with three plants in each (± SD). Different letters indicate significant differences between treatments ($p \leq 0.01$, Student's *t*-test)

	Sulfate (μM)	7 days	10 days	13 days
Biomass production				
Shoot	10	0.41±0.07a	1.34±0.22a	4.06±0.54a
	100	0.45±0.05a	1.59±0.20a	5.03±0.41b
Root	10	0.08±0.02a	0.24±0.06a	0.69±0.11a
	100	0.07±0.01a	0.26±0.05a	0.84±0.14a
S/R	10	5.6±0.9a	5.6±0.6a	5.9±0.5a
	100	6.8±0.7a	6.2±0.6a	6.1±0.8a
Sulfate content				
Shoot	10	12.0±1.4a	9.1±0.6a	7.6±0.3a
	100	15.2±1.2b	10.0±1.1a	7.5±0.8a
Root	10	7.4±0.6a	5.8±1.2a	6.6±0.2a
	100	9.3±0.5b	7.4±0.4b	7.5±0.3b
Sulfate uptake capacity	10	3.6±0.7b	2.5±0.3b	1.9±0.1b
	100	2.0±0.8a	1.4±0.1a	0.8±0.0a

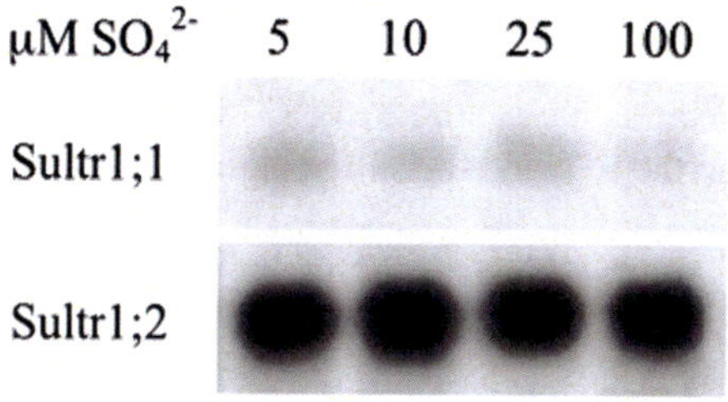

Fig. 1 Impact of external sulfate concentration (5, 10, 25 and 100 μM) on mRNA abundance of the sulfate transporters in Chinese cabbage roots (see for experimental details Table 1). Equal RNA loading was determined by ethidium bromide staining of gels (see Shahbaz et al. 2010 for more details)

expressed high affinity sulfate transporter in the root of Chinese cabbage, was hardly affected by the sulfate concentration in the root environment (Figs. 1 and 2). Sultr1;1 was hardly expressed in roots of Chinese cabbage, but its mRNA abundance was only slightly enhanced at ≤25 μM sulfate (Figs. 1 and 2). Moreover, the contents of sulfate and thiols in both root and shoot of Chinese cabbage were hardly affected at low sulfate concentrations in the root environment; there was only a slight decrease in the sulfate content of the root at ≤10 μM (Tables 1 and 2). From studies on prolonged sulfate deprivation it is evident that in addition to an increased uptake, the overall plant uptake capacity might also be enhanced by the change in biomass partitioning

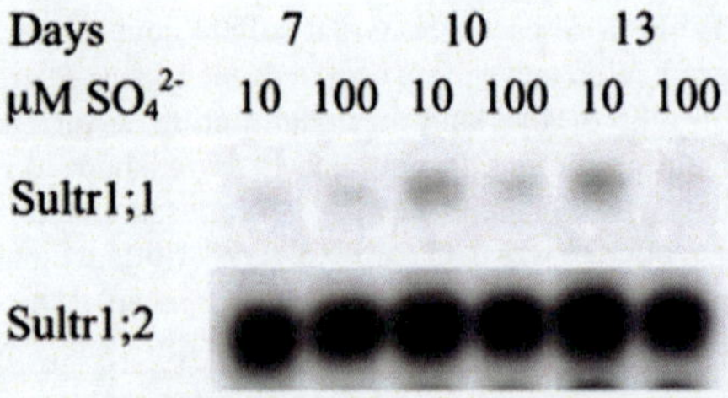

Fig. 2 Impact of external sulfate concentration (10 and 100 µM) on mRNA abundance of sulfate transporters in Chinese cabbage roots (see for experimental details Table 1)

in favor of the root (Hawkesford and De Kok 2006; Koralewska et al. 2007, 2008). However, in Chinese cabbage the shoot to root ratio remained unaffected at 5 and 10 µM sulfate, even upon more prolonged exposure (Tables 1 and 2).

The sulfate transporters may be regulated at a transcriptional, translational and/or post-translational level, which might be signaled or mediated by sulfate itself and/or by products of the assimilatory sulfate reduction pathway (*e.g.* sulfide, cysteine and/or glutathione; Hawkesford and De Kok 2006). However, on basis of the current knowledge and commonly applied research approaches, it is still unclear as to what extent measuring changes in concentrations of potential signal compounds and activity of the sulfate transporters, provides sufficient insight into the actual regulatory control of sulfate uptake at the root cellular level. Similar to other observations with *Brassica*, there was no clear relation between the overall tissue levels of sulfate and thiols and the activity of the sulfate transporters in Chinese cabbage upon variation in sulfate supply (Buchner et al. 2004; Koralewska et al. 2007, 2008, 2009a, b; Stuiver et al. 2009; Shahbaz et al. 2010). There was in Chinese cabbage no direct relation between the mRNA abundance of Sultr1;2, the primary sulfate transporter in the root and the sulfate uptake capacity.

References

Buchner P, Stuiver CEE, Westerman S, Wirtz M, Hell R, Hawkesford MJ, De Kok LJ (2004) Regulation of sulfate uptake and expression of sulfate transporter genes in *Brassica oleracea* as affected by atmospheric H_2S and pedospheric sulfate nutrition. Plant Physiol 136:3396–3408

Hawkesford MJ (2003) Transporter gene families in plants: the sulfate transporter gene family – redundancy or specialization? Physiol Plant 117:155–163

Hawkesford MJ (2007) Sulfur and plant ecology: a central role of sulfate transporters for responding to sulfur availability. In: Hawkesford MJ, De Kok LJ (eds) Sulfur in plants – an ecological perspective. Springer, Dordrecht, pp 1–15

Hawkesford MJ (2008) Uptake, distribution and subcellular transport of sulfate. In: Hell R, Dahl C, Knaff DB, Leustek T (eds) Sulfur metabolism in phototrophic organisms. Springer, Dordrecht, pp 15–30

Hawkesford MJ, De Kok LJ (2006) Managing sulphur metabolism in plants. Plant Cell Environ 29:382–395

Koralewska A, Posthumus FS, Stuiver CEE, Buchner P, Hawkesford MJ, De Kok LJ (2007) The characteristic high sulfate content in *Brassica oleracea* is controlled by the expression and activity of sulfate transporters. Plant Biol 9:654–661

Koralewska A, Stuiver CEE, Posthumus FS, Kopriva S, Hawkesford MJ, De Kok LJ (2008) Regulation of sulfate uptake, expression of the sulfate transporters Sultr1;1 and Sultr1;2, and APS reductase in Chinese cabbage (*Brassica pekinensis*) as affected by atmospheric H_2S nutrition and sulfate deprivation. Funct Plant Biol 35:318–327

Koralewska A, Buchner P, Stuiver CEE, Posthumus FS, Kopriva S, Hawkesford MJ, De Kok LJ (2009a) Expression and activity of sulfate transporters and APS reductase in curly kale in response to sulfate deprivation and re-supply. J Plant Physiol 166:168–179

Koralewska A, Stuiver CEE, Posthumus FS, Hawkesford MJ, De Kok LJ (2009b) The up-regulated sulfate uptake capacity, but not the expression of sulfate transporters is strictly controlled by the shoot sink capacity for sulfur in curly kale. In: Sirko A, De Kok LJ, Haneklaus S, Hawkesford MJ, Rennenberg H, Saito K, Schnug E, Stulen I (eds) Sulfur metabolism in plants: regulatory aspects, significance of sulfur in the food chain, agriculture and the environment. Margraf Publishers, Weikersheim, pp 61–68

Parmar S, Buchner P, Hawkesford MJ (2007) Leaf developmental stage affects sulfate depletion and specific sulfate transporter expression during sulfur deprivation in *Brassica napus* L. Plant Biol 9:647–653

Shahbaz M, Tseng MH, Stuiver CEE, Koralewska A, Posthumus FS, Venema JH, Parmar S, Schat H, Hawkesford MJ, De Kok LJ (2010) Copper exposure interferes with the regulation of the uptake, distribution and metabolism of sulfate in Chinese cabbage. J Plant Physiol 167:438–446

Stuiver CEE, Koralewska A, Posthumus FS, De Kok LJ (2009) The impact of sulfur deprivation on root formation, and activity and expression of sulfate transporters in Chinese cabbage. In: Sirko A, De Kok LJ, Haneklaus S, Hawkesford MJ, Rennenberg H, Saito K, Schnug E, Stulen I (eds) Sulfur metabolism in plants: regulatory aspects, significance of sulfur in the food chain, agriculture and the environment. Margraf Publishers, Weikersheim, pp 89–92

Characterization of Adenosine 5′-Phospho-Sulfate Kinase (*APSK*) Genes from Higher Plants

Susanna C.S. Leung, Diantha Smith, Ronan Chen, John A. McCallum, Marian McKenzie, and Michael T. McManus

Abstract Onion (*Allium cepa* L.) and broccoli (*Brassica oleracea*) are economically important crop plants in which differences in the S-assimilation pathway may operate. Onion is rich in reduced sulfur-containing compounds, which characterize its flavor while glucosinolates and their derivatives accumulate in Brassica species including broccoli. A significant branchpoint in the S-assimilation pathway is catalyzed by adenosine 5′-phosphosulfate (APS) kinase (APSK) which phosphorylates APS to form 3′-phosphoadenosine 5′-phosphosulfate (PAPS), one of the substrates used in the biosynthesis of glucosinolates. In comparison with Brassica species, there is no information on the significance or regulation of this enzyme in onion. Here we describe some preliminary characterization of an *APSK* gene and enzyme activity in onion, including some sequence and activity comparisons with Brassica species.

Onion (*Allium cepa* L.) accumulates reduced sulfur-containing compounds principally alk(en)yl-L-cysteine sulfoxides (ACSOs). Upon wounding these are hydrolyzed by the enzyme allinase (E.C. 4.4.1.4) and the derivatives produced characterize the flavor and health properties of onion (Lancaster and Boland 1990). In the reductive sulfur (S) assimilation pathway, the activities of enzymes in the pathway are

S.C.S. Leung • D. Smith • M.T. McManus (✉)
Institute of Molecular BioSciences, Massey University,
Private Bag 11 222, Palmerston North, New Zealand
e-mail: M.T.McManus@massey.ac.nz

R. Chen • M. McKenzie
New Zealand Institute for Plant and Food Research Ltd, Food Industry Science Centre,
Private Bag 11 600, Palmerston North, New Zealand

J.A. McCallum
New Zealand Institute for Plant and Food Research Ltd,
Canterbury Agriculture & Science Centre, Christchurch Mail Centre,
Private Bag 4704, Christchurch, New Zealand

L.J. De Kok et al. (eds.), *Sulfur Metabolism in Plants*, Proceedings of the International Plant Sulfur Workshop 1, DOI 10.1007/978-94-007-4450-9_8,

Table 1 APSK activity is determined in the crude plant extracts and is monitored by oxidation of NADH

Plant species and tissues	Specific activity (μmol NAD^+ mg protein^{-1} min^{-1})
Onion leaf	0.154
Broccoli florets	0.291
Broccoli leaf	0.584
Arabidopsis leaf	0.068
P. enysii leaf	0.086
P. fastigiata leaf	0.058

Change of absorbance was measured at 340 nm. For the calculation of the rate of NADH oxidation, the extinction coefficient of 6.2 mM^{-1} was used

influenced by S supply (McCallum et al. (2011); Thomas et al. 2011), with some (*in vitro*) evidence that ATPS and APR can form a complex to by-pass a branchpoint in the S-assimilation pathway catalyzed by adenosine 5′-phosphosulfate (APS) kinase (APSK, E.C. 2.7.1.25) (Cumming et al. 2007). APSK forms 3′-phosphoadenosine 5′-phosphosulfate (PAPS), an important substrate for the formation of the secondary S-containing metabolites including glucosinolates, and is therefore a significant enzyme in members of the Brassicaceae. Four *APSK* genes have been cloned from *Arabidopsis,* for example, with APSK1, APSK2 and APSK4 localised in the plastid, while APSK3 is a cytoplasmic isoform (Mugford et al. 2009). Mutant approaches have shown that APSK1 is able to produce sufficient PAPS to maintain normal plant growth (Mugford et al. 2010), while disruption of *APSK1* and *APSK2* expression reduces the biosynthesis of glucosinolates (Mugford et al. 2009) demonstrating that the expression of *APSK* genes are tightly linked with the biosynthesis of glucosinolates. However, in non-glucosinolate accumulating species, including for example onion, the secondary (APSK-mediated) pathway must also operate to generate important pools of sulfate esters. Thus it is of interest to us to determine how significant *APSK* expression and APSK activity is in onion given the importance of the reduced pathway to produce cysteine.

In onion, a full-length cDNA encoding *APSK* is not available and so initial studies have focused on the measurement of APSK activity in onion leaf extracts in comparison with glucosinolate-accumulating *Brassica* species such as broccoli and *Arabidopsis.* Activity is measured using a coupled assay in which excess pyruvate kinase and lactate dehydrogenase are added to utilize the ADP formed in the APSK reaction and the consumption of NADH measured (Burnell and Whatley 1975). However, as many plant enzymes in crude extracts can compete for NADH, particularly the inner membrane-localized mitochondrial NADH oxidase, soluble extracts are first centrifuged at 100,000 ×g to pellet plant membranes. Subsequent to this step, linear APSK activity was measured.

APSK activity in onion leaf extract was significantly higher than observed in *Arabidopsis* leaf extract and two closely related New Zealand native *Brassica* species of the *Pachycladon* genus, *P. enyii* and *P. fastigiata* (Table 1). It is known that expression of *APSK* is higher in *P. enyii* than in *P. fastigiata* (Voelckel et al. 2008) and in agreement with that, the APSK activity in leaf extracts of *P. enysii* was higher

```
Onion          -RDRDACRGLLPESNFIEVFMDMPIEICESRDPKGLYKLARAGKIKGFTGVDDPYEQPLN   59
AtAPSK2        -RDRDACRSLLPDGDFVEVFMDVPLHVCESRDPKGLYKLARAGKIKGFTGIDDPYEAPVN   258
A.lyrata       -RDRDACRSLLPEGDFVEVFMDVPLHVCESRDPKGLYKLARAGKIKGFTGIDDPYEAPVN   257
R.communis     RKDRDACRKILPNGDFIEVFMDIPLQVCESRDPKGLYKLARAGKIKGFTGIDDPYEPPLN   254
VvAPSK         -RDRDACRALVPEGSFIEVFMDVPLQVCEARDPKGLYKLARAGKIQGFTGIHDPYEPPLN   273
O.sativa       --DRSACRKLLPNSSFIEVFLNVPLEVCEERDPKGLYKLARAGKIKGFTGIDDPYETPSD   269
ZmASK1         --DRSACRDLLPKHSFIEVFLDVPLQVCEARDPKGLYKLARAGKIKGFTGIDDPYEPPSD   249
CrAPSK         ----DACRSLLPEGDFIEVFMDVPLKVCEARDPKGLYKLARAGKIKGFTGIDDPYEPPLK   278
                   .*** ::*. .*:***:::*:.:** ***************:****:.**** * .

Onion          CEIVMKPKDREEAMSPSAMAEQVISYLDMHGFLKA 94
AtAPSK2        CEVVLKHTGDDESCSPRQMAENIISYLQNKGYLEG 293
A.lyrata       CEVVLKYTGDDDSCSPRQMAENIISYLQNKGYLEG 292
R.communis     CEISLKLN-NGENASPCEMAEEVISYMEENGYLQA 288
VvAPSK         CEIVLKHS-EEVCTSPRDMAEKVIHYLEENGYL-- 307
O.sativa       CEIVIQCK-VGDCPSPKSMADQVVSYLEANGFFQ- 303
ZmASK1         CEIVIRCK-VGDCPSPESMAGHVVSYLETNGFLQ- 282
CrAPSK         SEIVLHQK-LGMCDSPCDLADIVISYLEENGYLKA 312
               .*: :: .      **  :*  :: *:: :*::
```

Fig. 1 CLUSTAL W alignment of APSK proteins from various plant species. The highlighted (*bold*) residues are 3 of 16 residues important for ligand binding within the P-loop NTPase domain conserved within the APSK genes characterized thus far. AtAPSK2, *A. thaliana* APSK2 (NP195704.1); *A. lyrata*, *Arabidopsis lyrata* subsp. APSK (XP_002866895.1); *R. communis*, *Ricinus communis* APSK (XP_002532415.1); VvAPSK, *Vitis vitifera* APSK (XP_002265518.1), *O. sativa*, *Oryzae sativa* APSK (ABA94897.1); ZmASK1, *Z. mays* APSK1 (ACG40162.1); CrAPSK, *Catharanthus roseus* APSK (O49204.1). * denotes an identical amino acid; **:** denotes a similar amino acid

than that measured in leaf extracts of *P. fastigiata* (Table 1). APSK activity was also measured in crude extracts of florets and leaves from broccoli, a known glucosinolate accumulator and activity was highest in these extracts. However, the significant activity measured in onion does suggest that the secondary pathway may also be of some physiological significance in onion.

A partial onion cDNA encoding APSK has been obtained and tBLASTx searching showed highest identity to *Arabidopsis APSK2* and *APSK2* from *A. lyrata* as well as high identity with *APSK* genes from other monocots including from *Zea mays* and *O. sativa*, as well as genes from *Ricinus communis*, *Vitus Vinifera* and *Catharanthus roseus* (data not shown). These sequences were then compared using a CLUSTAL W alignment (Fig. 1). A major feature of the APSK proteins is the C-terminal P (phosphate-binding)-loop NTPase domain, which spans residues 118–263 of AtAPSK2. Of the partial onion APSK sequence, residues 1–63 comprise the C-terminal region of the P-loop NTPase domain. Sixteen important residues have been identified within the P-loop for ligand binding (APS and ATP; Lansdon et al. 2002) and three of these, R_{228}, F_{246} and T_{247} (residue numbers from AtAPSK2) are also identified in the onion sequence (highlighted in Fig. 1). Phylogenetic analysis of the APSK proteins compared using CLUSTAL W in Fig. 1 shows that onion APSK forms a distinct clade with the other monocot sequences examined: one from *Oryzae sativa*, and *Zea mays* ASK1 (Fig. 2). The two Brassica species examined

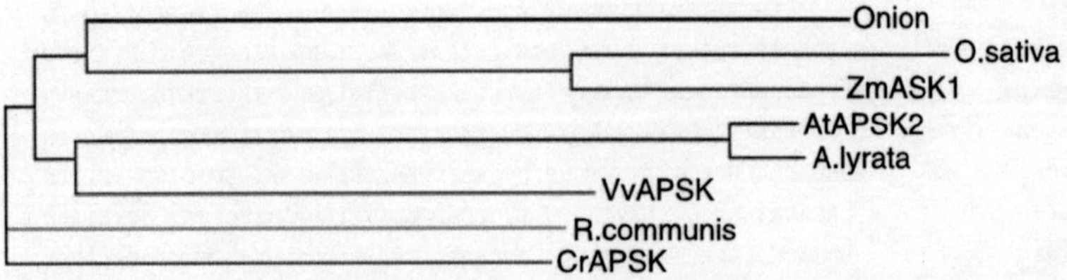

Fig. 2 Phylogenetic analysis of APSK proteins from different plant species, as indicated, as compared in Fig. 1

then group in a separate, but relatively closely related clade, with a sequence from *Vitis vitifera*, with sequences from *Ricunus communis* and *Catharanthus rosea* forming outliers.

In summary, APSK activity has been demonstrated in onion and a partial *APSK* gene isolated. To more fully determine the significance of APSK activity in onion, a full-length version of the *APSK* gene is now required which will provide an avenue for more detailed kinetic analysis of the recombinant APSK protein.

Acknowledgements Special thanks to Claudia Voelckel for providing the *P. enysii* and *P. fastigiata APSK* genomic DNA sequences. This research is partially funded by the FRST Future Vegetables Programme (C02X0701).

References

Burnell JN, Whatley FR (1975) A new, rapid, and sensitive assay for adenosine 5′-phosphosulphate (APS) kinase. Anal Biochem 68:281–288

Cumming M, Leung S, McCallum J, McManus MT (2007) Complex formation between recombinant ATP sulfurylase and APS reductase of *Allium cepa* (L.). FEBS Lett 581:4139–4147

Lancaster JE, Boland MJ (1990) Flavour biochemistry. In: Rabinowitch HD, Brewster JL (eds) Onions and allied crops, vol 3. CRC Press, Boca Raton, pp 33–72

Lansdon EB, Segel IH, Fisher AJ (2002) Light-induced structural changes in adenosine 5′-phosphosulfate kinase from *Penicillium chrysogenum*. Biochemistry 41:13672–13680

McCallum JA, Thomas LA, Shaw M, Pither-Joyce M, Leung S, Cumming M, McManus MT (2011) Genotypic variation in the sulfur assimilation and metabolism of onion (Allium cepa L.) I. Plant composition and transcript accumulation. Phytochemistry 72:882–887

Mugford SG, Yoshimoto N, Reichelt M, Wirtz M, Hill L, Mugford ST, Nakazato Y, Noji M, Takahashi H, Kramell R, Gigolashvili T, Flügge U-I, Wastemack C, Gershenzon J, Hell R, Saito K, Kopriva S (2009) Disruption of adenosine-5′-phosphosulfate kinase in *Arabidopsis* reduces levels of sulfated secondary metabolites. Plant Cell 21:910–927

Mugford SG, Matthewman CA, Hill L, Kopriva S (2010) Adenosine-5′-phosphosulfate kinase is essential for *Arabidopsis* viability. FEBS Lett 584:119–123

Thomas LA, Leung S, Shaw M, Albert N, McCallum JA, McManus MT (2011) Genotypic variation in sulfur assimilation and metabolism of onion (*Allium cepa* L.). II. Characterization of ATP sulfurylase activity. Phytochemistry 72:888–896

Voelckel C, Heenan PB, Janssen B, Reichelt M, Ford K, Hofmann R, Lockhart PJ (2008) Transcriptional and biochemical signatures of divergence in natural populations of two species of New Zealand alpine *Pachycladon*. Mol Ecol 17:4740–4753

Subcellular Compartmentation of Cysteine Synthesis in Plants – One Step More

Malgorzata Lewandowska, Frantz Liszewska, Markus Wirtz, and Rüdiger Hell

Abstract The cysteine synthase complex (CSC) is supposed to regulate the synthesis of cysteine (Cys). This complex is formed by two enzymes: serine acetyltransferase (SAT) and *O*-acetyl-serine (thiol) lyase (OAS-TL). The CSC occurs within plant cells in three compartments: cytoplasm, chloroplasts and mitochondria. Cys is one of the substrates for the synthesis of glutathione (GSH) – an important player in regulation of redox-state and stress-response in cells. Here we reported influence of light intensity on transgenic tobacco plants with overexpression of active and inactive forms of *Arabidopsis thaliana* SAT3 targeted to the cytosol.

Plants, with regard to their biomass, are the main producers of compounds containing organic sulfur. One of the key points in the sulfur metabolic pathway is the biosynthesis of cysteine (Cys), occurring after a series of reactions in which sulfate is reduced to sulfide. The following Cys synthesis is a two-step reaction. In the first step serine acetyltransferase (SAT) catalyzes synthesis of *O*-acetylserine (OAS) from serine (Ser) and acetyl-CoA, while in the next step *O*-acetyl-serine (thiol) lyase (OAS-TL) is responsible for synthesis of Cys by incorporation of sulfide into OAS. Both enzymes are able to form a complex named Cys synthase complex (CSC). The CSC consists of a dimer of homotrimers of SAT and two homodimers of OAS-TL (Wirtz et al. 2010). Activities of both enzymes are regulated by complex formation. Whereas OAS-TL is active as a free dimer, SAT is strongly activated by association

M. Lewandowska
Centre for Organismal Studies, University of Heidelberg, Heidelberg, Germany

Institute for Biochemistry and Biophysics, Polish Academy of Sciences, Warsaw, Poland

F. Liszewska • M. Wirtz • R. Hell (✉)
Centre for Organismal Studies, University of Heidelberg, Heidelberg, Germany
e-mail: Ruediger.hell@cos.uni-heidelberg.de

L.J. De Kok et al. (eds.), *Sulfur Metabolism in Plants*, Proceedings of the International Plant Sulfur Workshop 1, DOI 10.1007/978-94-007-4450-9_9,

with OAS-TL (Wirtz and Hell 2006). The CSC occurs within plant cells in three compartments: cytoplasm, chloroplasts and mitochondria. Synthesized Cys is used not only as a protein component, but also as substrate for many other sulfur-containing compounds such as glutathione (GSH), which plays a pivotal role in redox regulation and stress response (Droux et al. 1998; Lewandowska and Sirko 2008).

As we reported in Wirtz and Hell (2007) to investigate the role of the subcellular compartmentation of the CSC, transgenic tobacco plants overexpressing active (a) or inactive (i) form of *Arabidopsis thaliana* SAT3 (At-Sat3, At3g13110) in the cytosol (CC) were constructed. Inactivation o SAT was obtained by site-directed mutagenesis of His-309 to Ala, and did not influence interaction with OASTL. This strategy generates the competition of endogenous SAT with inactivated SAT in the CSC with the aim of deregulation of cysteine synthesis. Biochemical characterization of the lines showed increased level of At-SAT3 protein in both types of transgenic lines. As expected, the SAT activity was significantly increased only in lines overexpressing active form of At-SAT3, however, in both types of lines increased levels of non-protein thiols were observed.

Correlation between the overexpression of active SAT protein and increased levels of thiols was already well known (Lewandowska and Sirko 2008), but increased level of cysteine in plants overexpressing the inactive form of SAT was surprising. This interesting phenomenon was explained by compensation of the disrupted CSC in one compartment by CSC in other unaffected compartments, mitochondria and plastids (Wirtz and Hell 2007). Recently it was reported that the *Arabidopsis* chloroplastic cyclophilins CYP20-3 may interact with *Arabidopsis* SAT1 protein to link light and oxidative stress to SAT1 activity and synthesis of cysteine (Dominguez-Solis et al. 2008). We decided to investigate influence of the light intensity on our transgenic plants. Since deregulation of the CSC had no phenotypic effect on tobacco lines grown under 200 $\mu mol\ m^{-2}\ s^{-1}$ light (Wirtz and Hell 2007), CCa and CCi lines were grown in soil in a growth chamber for 6–7 weeks under lower (LL, 100 $\mu mol\ m^{-2}\ s^{-1}$) and higher (HL, 300 $\mu mol\ m^{-2}\ s^{-1}$) light intensity as compared to Wirtz and Hell (2007). Seedlings were grown for 1 week under long day condition at LL, at this time point the plants were divided into two groups growing either under LL, or under HL for 5 weeks. Other set of the plants were grown until 5 weeks under LL, next half of the plants was transfer to HL for 2 weeks. After this time the forth leaf from the top of the plants was used for determination of pigments and water content. For determination of the non-protein thiols grinded shoots were used. Phenotypic observation concerning lines growing under LL were confirmed using hydroponic system (half-strength Hoagland medium [2.5 mM $Ca(NO_3)_2$, 2.5 mM KNO_3, 0.5 mM $MgSO_4$, 0.5 mM KH_2PO_4, 40 mM Fe-EDTA, 25 mMH_3BO_3, 2.25 mM $MnCl_2$, 1.9 mM $ZnSO_4$, 0.15 mM, $CuSO_4$, and 0.05 mM $(NH_4)_6MO_7O_{24}$, pH 5.9]). *T*-test was used for statistical analysis. The statistically significant differences are marked with either a single (for $p<0.05$) or double (for $p<0.002$) asterisks.

Occurrence of the correct form of At-SAT3 was checked by PCR on genomic DNA isolated from leaves and with polyclonal serum against *Arabidopsis* SAT3 protein using total leaf protein of each transgenic line (as described in Wirtz and Hell 2007).

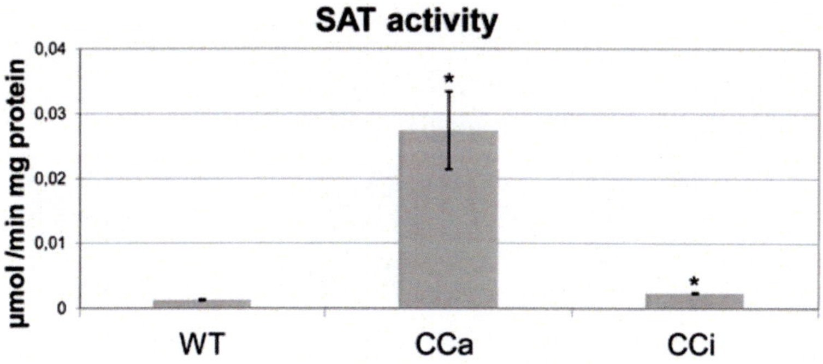

Fig. 1 SAT activity in WT and transgenic plants. Mean values and SD from two individual plants grown in two repetition (n=4). *Error bars* indicate SD, *asterisks* mark the statistically significant differences between the transformants and the wild-type line grown in the same conditions

Accordingly to results presented in Wirtz and Hell (2007), SAT activity in CCa plants was highly increased compared to wild-type (SNN) plants (Fig. 1.), while in CCi plants overexpressing the inactive form of At-SAT3 only slight increase of SAT activity was observed. No significant changes in OAS-TL activity were noticed (not shown).

The first remark concerning plants growing under LL was decreased biomass of plants from line CCa compared to WT, particularly visible in a hydroponic condition (Fig. 2a). In contrast, under HL condition plants from CCi showed reduction of biomass in comparison to WT and CCa lines. Growth retardation of CCi lines had already been observed under 600 µmol m^{-2} s^{-1} (Wirtz 2003). Spectrophotometrical measuring of the chlorophyll concentration (Lichtenthaler and Wellburn 1983) did not reveal differences between lines grown under LL conditions, which is in agreement with the results obtained by Wirtz and Hell (2007). In all plants grow under HL condition loss of pigments were measured, however in plants overexpressing inactive form of SAT3 the decrease of pigments concentration seemed to be slightly higher than in plants from other lines (Fig. 2b). Wild type and transgenic plants showed the characteristic phenotypes in response to light intensity (Lichtenthaler et al. 1981), such as a smaller leaf area, higher dry weight, lower water content and increased thickness of leaves (not shown).

As expected the determination of thiols according to Wirtz et al. (2004) in CCa and CCi lines revealed higher levels of thiols, particularly Cys, as compared to WT plants under LL and HL conditions (Fig. 3). However comparison of thiol levels in plants growing under both conditions revealed that in WT and CCa line, but not in CCi lines, Cys and GSH level significantly increased in plants growing under HL. The easiest explanation for the lack of increase of non-protein thiols in CCi lines would be that the concentration of these compounds is already high enough and there is no requirement for higher thiol contents. However, the slight decrease of biomass and amount of pigments may suggest elevated sensitivity to higher light condition in comparison to WT and CCa lines. The other possible solution is that the CCi plants are not able to regulate thiols synthesis or cannot additionally increase

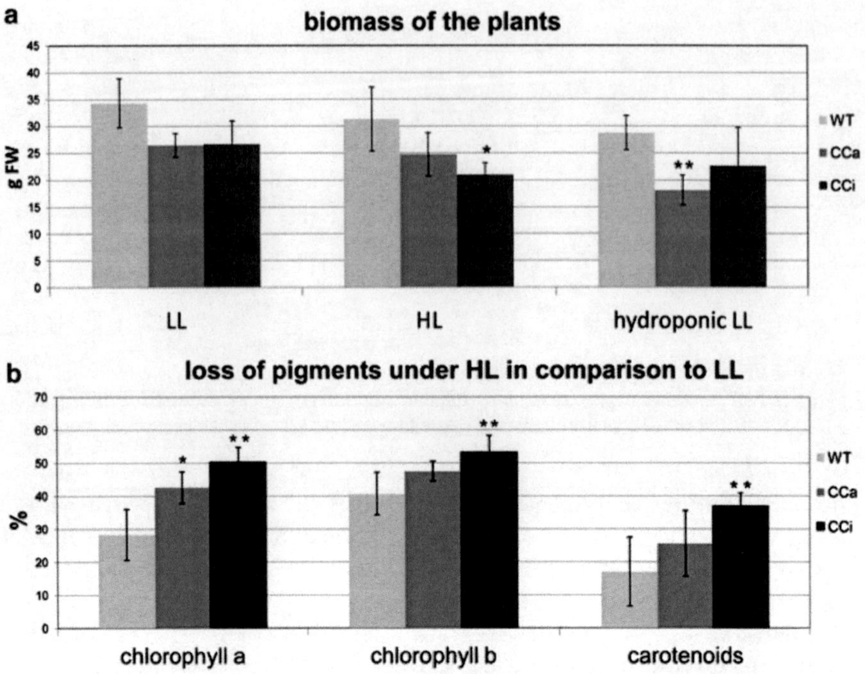

Fig. 2 Changes in biomass of the stems of the plants growing in LL and HL condition. Mean values and SD from three individual plants grown in two repetition (n=6), *error bars* indicate SD. *Asterisks* mark the statistically significant differences between the transformants grown under LL or HL and the wild-type line grown under LL (**a**). Decrease of photosynthetic pigments amounts depending on light conditions. *Asterisks* mark the statistically significant differences between the transformants and the wild-type line grown under HL (**b**)

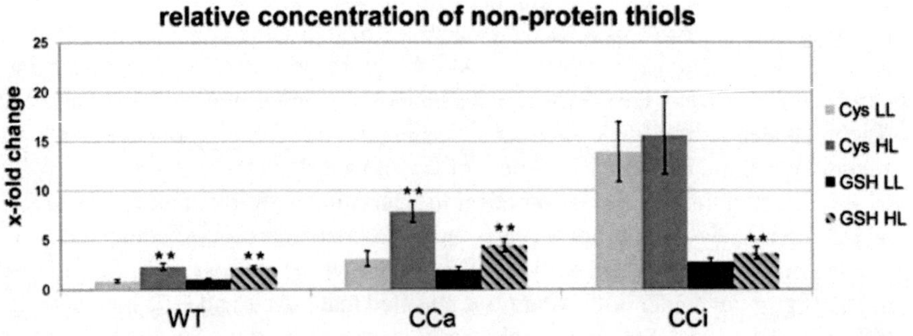

Fig. 3 Thiols concentration in plants growing in LL and HL condition was normalized to thiol levels in WT plants growing in LL. 1 unit – amount of Cys or GSH in WT plants under LL. Mean values and SD from three individual plants grown in two replication (n=6). *Error bars* indicate SD, *asterisks* mark the statistically significant differences between the same lines grown either under LL or under HL conditions

thiols level due to lack of SAT activity. Described by Dominguez-Solis et al. (2008) interaction between the *Arabidopsis* chloroplastic cyclophilins CYP20-3 and *Arabidopsis* SAT1 protein linking light and oxidative stress to SAT1 activity and synthesis of cysteine, gives as well other possibility. If overexpression of CCi would in fact deregulate level or activity of plastidic tobacco SAT isoform, and/or by this chloroplast redox system, theoretically there would be at least two ways of influencing of the inactive cytosolic SAT overexpression on chloroplasts: trough deregulation of redox-state or through interaction with the tobacco functional homolog of *Arabidopsis* CYP20-3. To resolve these questions more lines per one construct need to be analyzed. Nonetheless it is shown here that deregulation of CSC complex in the cytosol deregulates synthesis of thiols and putatively affects the adaptation to environmental changes. Cytosolic lines in combination with lines overexpressing inactive and active form of At-SAT3 in plastids and mitochondria would provide a very promising approach to investigate the puzzle of CSC compartmentation.

Acknowledgements ML was supported by German Academic Exchange Service (DAAD). Support by the German Science foundation (He 1848/13-1) is gratefully acknowledged.

References

Dominguez-Solis JR, He Z, Lima A, Ting J, Buchanan BB, Luan S (2008) A cyclophilin links redox and light signals to cysteine biosynthesis and stress responses in chloroplasts. Proc Natl Acad Sci USA 105:16386–16391

Droux M, Ruffet ML, Douce R, Job D (1998) Interactions between serine acetyltransferase and *O*-acetylserine (thiol) lyase in higher plants-structural and kinetic properties of the free and bound enzymes. Eur J Biochem 255:235–245

Lewandowska M, Sirko A (2008) Recent advances in understanding plant response to sulfur-deficiency stress. Acta Biochim Pol 55:457–471

Lichtenthaler HK, Wellburn AR (1983) Determinations of total carotenoids and chlorophyll *a* and *b* of leaf extracts in different solvents. Biochem Soc Trans 603:591–592

Lichtenthaler HK, Buschmann C, Döll M, Fietz H-J, Bach T, Kozel U, Meier D, Rahmsdorf U (1981) Photosynthetic activity, chloroplast ultrastructure, and leaf characteristics of high-light and low-light plants and of sun and shade leaves. Photosynth Res 2:115–141

Wirtz M (2003) Functionale Analyse des Cysteine-Synthasekomplex bei der Regulation des Schwefelmetabolismus in Pflanzen. PhD thesis, Martin-Luther-University Halle-Wittenberg

Wirtz M, Hell R (2006) Functional analysis of the cysteine synthase protein complex from plants: structural, biochemical and regulatory properties. J Plant Physiol 163:273–286

Wirtz M, Hell R (2007) Dominant-negative modification reveals the regulatory function of the multimeric cysteine synthase protein complex in transgenic tobacco. Plant Cell 19:625–639

Wirtz M, Droux M, Hell R (2004) *O*-acetylserine (thiol) lyase: an enigmatic enzyme of plant cysteine biosynthesis revisited in *Arabidopsis thaliana*. J Exp Bot 55:1785–1798

Wirtz M, Birke H, Heeg C, Throm C, Hosp F, Müller C, König S, Feldman-Salit A, Rippe K, Petersen G, Wade RC, Rybin V, Scheffzek K, Hell R (2010) Structure and function of the hetero-oligomeric cysteine synthase complex in plants. J Biol Chem 285:32810–32817

Preliminary Characteristics of a Tobacco Gene Down-Regulated by Sulfur Deprivation and Encoding a Cys-Rich Protein

Małgorzata Lewandowska, Katarzyna Zientara-Rytter, and Agnieszka Sirko

Abstract Tobacco cDNA clones representing genes differentially regulated by the short-term sulfur deficit were identified with the suppression subtractive hybridization (SSH) method. Many of them encoded proteins with established function, however for some of them not only their role in sulfur deficit response remained elusive but assigning them to any cellular processes was impossible. The down-regulated *D6* clone was representing one of such genes. The first steps of analysis included cloning of the full-length mRNA sequence and the sequence of its promoter region. Subsequent work, including monitoring of *D6* expression in various growth conditions, its intracellular localization and database screening for homologues lead to the hypothesis that *D6* may take part in differentiation and developing processes. However, its precise role and rationale for down-regulation during sulfur deficit remain unclear.

During screening of *Nicotiana tabacum* plants for genes regulated by short-term sulfur (S) deficit between those identified as encoding protein with well-defined function, a few were found for which it was not possible to assign a particular role (Lewandowska and Sirko 2009; Lewandowska et al. 2005; Wawrzyńska et al. 2005). The results of subsequent work on the most abundantly present up-regulated clones corresponding to the *UP9* gene encoding a member of UP9/LSU family allowed not only to add more information to the knowledge concerning regulation of S metabolism (Lewandowska et al. 2010) but also to link plant response to sulfur deficiency

M. Lewandowska
Institute for Biochemistry and Biophysics, Polish Academy of Sciences, Warsaw, Poland

Heidelberg Institute for Plant Sciences, University of Heidelberg, Heidelberg, Germany

K. Zientara-Rytter • A. Sirko (✉)
Institute for Biochemistry and Biophysics, Polish Academy of Sciences, Warsaw, Poland
e-mail: asirko@ibb.waw.pl

L.J. De Kok et al. (eds.), *Sulfur Metabolism in Plants*, Proceedings of the International Plant Sulfur Workshop 1, DOI 10.1007/978-94-007-4450-9_10,

a

```
                                                    M   A   V   K   G
  1 ATCAACAAGCTAAGATAAATCACTTACTATACTCTTAAGAAAAATCAATGGCAGTGAAAG
     F   A   V   L   L   F   V   A   M   L   L   M   S   T   E   Y   S   M   A   Q
 61 GTTTTGCGGTGTTGTTGTTTGTTGCAATGCTTCTCATGTCAACAGAGTATTCCATGGCTC
     Q   R   D   P   V   C   A   A   L   C   A   I   R   C   G   T   R   P   V   C
121 AACAACGTGATCCCGTATGCGCAGCATTATGCGCAATTAGGTGTGGAACAAGGCCTGTAT
     L   A   R   C   L   A   G   C   F   I   A   S   N   D   T   L   D   E   S   H
181 GCTTAGCTAGGTGCCTTGCTGGTTGCTTTATTGCAAGTAATGACACCTTAGACGAATCTC
     S   T   S   V   E   A   R   N   R   V   C   N   V   G   C   S   L   G   H   C
241 ATTCTACTTCAGTTGAAGCTAGAAACCGCGTCTGCAATGTTGGATGTTCCCTTGGCCATT
     S   Q   F   L   F   Q   Y   D   N   E   K   F   G   T   C   M   A   N   C   S
301 GTTCCCAATTCCTATTCCAATATGATAACGAGAAGTTTGGAACTTGCATGGCAAACTGCA
     E   N   Y   C   I   G   N   S   T   L   E   K   A   *
361 GTGAGAACTATTGCATTGGCAACAGTACTCTAGAGAAAGCTTAAGAAGCAAGCATTTGCT
421 CTAAGAAATAAGTAAACTTTGCCACAGTATGTGTCTGTGATAATGACACTGTGTACTCCT
481 ATAAATAAGTGGCAAACTCGTGTTTTGTGATCCAATTATCTAATAAGATCTTTTCAGTTC
541 AGCAAAAAAAAAAAAAAAAAAAA
```

b

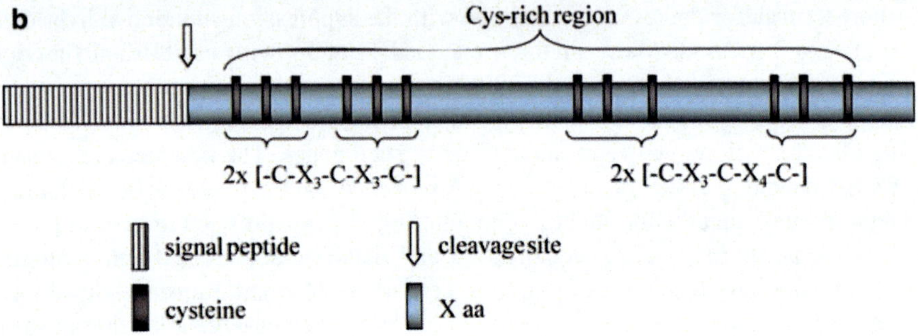

Fig. 1 *D6* cDNA and its protein product. (**a**) The cDNA sequence corresponding to the full-length mRNA. Putative coding region (underlined), non-coding mRNA, start and stop codons (*boxed*) are indicated. (**b**) Scheme of *D6* protein with indicated characteristic domains. The primers used for 5′RACE were: (5′-GGCCTTGTTCCACACCTAATTGCGCAT-3′) and (5′-CGGGATCACGTT GTTGAGCCATGGAAT-3′); while the primers used for 3′RACE were: (5′-ATGCGCAATTAGGTG TGGAACAAGGCC-3′) and (5′-GCTAGAAACCGCGTCTGCAATGTTGGA-3′)

with the process of selective autophagy (Zientara-Rytter et al. 2011). Here, we would like to summarize the results of investigations on the *D6* gene, which – in contrast to the previously characterized *UP9* – is down-regulated by S deficit.

The former results indicated that the *D6* gene is expressed mostly in leaves and that its mRNA level decreases significantly during S starvation (Wawrzyńska et al. 2005; Lewandowska et al. 2005). However, the previously established 374 bp sequence (based on five identified clones) covered only a part of *D6* mRNA with the fragment of a deduced open reading frame containing the potential START codon but no STOP codon (GenBank Acc No: AY547454). To obtain the complete sequence of *D6* mRNA the 5′- and 3′-RACE were performed. Both techniques allow to predict the complete *D6* mRNA sequence (GenBank Acc No: JN544241) consisting of an open reading frame of 118 amino acids flanked by non-coding regions (Fig. 1a). In silico analysis of the deduced open reading frame using the available on line programs, such as SignalP (http://www.cbs.dtu.dk/services/SignalP/),

PSORT (http://psort.hgc.jp) and TargetP (http://www.cbs.dtu.dk/services/TargetP/) revealed that *D6* contains N-terminal signal peptide of 24 amino acids and that it is probably targeted to the secretory pathway. The predicted molecular weight of the full-length protein is 12.8 kDa, while of the mature protein without the signal peptide – 10.2 kDa. Furthermore, the *D6* protein contains two pairs of characteristic Cys-motifs, the first pair with the [–C-X_3-C-X_3-C-] pattern and the second pair with the [–C-X_3-C-X_4-C-] pattern (Fig. 1b).

Screening of nucleotide and EST databases failed to reveal close homologues of *D6* in *Arabidopsis thaliana* and species from other families than Solanaceae. Concerning the homologous tobacco sequences, they were found in cDNA libraries obtained from multiple plant tissues, such as (i) mixture of leaves, apical buds and roots, (ii) early senescent leaves, (iii) trichomes, (v) flowers, (vi) seedling, (vii) anthers in early stage of development, (viii) pollen and (ix) floral nectar tissues. On the other hand, searching of protein databases showed that proteins with similar Cys-patterns occur in various plant species, including *Nicotiana glauca* x *Nicotiana langsdorffii* (tumor-related protein; BAA05479), *Ricinus communis* (hypotetical protein; XP_002528184), *Vitus vinifera* (unnamed protein product; CBI36344) *Antirrhinum majus* (TAP1 precursor; CAA40552) and in animals, including sea squirt *Ciona savignyi* (SCO-spondin protein; XP_002119345) and ant *Camponotus floridanus* (Tenascin-X; EFN73275). Interestingly, the cDNA encoding the first of these proteins (BAA05479) was identified by SSH method among the genetic tumor-related cDNA clones from interspecies hybrids between *N. glauca* and *N. langsdorffii*. A group of clones corresponding to this particular gene was called by the authors *TID771*. It was shown that *TID771* expression is restricted to genetic tumors, what could be direct or indirect result of stress response or hormone balance disruption. Further analysis showed that *TID771* is expressed at the late stage of genetic tumor formation, when the initiation of specific tumor organs starts. The authors failed to detect *TID771* mRNA in unorganized genetic tumor tissue, what led them to the statement that *TID771* is induced as a part of processes when unorganized tumor tissues are converted to the organized ones and that *TID771* could be putatively responsible for differentiation of some specific organs in teratomatous genetic tumors (Fujita et al. 1994). The expression of the other gene encoding the protein with the conserved [–C-X_3-C-X_{3-4}-C-] motifs, *TAP1*, is restricted to the transient period in flower development and to the tapetum of the anthers (Nacken et al. 1991). The *TAP1* was identified during search for putative target genes of *deficiens* (*def*) in *A. majus*. *Def* is involved in controlling petal and stamen organogenesis. The *TAP1* is a small protein with a putative N-terminal signal peptide (indicating that the protein may be secreted) and with Cys-patterns identical to these in *D6*. Tapetum functions as a kind of a feeder for the pollen mother cells and the developing pollen by delivering nutrients and it produces various secreted compounds. The authors suggested that *TAP1* is needed for differentiation and function of tapetal cells. On the other hand it might belong to a new class of thionins, taking part in defense response and jasmonic acid mediated signaling pathway. Indeed screening of *A. thaliana* protein databases for proteins with similar Cys-motifs allowed as to identify a group of proteins, some uncharacterized and some defined as thionins (among others THI2.1, AEE35295; THI2.2, AED94125; THI, AEE34462). However,

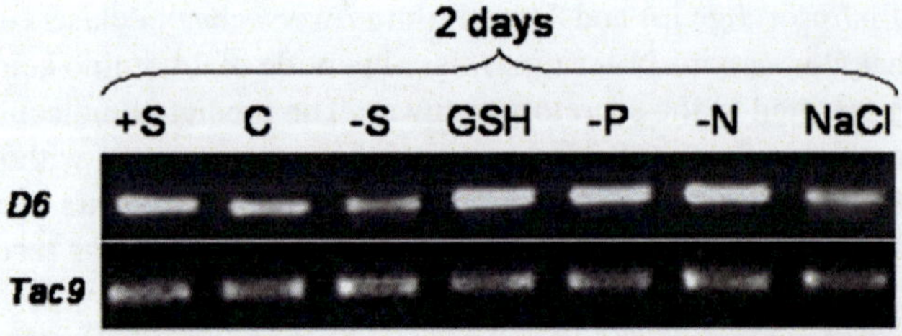

Fig. 2 The level of *D6* mRNA in various growth conditions. *D6* expression was monitored by RT-PCR using the following pair of primers: (5′-GAATTCATGGCAGTGAAAGGTT TTGCGG-3′) and (5′-CGGGATCACGTTGTTGAGCCATGGAAT-3′) and cDNA prepared from total RNA isolated from mature leaves of tobacco plants grown in the S-sufficient, control medium (C) or for 2 days in the following conditions: with 10-fold excess of sulfate (+S); without sulfate (–S); without phosphate (–P); without nitrate (–N); with 2 mM glutathione (GSH); with 100 mM NaCl (NaCl). Both, *D6* and *Tac9* (control gene, with expression presumably unaffected by the tested growth conditions) were amplified using 26 cycles

in thionins, the [–C-X_3-C-$X_{3\text{-}4}$-C-] motifs occur usually one- to twofold and localize in the central and/or the C-terminal part of the proteins. The other protein, SCO-spondin is a secreted big glycoprotein involved in neuronal cell differentiation (Meiniel et al. 2008), while tenascin-X is a member of a family of extracellular matrix glycoproteins. In mammals, this protein contributes to elasticity and strength of the dermis and is believed that it functions in matrix maturation during wound healing (Bristow et al. 2005; Egging et al. 2007). It was shown that tenascin-X, as well as its fragments derived from vascular endothelian cell extracellular matrices, stimulate wound healing (Egging et al. 2007; Demidova-Rice et al. 2011). In summary, information about the potential homologues did not bring us any closer to understanding of the *D6* function in tobacco, however all of these proteins seem to play a role in fast differentiating and/or developing cells.

In order to determine whether the transcriptional regulation of the *D6* gene is specific to S-deficiency stress, expression of the *D6* transcripts in tobacco plants grown in various conditions was determined (Fig. 2). A decreased level of *D6* mRNA was observed in mature leaves after 2 days of S deficit or 2 days of salt stress (100 mM NaCl). A small increase of *D6* mRNA amount could be observed after 2 days of N or P deficit, excess of S or in the presence of exogenous glutathione.

The genomic sequence of 1379-bp corresponding to the promoter of *D6* gene was identified using the Genome Walker technique (GenBank Acc No: JN544242). The promoter analysis resulted in identification of several motifs found also in promoters of genes which protein products take part in response to and/or which expression is regulated among others by light, dehydration, hormones, pathogens, salt, cold, diseases and wounding (results not shown). From the 'focused on primary response to S-deficit' point of view an interesting among many others are the potential EIN3 (Ethylene–insensitive 3) binding sites at the positions of 563–539 and 902–929 nucleotides from ATG, since they may be recognized by TEIL (tobacco EIN3-like) proteins. We have recently shown that some TEIL proteins are involved in S-deficiency induced up-regulation of the *UP9C* gene

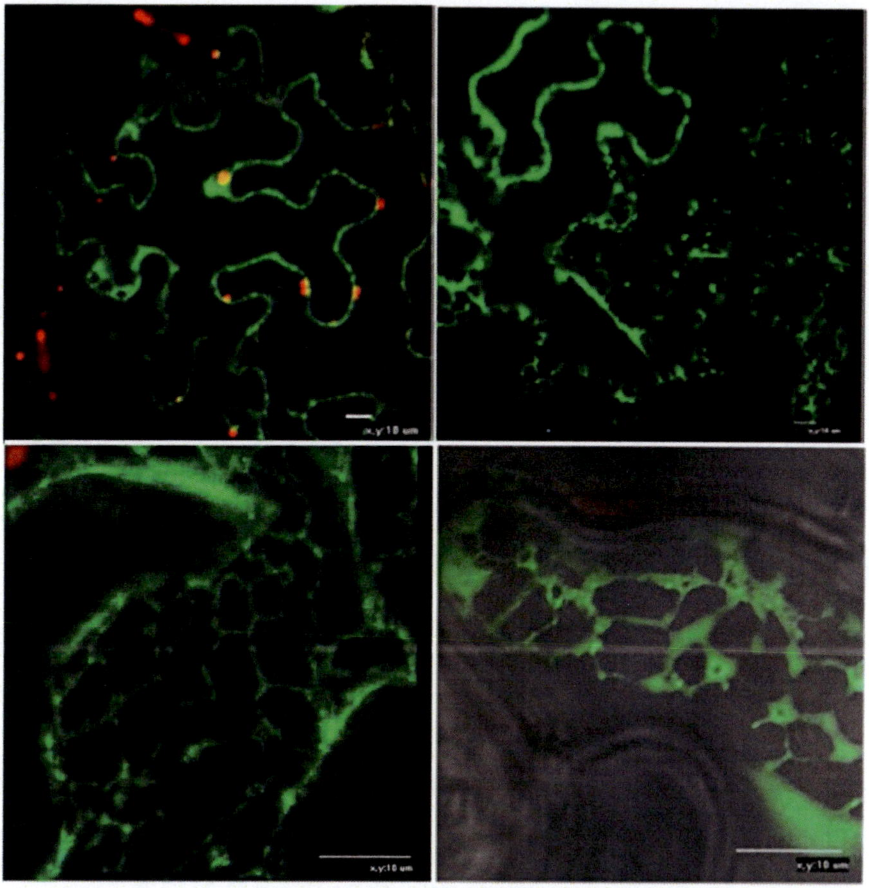

Fig. 3 Intracellular localization of the *D6*::YFP protein transiently expressed in *N. benthamiana* leaves. Several images of the *D6*::YFP protein are shown for comparison. The *scale bars* (10 μm) are indicated

(Wawrzyńska et al. 2010). Interestingly, the previously identified transcriptional factor SLIM1, which belongs to the same family and regulates gene expression in response to S-deficit in *A. thaliana* (Maruyama-Nakashita et al. 2006), can also bind to the *UP9C* promoter. However, contrary to TEIL, acting as an activator of *UP9C* transcription only in S-conditions, SLIM1 strongly activates the *UP9C* promoter regardless of S availability (Wawrzyńska et al. 2010).

In order to determine the subcellular localization, the cassette encoding *D6*::YFP fusion protein was transiently expressed in the leaves of *N. benthamiana* plants grown in the conditions of the optimal S-supply and its subcellular localization was analyzed. The transcription of the recombinant protein was under the control of the constitutive 35 S promoter from cauliflower mosaic virus (CaMV). The results indicated that *D6* localizes presumably in endoplasmic reticulum (ER) and possibly also in Golgi-like structures (Fig. 3).

The presence of the Cys-rich motifs might be one of the reasons for down-regulation of the *D6* gene by S-starvation resulting in a decreased amount of Cys pool. However, as shown in our previous paper, the genes encoding some other proteins with Cys-rich domains are up-regulated in such conditions (Wawrzyńska et al. 2005). An alternative explanation of the changes observed in *D6* expression could be linked to the putative involvement of *D6* protein in differentiation and/or development, however its association in plant response to S-deficit remains unclear. Perhaps the molecular function of *D6* can be established after analysis of the tobacco lines with 'sense' and 'antisense' expression of *D6* (in preparation). Additionally, information about *D6* promoter and the promoters of other genes down-regulated by S-deficit may allow in the future to identify specific *cis*- and *trans*-factors responsible for this effect. Despite numerous problems associated with work on genes of unknown function it is worth to make such efforts. Such results open new area of knowledge about processes and protein networks in plants.

Acknowledgements The work was supported by the Ministry of Science and Higher Education (N N302 119435). ML acknowledges support by German Academic Exchange Service (DAAD).

References

Bristow J, Carey W, Egging D, Schalkwijk J (2005) Tenascin-X, collagen, elastin, and the Ehlers–Danlos syndrome. Am J Med Genet C Semin Med Genet 139C:24–30

Demidova-Rice TN, Geevarghese A, Herman IM (2011) Bioactive peptides derived from vascular endothelial cell extracellular matrices promote microvascular morphogenesis and wound healing *in vitro*. Wound Repair Regen 19:59–70

Egging D, van Vlijmen-Willems I, van Tongeren T, van Tongeren T, Schalkwijk J, Peeters A (2007) Wound healing in tenascin-X deficient mice suggests that tenascin-X is involved in matrix maturation rather than matrix deposition. Connect Tissue Res 48:93–98

Fujita T, Kouchi H, Ichikawa T, Syono K (1994) Cloning of cDNAs for genes that are specifically or preferentially expressed during the development of tobacco genetic tumors. Plant J 5:645–654

Lewandowska M, Sirko A (2009) Identification of additional genes regulated by sulfur shortage in tobacco. In: Sirko A, De Kok LJ, Haneklaus S, Hawkesford MJ, Rennenberg H, Saito K, Schnug E, Stulen I (eds) Sulfur metabolism in plants. Regulatory aspects, significance of sulfur in the food chain, agriculture and the environment. Backhuys Publishers, Leiden, Margraf Publishers GmbH, Weikersheim, pp 73–77

Lewandowska M, Wawrzyńska A, Kamińska J, Liszewska F, Sirko A (2005) Identification of novel proteins of *Nicotiana tabacum* regulated by short term sulfur starvation. In: Saito K, De Kok LJ, Stulen I, Hawkesford MJ, Schnug E, Sirko A, Rennenberg H (eds) Sulfur transport and assimilation in plants in the postgenomic era. Backhuys Publishers, Leiden, pp 153–156

Lewandowska M, Wawrzyńska A, Moniuszko G, Łukomska J, Zientara K, Piecho M, Hodurek P, Zhukov I, Liszewska F, Nikiforova V, Sirko A (2010) A contribution to identification of novel regulators of plant response to sulfur deficiency: characteristics of a tobacco gene UP9C, its protein product and the effects of UP9C silencing. Mol Plant 3:347–360

Maruyama-Nakashita A, Nakamura Y, Tohge T, Saito K, Takahashi H (2006) *Arabidopsis* SLIM1 is a central transcriptional regulator of plant sulfur response and metabolism. Plant Cell 18: 3235–3251

Meiniel O, Meiniel R, Lalloue F, Didier R, Jauberteau M-O, Meiniel A, Petit D (2008) The lengthening of a giant protein: when, how, and why? J Mol Evol 66:1–10

Nacken WK, Huijser P, Beltran JP, Saedler H, Sommer H (1991) Molecular characterization of two stamen-specific genes, tap1 and fil1, that are expressed in the wild type, but not in the deficiens mutant of *Antirrhinum majus*. Mol Gen Genet 229:129–136

Wawrzyńska A, Lewandowska M, Hawkesford MJ, Sirko A (2005) Using a suppression subtractive library-based approach to identify tobacco genes regulated in response to short-term sulphur deficit. J Exp Bot 56:1575–1590

Wawrzyńska A, Lewandowska M, Sirko A (2010) *Nicotiana tabacum* EIL2 directly regulates expression of at least one tobacco gene induced by sulphur starvation. J Exp Bot 61:889–900

Zientara-Rytter K, Łukomska J, Moniuszko G, Gwozdecki R, Surowiecki P, Lewandowska M, Liszewska F, Wawrzyńska A, Sirko A (2011) Identification and functional analysis of Joka2, a tobacco member of the family of selective autophagy cargo receptors. Autophagy 7:1145–1158

Sulfate Uptake, Cysteine and GSH Contents Are Increased by 5-Aminolevulinic Acid in *Arabidopsis thaliana*

Akiko Maruyama-Nakashita

Abstract A key precursor of porphyrin biosynthesis, 5-aminolevulinic acid (ALA), is used as a fertilizer to promote plant growth and crop yields. Carbon fixation and nitrogen assimilation are promoted by ALA in plants, but the effect on other metabolic pathway has not been elucidated. In this study, we analyzed the effect of ALA on sulfur assimilation. Treatment with 0.3 and 1 mmol^{-1} ALA significantly increased the sulfate uptake and the transcript levels of key sulfur transport and assimilatory genes. The accumulation of cysteine and glutathione was increased in shoots but decreased in roots. These data demonstrated a new role of ALA that regulates positively transcript levels of some sulfur assimilatory genes, sulfate uptake and contents of cysteine and glutathione.

5-Aminolevulinic acid (ALA) is a key precursor of porphyrin biosynthesis including chlorophyll and heme. ALA treatment sometimes promotes the growth and yield of several crops (Hotta et al. 1997a, b; Tanaka et al. 2005). In these conditions, ALA promotes the carbon fixation through the stimulation of photosynthesis (Hotta et al.1997b) and nitrogen assimilation in plants through the induction of nitrate reductase and nitrite reductase activity (Mishra and Srivastava 1983). However, the effects of ALA on other metabolic pathways in plants have not been determined. Sulfur is an essential macronutrients required for plants. Plants use sulfate as the major sulfur source and synthesize the sulfur-containing amino acids cysteine and methionine (Crawford et al. 2000; Saito 2004). Sulfur assimilation starts from the uptake of external sulfate by the activity of sulfate transporter (SULTR) in roots. Sulfate is activated by ATP sulfurylase and then reduced by two-step reactions

A. Maruyama-Nakashita (✉)
Faculty of Agriculture, Laboratory of Plant Nutrition, Kyushu University, 6-10-1 Hakozaki, Higashi-ku, Fukuoka, 812-8581 Japan
e-mail: amaru@agr.kyushu-u.ac.jp

L.J. De Kok et al. (eds.), *Sulfur Metabolism in Plants*, Proceedings of the International Plant Sulfur Workshop 1, DOI 10.1007/978-94-007-4450-9_11,

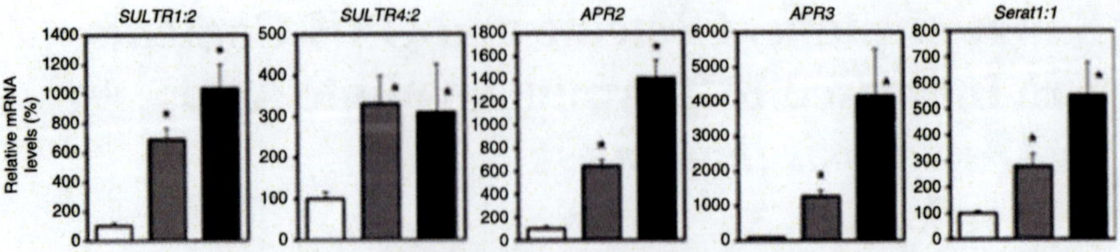

Fig. 1 Induction of mRNA accumulation of sulfur assimilatory genes by ALA treatment. *Arabidopsis thaliana* cv Columbia plants were grown for 7 days on agar medium, and then incubated for 2 days on the medium containing 0 (*white column*), 0.3 (*gray column*) and 1 (*black column*) μmol L^{-1} of ALA. The mRNA contents of *SULTR1;2*, *SULTR4;2*, *APR2*, *APR3*, *Serat2;1* and *UBQ2* in root tissues were determined by real-time RT-PCR. The relative values indicate comparisons with the control plants (0 mmol l^{-1} ALA). Statistically significant differences from the control with P-values less than 0.05 are shown as * (n=3)

catalyzed by APS reductase (APR) and sulfite reductase to produce sulfide. Then with cysteine synthase, sulfide reacts with O-acetyl-L-serine, which is produced from serine by serine acetyltransferase (Serat) activity, and turned into cysteine. Glutathione (GSH), methionine and many kinds of sulfur-containing compounds are produced from cysteine. The studies of the metabolic regulation of SULTR and APR have indicated that the regulatory network of sulfur assimilation is influenced by the nitrogen and carbon status of plants (Koprivova et al. 2000; Maruyama-Nakashita et al. 2004). In this study, we examined the effects of ALA on sulfur assimilation using *Arabidopsis thaliana*.

To determine whether ALA treatment increases the expression of sulfur assimilatory-related genes, the mRNA accumulation of genes of several key enzymes in the sulfur assimilatory pathway were analyzed (Fig. 2) with or without ALA treatment. Seven-day-old plants grown on agar media were transferred to the same media containing 0, 0.3 and 1 μmol L^{-1} of ALA. Two days after the transfer, root tissues of the plants were isolated and subjected to Real-Time RT-PCR analysis (Fig. 1; Maruyama-Nakashita et al. 2010). Genes analyzed their transcript levels were, *SULTR1;2*, a high affinity transporter responsible for sulfate uptake from root surface (Shibagaki et al. 2002; Yoshimoto et al. 2002), *SULTR4;2*, a sulfate exporter from vacuoles to the cytosol (Kataoka et al. 2004), *APR2* and *APR3*, catalyzing the reaction from APS to sulfite in plastids (Gutierrez-Marcos et al. 1996), and Serat2;1, catalyzing O-acetyl-L-serine synthesis from serine in plastids and the cytosol (Kawashima et al. 2005). Transcript levels of all tested genes were significantly increased by 0.3 and 1 mmol L^{-1} ALA (Fig. 1).

As ALA treatment increased the mRNA level of *SULTR1;2*, sulfate uptake experiments were carried out as described previously (Kataoka et al. 2004), to determine whether the ALA-mediated up-regulation of *SULTR1;2* results in modulation of the sulfate uptake activity (Fig. 2). Application of 0.3 or 1 mmol L^{-1} ALA significantly increased sulfate uptake activities. Then the cysteine and GSH contents in plants were analyzed to determine whether the up-regulation of the sulfur assimilatory genes by ALA reflects the levels of these compounds (Fig. 3).

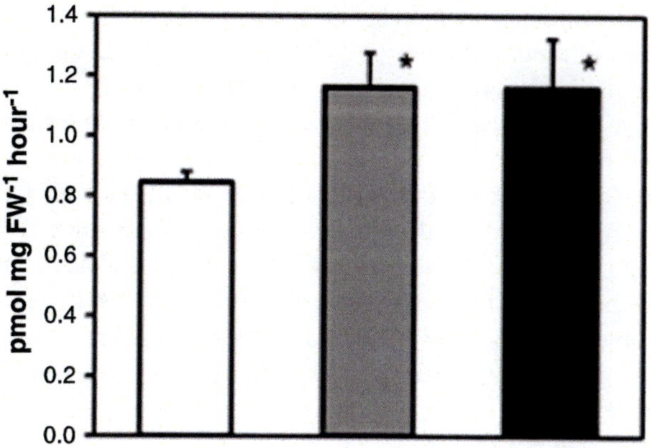

Fig. 2 Effects of ALA treatment on sulfate uptake activities in plants. Plants were treated as described in Fig. 1. Statistically significant differences from the control with P-values less than 0.05 are shown as * (n=8)

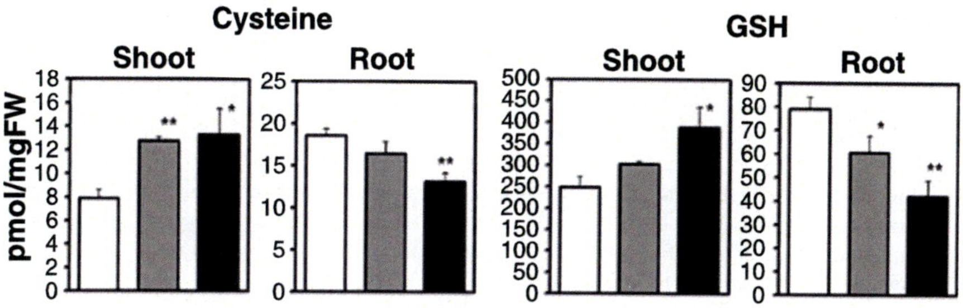

Fig. 3 Effects of ALA treatment on cysteine and GSH accumulation in plants. Plants were treated as described in Fig. 1. Statistically significant differences from the control with P-values less than 0.1 and 0.05 are shown as * and **, respectively (n=4)

Treatment with ALA increased both cysteine and GSH contents in shoots, whereas those in roots were decreased. Cysteine contents in shoots and roots of the ALA-treated plants were 5.42 pmol mg FW^{-1} more and 5.44 pmol mg FW^{-1} less, respectively, than those of non-treated plants (Fig. 3). GSH contents in shoots and roots of the ALA-treated plants were 139.9 pmol mg FW^{-1} more and 36.8 pmol mg FW^{-1} less, respectively, than those of non-treated plants (Fig. 3). Since the fresh weights of the shoots were twofold to threefold more than those of roots (data not shown), the ALA treatment increased cysteine and GSH contents in plants.

The different patterns of cysteine and GSH accumulation between shoots and roots (Fig. 3) were unexpected results considering the transcript levels of sulfur assimilatory genes in roots (Fig. 1). Cysteine and GSH are synthesized both in photosynthetic and non-photosynthetic tissues (Saito 2004; Noctor et al. 2002), and are translocated from roots to shoots through xylem flow and *vice versa* from shoot to root through phloem (Noctor et al. 2002; Tausz et al. 2004; Li et al. 2006).

The ALA treatment possibly induced the biosynthesis of cysteine and GSH in roots and the transfer of them to shoots, probably because the ALA treatment activate the translocation mechanism only from roots to shoots.

The positive effects of ALA on the transcript levels of sulfur assimilatory genes were revealed in this study (Fig. 1). ALA stimulates carbon fixation and nitrogen assimilation (Hotta et al. 1997b; Mishra and Srivastava 1983), and the positive effects of carbon and nitrogen availability on the mRNA levels of sulfur assimilatory genes have been reported (Koprivova et al. 2000; Maruyama-Nakashita et al. 2004). It is not clear whether the effects of ALA on mRNA levels of sulfur assimilatory genes and cysteine and GSH contents are due to the increased level of carbon and nitrogen assimilation or the direct influence of ALA on the sulfur assimilatory genes, which is an interesting question concerning the regulatory mechanism of sulfur assimilatory pathway. The tested concentrations of ALA in this study did not increase the plant growth of *A. thaliana,* however, the new roles of ALA demonstrated in this study possibly contributes to the ALA-mediated stimulation of plant growth.

References

Crawford NM, Kahn ML, Leustek T, Long SR (2000) Nitrogen and sulfur. In: Buchanan BB, Gruissem W, Jones RL (eds) Biochemistry and molecular biology of plants. American Society of Plant Biologists, Rockville, pp 824–849

Gutierrez-Marcos JF, Roberts MA, Campbell EI, Wray JL (1996) Three members of a novel small gene-family from *Arabidopsis thaliana* able to complement functionally an *Escherichia coli* mutant defective in PAPS reductase activity encode proteins with a thioredoxin-like domain and "APS reductase" activity. Proc Natl Acad Sci USA 93:13377–13382

Hotta Y, Tanaka T, Takaoka H, Takeuchi Y, Konnai M (1997a) New physiological effects of 5-aminolevulinic acid in plants: the increase of photosynthesis, chlorophyll content, and plant growth. Biosci Biotech Biochem 61:2025–2028

Hotta Y, Tanaka T, Takaoka H, Takeuchi Y, Konnai M (1997b) Promotive effects of 5-aminolevulinic acid on the yield of several crops. Plant Growth Regul 22:109–114

Kataoka T, Watanabe-Takahashi A, Hayashi N, Ohnishi M, Mimura T, Buchner P, Hawkesford MJ, Yamaya T, Takahashi H (2004) Vacuolar sulfate transporters are essential determinants controlling internal distribution of sulfate in *Arabidopsis*. Plant Cell 16:2693–2704

Kawashima CG, Berkowitz O, Hell R, Noji M, Saito K (2005) Characterization and expression analysis of a serine acetyltransferase gene family involved in a key step of the sulfur assimilation pathway in *Arabidopsis*. Plant Physiol 137:220–230

Koprivova A, Suter M, Op den Camp R, Brunold C, Kopriva S (2000) Regulation of sulfur assimilation by nitrogen in *Arabidopsis*. Plant Physiol 122:737–746

Li Y, Dankher OP, Carreira L, Smith AP, Meagher RB (2006) The shoot-specific expression of gamma-glutamylcysteine synthetase directs the long-distance transport of thiol-peptides to roots conferring tolerance to mercury and arsenic. Plant Physiol 141:288–298

Maruyama-Nakashita A, Nakamura Y, Yamaya T, Takahashi H (2004) Regulation of high-affinity sulfate transporters in plants: towards systematic analysis of sulfur signaling and regulation. J Exp Bot 55:1843–1849

Maruyama-Nakashita A, Hirai MY, Funada S, Fueki S (2010) Exogenous application of 5-aminolevulinic acid increases transcript levels of sulfur transport and assimilatory genes, sulfate uptake, and cysteine and glutathione contents in *Arabidopsis thaliana*. Soil Sci Plant Nutr 56:281–288

Mishra SN, Srivastava HS (1983) Stimulation of nitrate reductase activity by delta amino levulinic acid in excised maize leaves. Cell Mol Life Sci 39:1118–1120

Noctor G, Gomez L, Vanacker H, Foyer CH (2002) Interactions between biosynthesis, compartmentation and transport in the control of glutathione homeostasis and signaling. J Exp Bot 53:1283–1304

Saito K (2004) Sulfur assimilatory metabolism. The long and smelling road. Plant Physiol 136:2443–2450

Shibagaki N, Rose A, McDermott JP, Fujiwara T, Hayashi H, Yoneyama T, Davies JP (2002) Selenate-resistant mutants of *Arabidopsis thaliana* identify *Sultr1;2*, a sulfate transporter required for efficient transport of sulfate into roots. Plant J 29:475–486

Tanaka T, Iwai K, Watanabe K, Hotta Y (2005) Development of 5-aminolevulinic acid for agricultural uses. Regul Plant Growth Dev 40:22–29

Tausz M, Pilcha B, Rennenberg H, Grilla D, Herschbach C (2004) Root uptake, transport, and metabolism of externally applied glutathione in *Phaseolus vulgaris* seedlings. J Plant Physiol 161:347–349

Yoshimoto N, Takahashi H, Smith FW, Yamaya T, Saito K (2002) Two distinct high-affinity sulfate transporters with different inducibilities mediate uptake of sulfate in *Arabidopsis* roots. Plant J 29:465–473

Partitioning of Sulfur Between Primary and Secondary Metabolism

Sarah G. Mugford, Colette Matthewman, Bok-Rye Lee, Ruslan Yatusevich, Naoko Yoshimoto, Markus Wirtz, Lionel Hill, Ruediger Hell, Hideki Takahashi, Kazuki Saito, Tamara Gigolashvili, and Stanislav Kopriva

Abstract The essential macronutrient sulfur enters plants as inorganic sulfate, which is assimilated by reduction to sulfide and incorporation into cysteine or by transfer to various metabolites as organic sulfo-group. The two pathways share the activation of sulfate by adenylation to adenosine 5′-phosphosulfate (APS). The further fate of APS is decided in interplay of two enzymes, APS reductase and APS kinase. Here we summarize recent progress in analysis of APS kinase and its role in plant sulfur metabolism as well as the contribution of the two enzymes to control of flux through sulfate assimilation and partitioning of sulfur between primary and secondary metabolism.

Plants cover their need for sulfur by taking up sulfate from the soil. The sulfate has to be assimilated into bioorganic compounds. The first step of such assimilation is activation to adenosine 5′-phosphosulfate (APS) by ATP sulfurylase (ATPS).

S.G. Mugford • C. Matthewman • B.-R. Lee • L. Hill • S. Kopriva (✉)
John Innes Centre , Norwich NR4 7UH, UK
e-mail: stanislav.kopriva@jic.ac.uk

R. Yatusevich • T. Gigolashvili
Botanisches Institut der Universität zu Köln, Gyrhofstrasse 15, Köln D-50931, Germany

N. Yoshimoto • K. Saito
RIKEN Plant Science Center, 1-7-22 Suehiro-cho, Tsurumi-ku, Yokohama 230-0045, Japan

Department of Molecular Biology and Biotechnology, Graduate School of Pharmaceutical Sciences, Chiba University, Yayoi-cho 1-33, Inage-ku, Chiba 263-8522, Japan

M. Wirtz • R. Hell
Heidelberg Institute for Plant Sciences, Im Neuenheimer Feld 360, Heidelberg D-69120, Germany

H. Takahashi
RIKEN Plant Science Center, 1-7-22 Suehiro-cho, Tsurumi-ku, Yokohama 230-0045, Japan

L.J. De Kok et al. (eds.), *Sulfur Metabolism in Plants*, Proceedings of the International Plant Sulfur Workshop 1, DOI 10.1007/978-94-007-4450-9_12,

In the primary sulfate assimilation APS is reduced to sulfide and incorporate into O-acetylserine to form cysteine (Kopriva 2006). APS can also be phosphorylated by APS kinase and used as a source of active sulfate for the synthesis of various secondary metabolites. In order to dissect the control of sulfur partitioning between primary and secondary metabolism we have analyzed the APS kinase gene family in *Arabidopsis*. We generated plants with modified expression of the two enzymes metabolizing APS, APS reductase and APS kinase, and analyzed the fluxes of sulfur. We found that, at least in *Arabidopsis*, the interplay between APS reductase and APS kinase is important for sulfur partitioning between primary and secondary metabolism.

APS kinase (APK) is encoded by a small gene family with four members. We showed by analysis of recombinant proteins that all four genes encode functional enzymes (Mugford et al. 2009). Three APK isoforms are localized in the plastids, while one, APK3, is cytosolic. To elucidate functions of the individual APK isoforms, T-DNA insertions lines disrupting the *APK* genes have been obtained from the Nottingham *Arabidopsis* Stock Centre and analyzed. The lines with a single *APK* gene disrupted did not show any phenotypes, neither in growth, development, or levels of key sulfur containing metabolites (Mugford et al. 2009). Therefore, double and triple mutants were obtained by crossing. All six combinations of double mutants were recovered, while only three out of four combinations of triple *apk* mutants were viable (Mugford et al. 2010). We could not obtain plants with APK2 as a single active isoform. Selfing of *apk1 APK2 apk3/APK3 apk4* plants, which should result in 25% of triple mutants, produced 1:1 ratio of WT and heterozygous plants at APK3 locus, which indicates pollen lethality. Indeed, APK2 is the only APK isoform not expressed in the pollen indicating a necessity of availability of PAPS for pollen development. Interestingly, *apk2 apk3 apk4* plants were not distinguishable from WT plants (Fig. 1), showing that APK1 is the major isoform in *Arabidopsis*, which on its own is capable to support growth under normal conditions (Mugford et al. 2010). On the other hand, *apk1 apk2 apk3* and *apk1 apk2 apk4* mutants are much smaller that WT and smaller that *apk1 apk2*, the only double mutant that showed a growth phenotype (Fig. 1; Mugford et al. 2009, 2010).

Apart from smaller size, the *apk1 apk2* mutant showed significant alteration in the levels of glucosinolates (GLS) and their desulfo- precursors (ds-GLS). The levels of each individual GLS were reduced in the leaves of the mutant so that total GLS levels reached only ca. 15% of that in Col-0 (Fig. 2). The reduction in GLS was accompanied by a massive increase in ds-GLS, which reached a tenfold higher concentration than the mature GLS in WT leaves. A similar reduction in GLS levels was detected in the seeds of the mutant plants, however, the ds-GLS did not accumulate in the seeds. These results confirm that GLS are not synthesized in seeds but are transported there in the mature sulfate form (Magrath and Mithen 1993). The high accumulation of ds-GLS in leaves of *apk1 apk2* also indicated that the GLS biosynthetic pathway is upregulated in the mutant. Indeed, both microarray analysis and qPCR revealed that the transcript levels for genes involved in synthesis of both aliphatic and indolic GLS were three- to sixfold higher in the mutant than in Col-0 (Mugford et al. 2009).

Fig. 1 Phenotypes of APK triple mutants compared with *apk1 apk2* double mutant and the WT Col-0. The plants were grown for 5 weeks at long days in a controlled environment room (Adapted from Mugford et al. (2010))

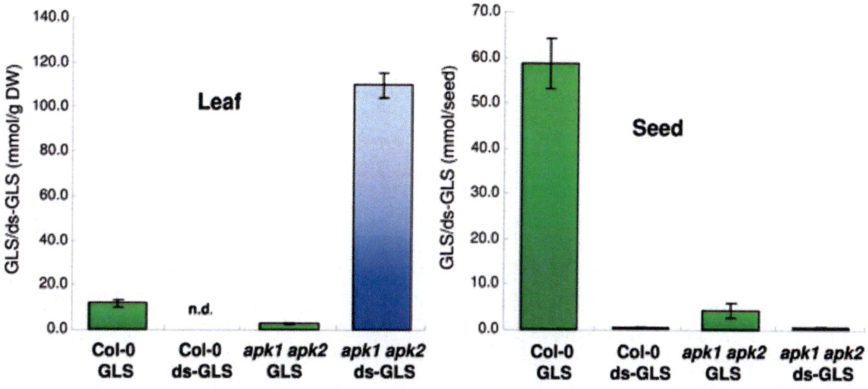

Fig. 2 Levels of total glucosinolates (GLS) and desulfo-glucosinolates (ds-GLS) in 5 weeks old rosette leaves and dry seeds of Col-0 and *apk1 apk2* mutants (From Mugford et al. (2009))

Clearly, reduction in APS kinase activity limited synthesis of GLS. In addition, among the genes upregulated in *apk1 apk2* mutants were genes encoding three isoforms of ATPS, *ATPS1*, *-2*, and *-3*. Thus, we hypothesized that genes involved in PAPS synthesis may be part of the GLS biosynthetic network controlled by two groups of MYB transcription factors (Hirai et al. 2007; Gigolashvili et al. 2007a, b, 2008).

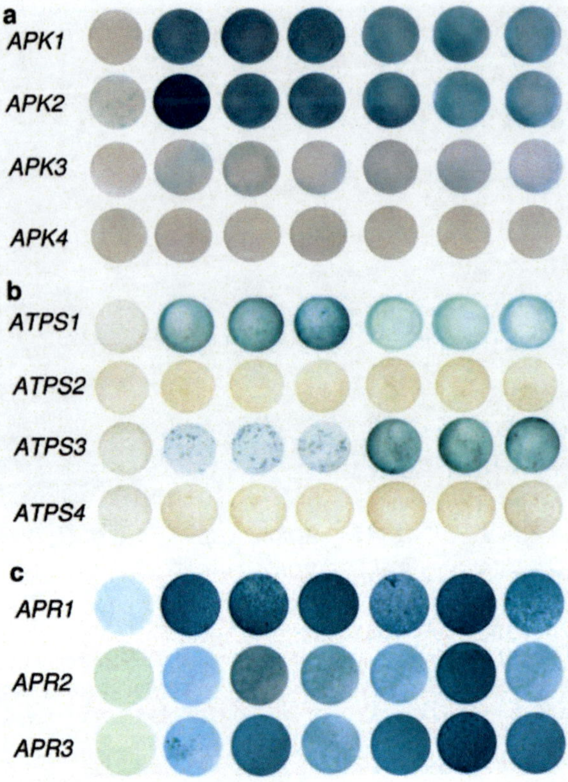

Fig. 3 Transactivation assay of (**a**) APK (**b**) ATPS and (**c**) APR promoters with MYB transcription factors. The promoters of all four APK, four ATPS, and three APR isoforms were fused to the uidA (GUS) reporter gene. Cultured *A. thaliana* Col-0 cells were inoculated with the supervirulent Agrobacterium strain LBA4404.pBBR1MCS. virGN54D containing either only the reporter construct or the reporter construct and, in addition, Pro35S:MYB effector constructs. GUS staining indicates trans-activation of a promoter by an effector

To test the hypothesis, the promoters of all ATPS and APK isoforms were isolated and used for transactivation assays to detect binding of the MYB factors (Yatusevich et al. 2010). This analysis revealed that promoters of *APK1* and *APK2* strongly interact with all 6 MYB factors controlling GLS synthesis (Fig. 3a). *ATPS1* and *ATPS3* are also under control of these factors. However, *ATPS1* seems to interact stronger with the MYB factors controlling aliphatic GLS, while *ATPS3* is preferentially controlled by the factors affecting indolic GLS (Fig. 3b). Interestingly, promoters of three APS reductase isoforms, encoding the key step in the reductive primary sulfate assimilation, also interact with the MYB factors. The results of the transactivation assays were confirmed by qPCR analysis of plants overexpressing these MYB factors (Yatusevich et al. 2010). Thus, PAPS synthesis is an integral part of GLS biosynthesis network, which also includes genes of primary sulfate assimilation.

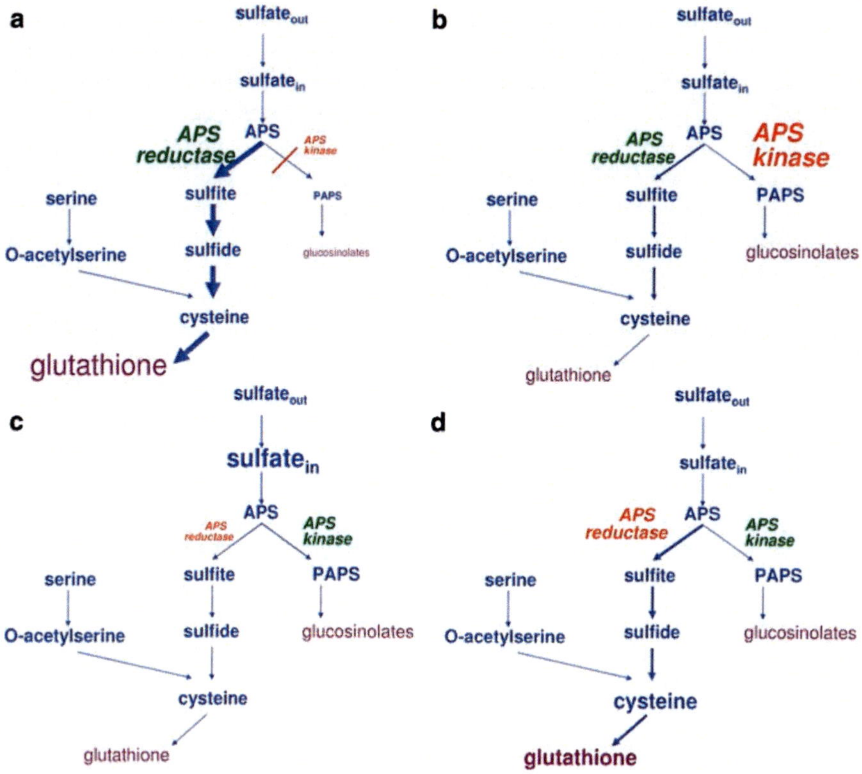

Fig. 4 Effects of manipulation of APS kinase and APS reductase on levels of S-containing compounds and flux through the pathway. The size of script symbolizes the size of the pool or enzyme activity, the *thickness of the arrows* corresponds to the flux, the manipulated enzyme is colored *red*. (**a**) *apk1 apk2*, (**b**) EcAPK targeted to plastids, (**c**) *apr2*, (**d**) PpAPR-B expressed in plastids

To determine the effects of manipulation of APS kinase and APS reductase on the partitioning of sulfur between primary and secondary assimilation we created transgenic plants overexpressing bacterial APS kinase in the cytosol and in the plastids as well as lines overexpressing the APR-B form of APS reductase from *Physcomitrella patens* (Mugford et al. 2011). These plants were analysed together with *apk1 apk2* and *apr2* mutants (Mugford et al. 2009; Loudet et al. 2007) for the accumulation of thiols and GLS and for the flux through the sulfate assimilation pathway. As summarized in Fig. 4, the *apk1 apk2* plants accumulate glutathione and show increased flux through assimilatory sulfate reduction partly due to elevated APR activity (Fig. 4a). Overexpression of APS kinase does not affect GLS accumulation but increases slightly the flux through primary assimilation (Fig. 4b). Reduction of APR activity by 75% in apr2 mutant has surprisingly no effect on the metabolite pools or flux (Fig. 4c), while overexpression of APR-B, despite only ca. 10% increase in the enzyme activity results in increased flux and increased levels of glutathione (Fig. 4d; Mugford et al. 2011).

The analysis reported here shows that APS kinase is an essential enzyme. However, the activity can be located either in plastids or in the cytosol. Strong reduction in APK results in semi-dwarf phenotype and in reduced accumulation of sulfated compounds such as glucosinolates. APK as well as other components of sulfate assimilation ATPS and APR are part of glucosinolate biosynthesis network controlled by two groups of MYB factors. Limitation of PAPS production leads to upregulation of GLS biosynthesis and to redirection of sulfur flow to primary assimilation. The partitioning of sulfur is controlled by a complex interplay of APS kinase and APS reductase.

Acknowledgements The research in S. Kopriva's laboratory is supported by Biotechnology and Biological Science Research Council (BBSRC) of the UK.

References

Gigolashvili T, Berger B, Mock HP, Müller C, Weisshaar B, Flügge UI (2007a) The transcription factor HIG1/MYB51 regulates indolic glucosinolate biosynthesis in *Arabidopsis thaliana*. Plant J 50:886–901

Gigolashvili T, Yatusevich R, Berger B, Müller C, Flügge UI (2007b) The R2R3-MYB transcription factor HAG1/MYB28 is a regulator of methionine-derived glucosinolate biosynthesis in *Arabidopsis thaliana*. Plant J 51:247–261

Gigolashvili T, Engqvist M, Yatusevich R, Müller C, Flügge UI (2008) HAG2/MYB76 and HAG3/MYB29 exert a specific and coordinated control on the regulation of aliphatic glucosinolate biosynthesis in *Arabidopsis thaliana*. New Phytol 177:627–642

Hirai MY, Sugiyama K, Sawada Y, Tohge T, Obayashi T, Suzuki A, Araki R, Sakurai N, Suzuki H, Aoki K, Goda H, Nishizawa OI, Shibata D, Saito K (2007) Omics-based identification of *Arabidopsis* Myb transcription factors regulating aliphatic glucosinolate biosynthesis. Proc Natl Acad Sci USA 104:6478–6483

Kopriva S (2006) Regulation of sulfate assimilation in *Arabidopsis* and beyond. Ann Bot 97:479–495

Loudet O, Saliba-Colombani V, Camilleri C, Calenge F, Gaudon V, Koprivova A, North KA, Kopriva S, Daniel-Vedele F (2007) Natural variation for sulfate content in *Arabidopsis* is highly controlled by adenosine 5′-phosphosulfate reductase. Nat Genet 39:896–900

Magrath R, Mithen R (1993) Maternal effects on the expression of individual aliphatic glucosinolates in seeds and seedlings of *Brassica napus*. Plant Breed 111:249–252

Mugford SG, Yoshimoto N, Reichelt M, Wirtz M, Hill L, Mugford ST, Nakazato Y, Noji M, Takahashi H, Kramell R, Gigolashvili T, Flügge UI, Wasternack C, Gershenzon J, Hell R, Saito K, Kopriva S (2009) Disruption of adenosine-5′-phosphosulfate kinase in *Arabidopsis* reduces levels of sulfated secondary metabolites. Plant Cell 21:910–927

Mugford SG, Matthewman CA, Hill L, Kopriva S (2010) Adenosine 5′ phosphosulfate kinase is essential for *Arabidopsis* viability. FEBS Lett 584:119–123

Mugford SG, Lee B-R, Koprivova A, Matthewman C, Kopriva S (2011) Control of sulfur partitioning between primary and secondary metabolism. Plant J 65:96–105

Yatusevich R, Mugford SG, Matthewman C, Gigolashvili T, Frerigmann H, Delaney S, Koprivova A, Flügge U-I, Kopriva S (2010) Genes of primary sulfate assimilation are part of the glucosinolate biosynthetic network in *Arabidopsis thaliana*. Plant J 62:1–11

Interaction Between Atmospheric and Pedospheric Sulfur Nutrition in *Eucalyptus camaldulensis*

Freek S. Posthumus, C. Elisabeth E. Stuiver, Michael Tausz, and Luit J. De Kok

Abstract Plants are able to use gaseous sulfurous air pollutants as a sulfur source for growth, especially under circumstances where the sulfur supply to the roots is limited. In *Eucalyptus camaldulensis* seedlings there was a direct interaction between atmospheric H_2S exposure and the uptake of sulfate by the root. At an ample sulfate supply, exposure of *E. camaldulensis* to 0.2 μl l^{-1} H_2S for 1 week resulted in a decrease of the sulfate uptake capacity of the root (up to 70%). Sulfate deprivation resulted in an up to sixfold increase in the sulfate uptake capacity of the root, however, it was completely alleviated upon H_2S exposure. At all conditions, the sulfate uptake capacity was strongly related to the sulfate concentration in the root. Evidently, upon H_2S exposure *E. camaldulensis* transferred from sulfate taken up by the root to sulfide absorbed by the shoot as sulfur source for growth. Despite that down-regulation of the sulfate uptake capacity occurred upon H_2S exposure there was no direct metabolic control of the uptake and utilization of foliarly absorbed H_2S. H_2S exposure resulted in enhanced levels of total and organic sulfur (and nitrogen) in the shoot, whereas the N/S ratio strongly decreased.

F.S. Posthumus • C.E.E. Stuiver
Laboratory of Plant Physiology, University of Groningen,
P.O. Box 11103, Groningen 9700 CC, The Netherlands

M. Tausz
Department of Forest and Ecosystem Science, University of Melbourne,
Water Street, Creswick, VIC 3363, Australia

L.J. De Kok (✉)
Laboratory of Plant Physiology, University of Groningen,
P.O. Box 11103, Groningen 9700 CC, The Netherlands

Department of Forest and Ecosystem Science, University of Melbourne,
Water Street, Creswick, VIC 3363, Australia
e-mail: l.j.de.kok@rug.nl

L.J. De Kok et al. (eds.), *Sulfur Metabolism in Plants*, Proceedings of the International Plant Sulfur Workshop 1, DOI 10.1007/978-94-007-4450-9_13,

Plants are able to use foliarly absorbed H_2S as sulfur source for growth, especially under conditions where the sulfur supply to the roots is limited (De Kok et al. 2007, 2009; Koralewska et al. 2008). In *Brassica*, there was direct interaction between atmospheric and pedospheric sulfur utilization and H_2S exposure hardly affected the total plant sulfur content, whereas the expression and activity of the sulfate transporters and APS reductase were reduced (Westerman et al. 2000, 2001; Durenkamp et al. 2007; Koralewska et al. 2008). However, in other species *viz. Allium cepa* and *Picea abies*, H_2S exposure hardly affected the uptake of sulfate, whereas it resulted in an accumulation of organic sulfur compounds in the shoot (Tausz et al. 2003; Durenkamp and De Kok 2004; Durenkamp et al. 2007). Apparently, the shoot to root signaling in the regulation of the uptake and assimilation of sulfate differs between species. In the current study the interaction between atmospheric and pedospheric sulfur nutrition was studied in seedlings of the Australian native trees species *Eucalyptus camaldulensis* (River Red Gum).

Seeds of *Eucalyptus camaldulensis* var. *camaldulensis* were germinated in vermiculite for 14 days and then transferred to a 25% Hoagland solution for 14 days. Subsequently, plants were exposed to 0 or 0.2 $\mu l\ l^{-1}$ H_2S at sulfate-sufficient (0.5 mM sulfate) and sulfate-deprived (0 mM sulfate) conditions for 7 and 14 days. Day and night temperature were 21°C and 18°C, respectively, the relative humidity was 60–70% and the photoperiod 14 h at a photon fluence rate of 350 $\mu mol\ m^{-2}\ s^{-1}$ (for more experimental details see Koralewska et al. 2008). Biomass production and the contents of pigments, total sulfur and nitrogen, sulfate and nitrate, and the sulfate uptake capacity were determined as described by Koralewska et al. (2008).

A 14-day exposure of *E. camaldulensis* to sulfate deprivation resulted in a decreased plant biomass production (by 35%) and an altered shoot to root biomass partitioning in favor of the root such that the shoot to root ratio decreased from 1.1 in sulfate-sufficient (+S) to 0.74 in sulfate-deprived plants (–S). However, after 7 days of sulfate deprivation, plant biomass production (on average growth rate was 14% day^{-1}), shoot to root ratio and pigment content were not yet affected, nevertheless the total sulfur and sulfate content of both root and shoot were diminished, whereas the nitrate content in the shoot was substantially enhanced (Table 1).

A 7 day exposure of *Eucalyptus* seedlings to 0.2 $\mu l\ l^{-1}$ H_2S at sulfate-sufficient conditions (+S, H_2S) did not affect shoot and root biomass production and shoot pigment content, but did not result in an increase in total sulfur and sulfate content in both root and shoot and total nitrogen in the shoot (Table 1). The sulfate uptake capacity of sulfate-sufficient plants was partly down-regulated (by 67%) upon H_2S exposure, indicating that plants had partially transferred from sulfate taken up by the root to H_2S taken by the shoot as sulfur source for growth (Fig. 1). H_2S exposure of sulfate-deprived plants (–S, H_2S) did neither affect plant biomass, nor shoot to root ratio and pigment content, but there was a substantial increase in total sulfur and sulfate content in both root and shoot and total nitrogen in the shoot (Table 1).

Sulfate deprivation resulted in a more than sixfold increase of the sulfate uptake capacity (Fig. 1). If sulfate-deprived plants were simultaneously exposed to H_2S, then the sulfate uptake capacity was not up-regulated, but it was even lower than

Table 1 Impact of H_2S exposure and sulfate deprivation on biomass and pigment content of *E. camaldulensis*

	+S	+S, H_2S	−S	−S, H_2S
Shoot				
Biomass production (g FW)	0.25±0.08a	0.18±0.09a	0.28±0.08a	0.22±0.06a
DMC (%)	24±4a	19±4a	22±2a	19±1a
Chlorophyll a+b (mg g^{-1} FW)	1.76±0.04a	1.72±0.07a	1.76±0.05a	1.69±0.60a
Chlorophyll a/b	3.22±0.33a	3.28±0.06a	3.49±0.05a	3.30±0.20a
Total N (mmol g^{-1} DW)	1.43±0.21a	2.29±0.01b	1.10±0.12a	2.15±0.08b
Total S (μmol g^{-1} DW)	74±9a	151±17b	45±19a	141±11b
Nitrate (μmol g^{-1} FW)	10.2±1.7a	10.2±1.3a	19.0±11.5a	11.2±1.2a
Sulfate (μmol g^{-1} FW)	7.5±0.5b	11.3±1.1d	3.3±0.9a	9.1±1.4c
Root				
Biomass production (g FW)	0.23±0.06a	0.17±0.08a	0.27±0.06a	0.22±0.06a
Root DMC (%)	9.0±0.5a	9.1±0.6a	9.6±1.1a	8.8±0.8a
Total N (mmol g^{-1} DW)	3.05±0.06a	2.83±0.13a	2.47±0.30a	3.01±0.13a
Total S (μmol g^{-1} DW)	114±8bc	130±12c	38±12a	104±6b
Nitrate (μmol g^{-1} FW)	31.8±1.8a	32.2±5.8a	27.8±1.0a	28.0±2.4a
Sulfate (μmol g^{-1} FW)	7.9±0.8c	6.4±1.0b	2.5±1.5a	5.0±0.6b

Plants were exposed for 7 days and data on biomass production and dry matter content (DMC) represent the mean of 18 and 6 measurements with 3 plants in each (±SD), respectively. The initial shoot and root fresh weight were 0.14±0.01 g and 0.13±0.02 g, respectively. Data on pigment content represent the mean of three measurements and on total nitrogen, total sulfur, nitrate and sulfate content the mean of three measurements with three plants in each (±SD), respectively. Different letters indicate significant differences between treatments ($p<0.01$, Student's t-test)

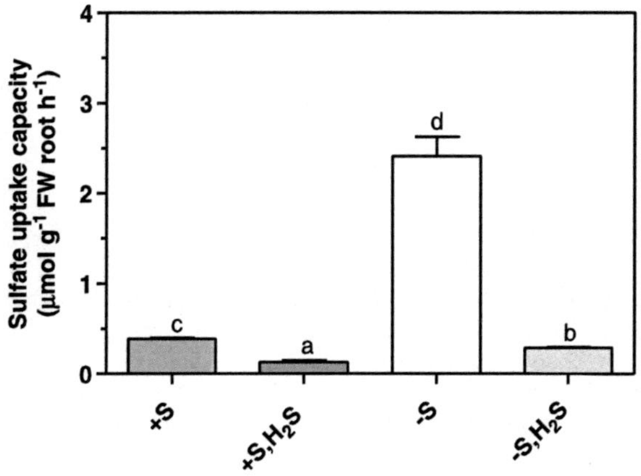

Fig. 1 Effect of H_2S exposure and sulfate deprivation on the sulfate uptake capacity of *E. camaldulensis*. Plants were exposed for 7 days and data represent the mean of three measurements with three plants in each (±SD). Different letters indicate significant differences between treatments ($p<0.01$, Student's t-test)

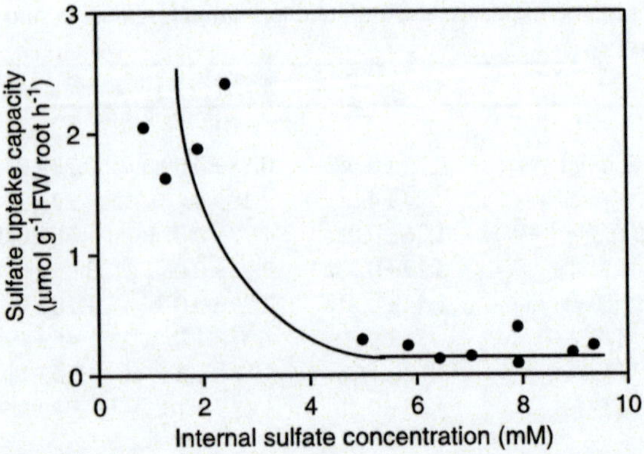

Fig. 2 The relation between the sulfate uptake capacity and the overall sulfate concentration in roots of *E. camaldulensis* at various levels of atmospheric and pedospheric sulfur nutrition. Derived from data presented in Table 1

that of sulfate-sufficient non-exposed plants (Fig. 1). For all treatments there was a direct relation between the sulfate concentration in the root and the sulfate uptake capacity (Fig. 2).

Evidently, *Eucalyptus* was able to utilize atmospheric H_2S as sulfur source, resulting in a down-regulation of the sulfate uptake capacity by the root both in the presence and absence of sulfate in the root environment. However, in contrast to other plant species (*e.g. Brassica;* Koralewska et al. 2008), H_2S exposure resulted in an increase in sulfur content of both root and shoot, indicating the absence of a strict regulatory control between shoot and root in atmospheric H_2S and pedospheric sulfate utilization. Moreover, H_2S exposure also resulted in enhanced nitrogen content in the shoot, which could be (in common with sulfur) ascribed to an enrichment of the organic nitrogen fraction. The nature of both the enhancement of organic S and N accumulation upon H_2S exposure needs further to be investigated.

References

De Kok LJ, Durenkamp M, Yang L, Stulen I (2007) Atmospheric sulfur. In: Hawkesford MJ, De Kok LJ (eds) Sulfur in plants – an ecological perspective. Springer, Dordrecht, pp 91–106

De Kok LJ, Yang L, Stuiver CEE, Stulen I (2009) Negative vs. positive functional plant responses to air pollution: a study establishing cause-effect relationships of SO_2 and H_2S. In: Legge AH (ed) Air quality and ecological impacts: relating sources to effects, vol 9, Developments in environmental sciences. Elsevier B.V, Amsterdam, pp 121–134

Durenkamp M, De Kok LJ (2004) Impact of pedospsheric and atmospheric sulphur nutrition on sulphur metabolism of *Allium cepa* L., a species with a potential sink capacity for secondary sulphur compounds. J Exp Bot 55:1821–1830

Durenkamp M, De Kok LJ, Kopriva S (2007) Adenosine 5′ phosphosulphate reductase is regulated differently in *Allium cepa* L. and *Brassica oleracea* L. upon exposure to H_2S. J Exp Bot 58: 1571–1579

Koralewska A, Stuiver CEE, Posthumus FS, Kopriva S, Hawkesford MJ, De Kok LJ (2008) Regulation of sulfate uptake, expression of the sulfate transporters Sultr1;1 and Sultr1;2, and APS reductase in Chinese cabbage (*Brassica pekinensis*) as affected by atmospheric H2S nutrition and sulfate deprivation. Funct Plant Biol 35:318–327

Tausz M, Wonisch W, Weidner W, De Kok LJ, Grill D (2003) Modification of S-metabolism in spruce trees by H_2S studied by radiolabelled sulfate uptake. In: Davidian J-C, Grill D, De Kok LJ, Stulen H, Hawkesford MJ, Schnug E, Rennenberg H (eds) Sulfur transport and assimilation in plants: regulation, interaction and signaling. Backhuys Publishers, Leiden, pp 341–343

Westerman S, De Kok LJ, Stulen I (2000) Interaction between metabolism of atmospheric H_2S in the shoot and sulfate uptake by the roots of curly kale (*Brassica oleracea* L.). Physiol Plant 109:443–449

Westerman S, Stulen I, Suter M, Brunold C, De Kok LJ (2001) Atmospheric H_2S as sulfur source for *Brassica oleracea*: consequences for the activity of the enzymes of the assimilatory sulfate reduction pathway. Plant Physiol Biochem 39:425–432

Comparative Analysis of Amino Acids from *Arabidopsis* Wild-Type and Mutant *sir1-1* Leaves by Reverse-Phase High-Pressure Liquid Chromatography (HPLC) and Gas Chromatography-Mass Spectrometry (GC-MS)

Arman Allboje Samami, Adriano Nunes-Nesi, Markus Wirtz, and Rüdiger Hell

Abstract The accepted gold standard for determination of amino acids levels in eukaryotic cells is derivatization of extracted amino acids with a fluorescent dye followed by separation of the respective derivates by high performance liquid chromatography (HPLC). The method is highly sensitive but has a narrow spectrum of metabolites that can be analyzed. The coupling of gas chromatography to a sensitive mass-spectrometer (GC-MS) allows analyzing a wide range of metabolites in one run. In this study both methods for determination of amino acid levels were compared in wild-type and *sir1-1*, an *Arabidopsis thaliana* T-DNA knock in mutant in the promoter of sulfite reductase. Sulfite reductase is a key enzyme in the sulfate reduction pathway of plants and is exclusively localized in plastids. Both methods confirm independently a significant increase of total amino acids level in leaves of *sir1-1* plants in comparison to the wild-type. The degree of increase for each amino acid detected by GC-MS could be confirmed in almost all cases by HPLC. The GC-MS method is suitable for screening of mutants disturbed in amino acid composition and the assessment of adaptations in primary metabolism of various mutants.

A.A. Samami • M. Wirtz • R. Hell (✉)
Centre for Organismal Studies, University of Heidelberg,
Im Neuenheimer Feld 360, Heidelberg 69120, Germany
e-mail: Ruediger.hell@cos.uni-heidelberg.de

A. Nunes-Nesi
Department of Metabolic Networks Max-Planck-Institut für
Molekulare Pflanzenphysiologie,
Am Mühlenberg 1, Golm 14476, Germany

L.J. De Kok et al. (eds.), *Sulfur Metabolism in Plants*, Proceedings of the International Plant Sulfur Workshop 1, DOI 10.1007/978-94-007-4450-9_14,

Sulfur is an important macronutrient in higher plants. Reduced sulfur can be found in proteinogenic amino acids (cysteine and methionine), cofactors (biotin, thiamine, Coenzyme A and B1) and iron-sulfur clusters and in a broad range of secondary metabolites, *e.g.* glucosinolates. Unlike humans and animals, plants and most micro-organisms are able to take up, reduce and assimilate inorganic sulfate into cell metabolism. Regulation of primary sulfur metabolism in plants is tightly connected in a network with assimilation of nitrogen and carbon (Hesse et al. 2004; Howarth et al. 2008). The discovery of a knock-down mutant of the single copy gene encoding sulfite reductase (SiR) recently strongly supported the intricate correspondence of sulfate reduction with carbon and nitrogen metabolism (Khan et al. 2010). T-DNA knockdown of SiR (*sir1-1*) displayed lower amounts of SiR transcript and protein, lower SiR activity in leaf tissue, and a severe retardation of vegetative growth (Fig. 1). Gene expression analyses indicated strong compensatory reactions in primary metabolism in leaves. The flux of sulfate through the reduction pathway and APS reductase (*APR2*) transcript levels were reduced and, as a consequence, sulfate accumulated significantly. The ratio of total C and N was shifted to lower values, free sugar and starch contents were reduced (Khan et al. 2010). Based on these observations the question arises if lowered flux and reduction of sulfate can lead to a change in N metabolite composition? Since the sulfur-containing amino acids cysteine and methionine are present in proteins and the flux of sulfate into cysteine is down-regulated in *sir1-1* (Khan et al. 2010), it can be hypothesized that global translation is affected, hence, free amino acids would accumulate.

Two techniques are currently accepted as state-of-the-art for determination of amino acid levels from tissue samples: gas chromatography coupled mass-spectrometry (GC-MS) and fluorescence based detection after separation by high performance liquid chromatography (HPLC). Here, we performed a comparative analysis of extracted amino acids from leaves of 7-week-old *Arabidopsis* wild-type and *sir1-1* mutant grown under short day conditions (8.5 h light) with a light intensity of 100 μmol m^{-2} s^{-1}, and relative humidity of 50%. The temperature during the day and night was 22°C and 18°C, respectively. The extraction protocol for the GC-MS method used methanol to isolate soluble and insoluble cell constituents (Lisec et al. 2006). Extracted metabolites were derivatized with *N*-methyl-*N*-[trimethylsilyl] trifluoroacetamide and methoxyamine hydrochloride to allow separation by GC. Ribitol was used as an internal standard to allow normalization of the extraction procedure. Metabolites were separated on a MDN-35 capillary column (30 m length, 0.32 mm inner diameter, 0.25 mm film thickness) connected to 6890 N gas chromatograph (Agilent) with a flow of 1 ml min^{-1} argon and detected in Pegasus III time-of-flight mass spectrometer (Leco Instruments). Settings for GC-MS were controlled with the PAL cycle composer software version 1.5 (CTC Analytics) as described by Lisec et al. (2006). This technique gave an average error of 39% based on 5 biological replicas and including technical errors.

For the second technique metabolites were extracted with 0.1M hydrochloric acid prior derivatization with the fluorescent dye AccQ-Tag™ (Waters) and separated by high-performance liquid chromatography using a Nova-PakTM C18, 3.9 × 150 mm column (Waters) as described in Hartmann et al. (2004). Separated

Fig. 1 Phenotype of *sir1-1 Arabidopsis*. 7-week-old wild-type and *sir1-1* plants were grown on soil in a growth chamber under short day conditions

amino acid AccQ-Tag™ derivates were detected with a fluorescence detector (Jasco FP-920, Germany) at 395 nm upon excitation with 250 nm. Quantitation was performed using the Waters LC control- and analysis software Millenium 32 (Waters, USA), using external standard calibration curves for each amino acid. Here, the average error was 29% based on five biological replicas and including technical errors.

Derivatization and detection of amino acids by HPLC was found to be an appropriate technique to quantify all proteinogenic amino acids. In contrast, only 13 proteinogenic amino acids could be detected in leaf extracts form *Arabidopsis* with the here applied GC-MS method. These amino acids are: aspartic acid (Asp), glutamic acid (Glu), serine (Ser), asparagine (Asn), glycine (Gly), glutamine (Gln), threonine (Thr), alanine (Ala), proline (Pro), tyrosine (Tyr), valine (Val), isoleucine (Ile), and phenylalanine (Phe) (Fig. 2). All amino acids, measured by both techniques, accumulated significantly in leaves of *sir1-1* as compared to wild-type. This result is in line with the assumption drawn from the reduced flux rates of sulfur into cysteine observed in leaves of *sir1-1* plants. The production of cysteine by the assimilatory sulfate reduction pathway limits efficient translation in leaves of *sir1-1*. The potential

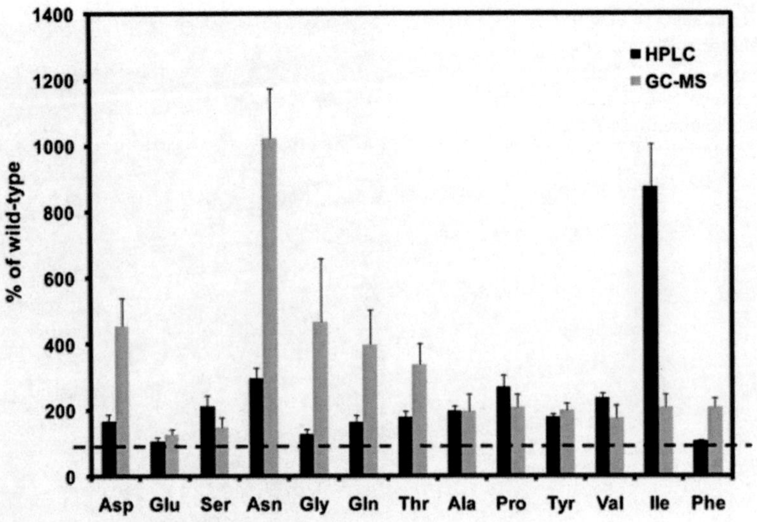

Fig. 2 Comparative spectrum of amino acids in leaves of *sir1-1 Arabidopsis* measured via GC-MS and HPLC. Amino acid values of wild-type were defined as 100%. Levels of amino acids from 7 week-old *sir1-1* leaves as percentage of wild-type plants grown on soil in a growth chamber under short day conditions measured by HPLC (*black*) and GC-MS (*grey*). Means ± SE (n = 5) are shown. *Asp* aspartic acid, *Glu* glutamic acid, *Ser* serine, *Asn* asparagine, *Gly* glycine, *Gln* glutamine, *Thr* threonine, *Ala* alanine, *Pro* proline, *Tyr* tyrosine, *Val* valine, *Ile* isoleucine, *Phe* phenylalanine

decrease of global translation in *sir1-1* is most likely the cause for the accumulation of non-sulfur containing amino acids, since they are not incorporated into proteins.

Remarkably, differences in the estimated fold-changes between *sir1-1* and wild-type for some amino acids were massive, *e.g.* for Asp, Asn, Gly, Gln, Thr, and Ile. The relative differences between the amino acid contents that had been determined by HPLC or GC-MS was largest for Asp, Asn, Gly, Gln, and Thr. These belong to the charged amino acids but since detection differences were very low for Glu or Pro this property may not be the cause for the observed discrepancies. Moreover, the GC-MS method was found to be only useful for comparative and relative quantitation where an internal standard is necessary, conversely, HPLC technique can provide absolute quantification. However, the large number of additional metabolites that can be detected by GC-MS may add extremely useful information *e.g.* for comparisons of multiple genotypes. With respect to standard errors, both methods were comparable, but HPLC showed better performance. Similarly, both methods were reproducible.

In this study we could show that both methods were suitable to extract, derivatize, and separate amino acids. The increase of most free amino acids contents in leaves of *sir1-1* plants as compared to wild-type was evident independent from the method applied. While HPLC method provides the possibility to detect all amino acids and to quantify them, GC-MS is not useful for the quantitation of total amino acids spectrum, and neither for absolute quantitation of their amount in relation to fresh weight. However, GC-MS is able to detect a huge variety of metabolites besides amino acids in the same injection run, *e.g.* sugars, which can be very useful depending on researcher's demand. On the other hand, for amino acid determination from *Arabidopsis* leaf tissue, the HPLC method is highly suitable.

References

Hartmann T, Honicke P, Wirtz M, Hell R, Rennenberg H, Kopriva S (2004) Regulation of sulphate assimilation by glutathione in poplars (*Populus tremula* x *P. alba*) of wild type and overexpressing γ-glutamylcysteine synthetase in the cytosol. J Exp Bot 55:837–845

Hesse H, Nikiforova V, Gakiere B, Hoefgen R (2004) Molecular analysis and control of cysteine biosynthesis: integration of nitrogen and sulphur metabolism. J Exp Bot 55:1283–1292

Howarth JR, Parmar S, Jones J, Shepherd C, Carol C, Galster A, Hawkins N, Miller S, Baker J, Ward J, Beale M, Verrier P, Barraclough P, Hawkesford MJ (2008) Co-ordinated expression of amino acid metabolism in response to N and S deficiency during wheat grain filling. J Exp Bot 59:3675–3689

Khan MS, Haas FH, Allboje Samami A, Gholami AM, Bauer A, Fellenberg K, Reichelt M, Hansch R, Mendel RR, Meyer AJ, Wirtz M, Hell R (2010) Sulfite reductase defines a newly discovered bottleneck for assimilatory sulfate reduction and is essential for growth and development in *Arabidopsis thaliana*. Plant Cell 22:1216–1231

Lisec J, Schauer N, Kopka J, Willmitzer L, Fernie AR (2006) Gas chromatography mass spectrometry-based metabolite profiling in plants. Nat Protoc 1:387–396

Homomeric Interaction of the STAS Domain in Sultr1;2

Nakako Shibagaki and Arthur R. Grossman

Abstract We recently reported the interaction of the STAS domain of SULTR1;2 and O-acetylserine(thiol)lyase (OASTL). This inter-protein interaction was initially identified in the yeast two hybrid system and was confirmed using other methods. Interestingly, we also found that the STAS domain (comprised of both the L and STAS' regions) of SULTR1;2 undergoes a homomeric interaction; an association between two STAS domains occurs as a consequence of a specific interaction between the L and STAS' regions of the domain. The strength of this interaction depends on whether one or both of the interacting STAS partners contains an L region. A similar interaction was observed for the L and STAS' regions of SULTR3;1 but not of SULTR1;1. Lesions that alter the SULTR1;2 L and STAS' interactions do not appear to markedly impact sulfate transport activity, at least in yeast cells, although they may impact transport activity of SULTR1;2 in *Arabidopsis*. Hence, while regulation of sulfate transporter activity occurs through interactions of the transporter with OASTL, transporter activity may also be modulated by homomeric interactions in the STAS domain.

We recently proposed a novel mode of regulation in which SO_4^{2-} uptake to *Arabidopsis* roots is controlled by the level of reductant in the root; the reductant required for SO_4^{2-} assimilation indirectly controls the physical interaction

N. Shibagaki (✉)
Department of Plant Biology, The Carnegie Institution for Science,
260 Panama Street, Stanford, CA 94305, USA

Research and Innovation Center, Nihon L'Oreal,
3-2-1 Sakado, Takatsu, Kawasaki, Kanagawa 213-0012, Japan
e-mail: snakako@gmail.com

A.R. Grossman
Department of Plant Biology, The Carnegie Institution for Science,
260 Panama Street, Stanford, CA 94305, USA

L.J. De Kok et al. (eds.), *Sulfur Metabolism in Plants*, Proceedings of the International Plant Sulfur Workshop 1, DOI 10.1007/978-94-007-4450-9_15,

between the STAS domain of SULTR1;2, a high affinity sulfate transporter, and *O*-acetylserine(thio)lyase (OASTL, also called cysteine synthase) (Shibagaki and Grossman 2010). This interaction may impact the activity of SULTR1;2 under sulfate rich conditions and coordinate sulfate uptake by *Arabidopsis* roots with the ability of the plant to reduce the sulfate anion. The recently solved crystal structure of the STAS domain of prestin (SLC26A5), a member of sulfate transporter superfamily, suggests that a linker (L) region, connecting the transmembrane domains to the region previously designated STAS (STAS'), is closely connected to the STAS' (Pasqualetto et al. 2010); for the results described in this manuscript we divide the STAS domain in the L and the STAS' regions. During experiments in which we defined interactions of SULTR1;2 and OASTL, we also observed a physical interaction between STAS domains and demonstrated that both the L and the STAS' regions were required for this interaction. We also compared homomeric interactions of STAS domains of SULTR1;1 and SULTR3;1.

The yeast two-hybrid system was used to examine potential homomeric interactions between STAS domains of the SULTR transporters of *A. thaliana*. Initially, interactions between the LexA-based bait (DNA binding domain) plasmid from pEG202 and prey (activation domain) plasmid from pJG4-5, both containing DNA fragments encoding either the L or STAS' regions of SULTR1;2 (from ser 490 to val 653; presented as 1;2LSTAS') or STAS' without the L region (from ile 520 to val 653; presented as 1;2STAS'). These plasmids were introduced into *Saccharomyces cerevisiae* strain EGY48 along with the reporter plasmid pSH18-34, and the cells were grown at 30°C for 2 days in synthetic galactose/raffinose medium containing X-gal. In no case did we detect the formation of blue colonies, even though the fusion protein could be detected by immunoblots (data not shown). Both conformation and steric constraints associated with fused domains in bait/prey constructs could diminish protein-protein interactions. Therefore, we examined the same interactions as noted above but included a DNA sequence encoding a 10 alanine (10 ala) separation sequence (Doyle and Botstein 1996) between the bait insertion and the LexA coding region, and also between the prey insertion and the activation domain sequence. Using the 1;2L-STAS' insert in both the bait and prey constructs with the 10 ala separation sequence yielded blue colonies in the yeast two hybrid assay (Fig. 1a), suggesting the occurrence of L-STAS' interactions, and that a flexible separation sequence is required to observe such interactions.

The level of β-galactosidase activity measured in *S. cerevisiae* cells used for the yeast two hybrid assay reflects the strength of the interaction between the bait and prey fusion proteins. This activity was assessed using both qualitative and quantitative assays (Fig. 1 *right* and *left panel*, respectively). Expression of 1;2L-STAS' fusion proteins by both bait and prey constructs in *S. cerevisiae* cells was confirmed by Western blot analyses using antibodies against the LexA and HA epitopes (data not shown). When the sequence encoding the L region that precedes the STAS' region in the native protein was omitted from the construct (1;2STAS'), no interaction between fusion proteins was observed (Fig. 1c). This result suggests that the L region that separates the STAS' from the catalytic moiety of SULTR1;2 is critical for establishing interactions between L-STAS' domains. A heterogenous combination

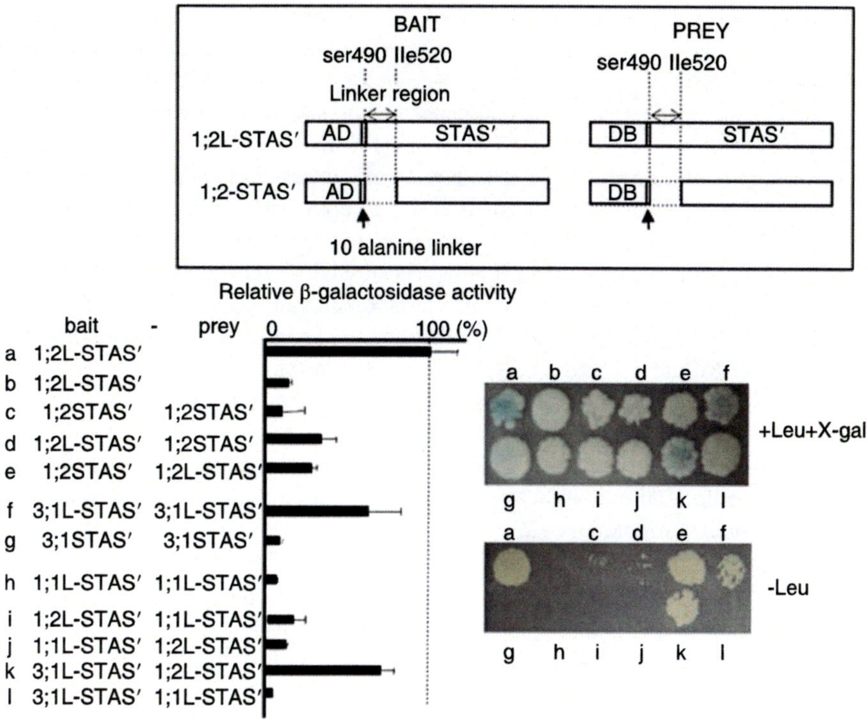

Fig. 1 Interactions among the STAS domains in yeast two hybrid system. The LexA-based bait plasmid derived from pEG202 and prey plasmid from pJG4-5 were introduced into EGY48 with the reporter plasmid pSH18-34. The bait and prey constructs used are shown in the *box*. Homomeric and heteromeric interactions between STAS domains of SULTR1;1, SULTR1;2 and SULTR3;1. The yeast transformants expressing indicated bait and prey fusion proteins were grown on agar-solidified synthetic galactose/raffinose medium lacking for uracil, histidine and tryptophan containing X-gal or medium lacking also for leucine as well as the above three amino acids without X-gal at 30°C for 2 days before pictured. *Blue* colonies on X-gal and leucine auxotroph indicate the active transcription of lacZ gene and Leu2 gene as a result of bait and prey interaction. β-galactosidase activity indicative of the strength of interaction was determined and calculated as relative value to that from the homogenous interaction between the L-STAS' from Sultr1;2. Average values from three independent experiments are shown with standard deviations. AD, activation domain; DB, DNA binding domain

of a 1;2L-STAS' bait construct with a 1,2STAS' (deleted for L region) prey construct yielded significantly more β-galactosidase activity than the homogenous 1,2STAS' bait/1,2STAS' prey combination, but less than the homomeric 1;2L-STAS'/1;2L-STAS' prey combination (Fig. 1a, c, d). A similar intermediate level of interaction, based on the level of β-galactosidase activity, was obtained if the bait construct contained 1;2STAS' and the prey construct contained 1;2L-STAS' (Fig. 1e). These findings suggest that both the L and STAS regions are important for the interactions observed in the yeast two hybrid system and that the L may directly interact with the STAS' region, as supported by preliminary results from *in vitro* binding assays which showed enhanced binding of the GST-L to MBP (maltose binding protein)-STAS' when sulfate and methionine were added to the assay buffer (data not shown).

Analogous constructs were prepared using sequences encoding the STAS domains of SULTR1;1 and SULTR3;1. As with SULTR1;2, the STAS domain of SULTR3;1 (481A-658V) exhibited a homomeric interaction in the yeast two hybrid system (Fig. 1f). Also similar to SULTR1;2, both the L and STAS' regions of SULTR3;1 were required for this interaction (Fig. 1g). In contrast, no interactions were observed if STAS domain constructs of SULTR1;1 were similarly assayed (Fig. 1h). Heterogenous interactions of STAS domains among SULTR1;1, SULTR1;2 and SULTR3;1 were also tested. We only observed an interaction between the STAS domains of SULTR3;1 and SULTR1;2 (Fig. 1k); no interaction was ever detected when the SULTR1;1 STAS domain was used in either the bait or prey construct (Fig. 1i, j, l). It is not clear whether the heterogenous interaction between the STAS domains of SULTR1;2 and SULTR3;1 has biological meaning since at this time there is no information to indicate that these two transporters are ever present in the same *Arabidopsis thaliana* cell.

We also examined the effect of mutations in the STAS domain of SULTR1;2, previously reported to affect the transport activity (Shibagaki and Grossman 2006), on homomeric SULTR1;2 STAS interactions (Shibagaki and Grossman, unpublished data). The capability of the STAS domain to undergo L-STAS' interactions varied in strains with lesions that alter the activity of SULTR1;2. Some of the mutations that depress SULTR1;2 activity without altering accumulation of the SULTR1;2 protein did not impact the STAS domain homomeric interaction, while all mutations that lead to SULTR1;2 instability diminish the interaction. As an example, the T587A lesion in the STAS' region, which eliminates the potential phosphorylation site on SULTR1;2 as well as transporter activity (Rouached et al. 2005), did not show a decrease in the ability of the SULTR1;2 STAS domain to participate in a homomeric interaction. Therefore, there does not appear to be a simple correlation between altered functionality of SULTR1;2 and the ability of the transporter to participate in STAS domain-dependent homomeric interactions. One possibility is that interactions between STAS domains impact transporter activity negatively; at least such a possibility cannot be excluded from the observations discussed above. Although not examined in SULTR1;2, prestin has been shown to form a stable dimer that serves as a building block for establishing higher order oligomers (Zheng et al. 2006). Therefore it is very much likely that SULTR1;2 also forms a dimer and possibly multimeric assemblies and that interactions through the cytosolic STAS domain could potentially impact transporter function (Fig. 2 *left*). This type of regulation has been reported for an ammonium transporter in which a cytosolic domain triggers trans-activation for functional transport (Loque et al. 2007). Potential regulation of transporter function through homomeric interactions in the STAS domain is likely to be specific since such interactions are observed with SULTR1;2, but not with SULTR1;1 (Fig. 2).

Since SULTR1;1 is not expressed in sulfate-rich conditions, there is no need for repressive control of its activity. In contrast, SULTR1;2 is present even when

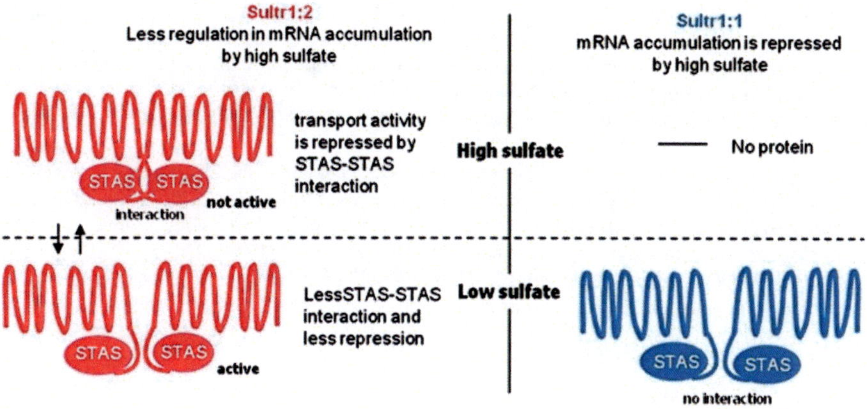

Fig. 2 Hypothetical regulation mode of SULTR1;2 activity through the interaction in the STAS domain. Under sulfur sufficient condition, higher cellular concentration of sulfate may promote the interaction among two SULTR1;2 proteins between the L and STAS' regions, as preliminary results in *in vitro* binding assay showed higher interaction when sulfate and methionine are added to the assay. This interaction is likely to repress the activity of SULTR1;2, indirectly suggested from the analyses of the effects of mutations in L-STAS' on both sulfate transporter activity and homomeric interaction as described in the text. Under sulfur deficient condition, SULTR1;2 may be released from this repressive regulation of the activity and SULTR1;1, which is expressed exclusively during sulfur deficiency, does not show the homomeric STAS interaction or repressive regulatory mechanism of its activity

there is an abundance of sulfate in the environment, which likely required the evolution of various mechanisms to coordinate SULTR1;2 activity with the metabolic state of the plant; this regulation would be in addition to the hypothesized regulation that occurs as a consequence of the interaction of SULTR1;2 with OASTL (Shibagaki and Grossman 2010). In this context, it is intriguing that the chimeric SULTR1;2-STAS1;1 protein, which has the catalytic domain from SULTR1;2 fused to the SULTR1;1 STAS domain, exhibited a higher V_{max} than native SULTR1;2 in *S. cerevisiae* cells (Shibagaki and Grossman 2004).

Acknowledgements This work was supported by NSF Grant MCB-0824469 awarded to ARG.

References

Doyle T, Botstein D (1996) Movement of yeast cortical actin cytoskeleton visualized *in vivo*. Proc Natl Acad Sci U S A 93:3886–3891

Loque D, Lalonde S, Looger LL, von Wiren N, Frommer WB (2007) A cytosolic trans-activation domain essential for ammonium uptake. Nature 446:195–198

Pasqualetto E, Aiello R, Gesiot L, Bonetto G, Bellanda M, Battistutta R (2010) Structure of the cytosolic portion of the motor protein prestin and functional role of the STAS domain in SLC26/SulP anion transporters. J Mol Biol 400:448–462

Rouached H, Berthomieu P, El Kassis E, Cathala N, Catherinot V, Labesse G, Davidian JC, Fourcroy P (2005) Structural and functional analysis of the C-terminal STAS (sulfate transporter and anti-sigma antagonist) domain of the *Arabidopsis thaliana* sulfate transporter SULTR1.2. J Biol Chem 280:15976–15983

Shibagaki N, Grossman AR (2004) Probing the function of STAS domains of the *Arabidopsis* sulfate transporters. J Biol Chem 279:30791–30799

Shibagaki N, Grossman AR (2006) The role of the STAS domain in the function and biogenesis of a sulfate transporter as probed by random mutagenesis. J Biol Chem 281:22964–22973

Shibagaki N, Grossman AR (2010) Binding of cysteine synthase to the STAS domain of sulfate transporter and its regulatory consequences. J Biol Chem 285:25094–25102

Zheng J, Du GG, Anderson CT, Keller JP, Orem A, Dallos P, Cheatham M (2006) Analysis of the oligomeric structure of the motor protein prestin. J Biol Chem 281:19916–19924

Simultaneous Determination of Sulfur Metabolites in Plant Tissue by LC-ESI-MS/MS

Mei-Hwei Tseng, Luit J. De Kok, and Yao-Moan Huang

Abstract An analytical method was developed for the accurate quantification of the contents of sulfur metabolites of plant tissues based on liquid chromatography – electrospray ionization – tandem mass spectrometry (LC-ESI-MS/MS). ^{34}S-Labeled sulfur metabolites were biosynthesized in *Arabidopsis thaliana* from $^{34}S\text{-}SO_4^{2-}$ and used as internal standards for the analysis and quantification of these sulfur-containing metabolites in *A. thaliana* and *O. banksiaefolia* fern spores.

Sulfur is an essential macronutrient for all living organisms. There has been substantial progress on the elucidation of the pathways involved in sulfur uptake and assimilation and the significance of sulfur metabolites in plant functioning and adaptation to abiotic and biotic stress (Hawkesford and De Kok 2006; Koralewska et al. 2009; Shahbaz et al. 2010). Sulfur-containing natural products have been investigated to determine their therapeutic potential including antioxidant, antibacterial, antimicrobial, antifungal and anticancer properties (Jacob 2006). However, there is still the need for accurate analyses of sulfur metabolites in plant tissue.

Recently advances in mass spectrometry (MS) enable improved analysis of metabolites in complex biological samples (Bajad and Shulaev 2007; Lu et al. 2008).

M.-H. Tseng (✉)
Department of Applied Physics and Chemistry, Taipei Municipal University of Education, Taipei 10048, Taiwan
e-mail: biomei@tmue.edu.tw

L.J. De Kok
Laboratory of Plant Physiology, University of Groningen, P.O. Box 11103, Groningen 9700 CC, The Netherlands
e-mail: l.j.de.kok@rug.nl

Y.-M. Huang
Division of Silviculture, Taiwan Forestry Research Institute, 53 Nanhai Rd., Taipei 10066, Taiwan

L.J. De Kok et al. (eds.), *Sulfur Metabolism in Plants*, Proceedings of the International Plant Sulfur Workshop 1, DOI 10.1007/978-94-007-4450-9_16,

Combined with effective sample preparation, LC-ESI-MS has been used in simultaneous identification and quantification of selected metabolites in biological samples with high selectivity, adequate sensitivity, and minimal material (Harwood et al. 2009). However, matrix effects interfere with detections in MS analysis (Annesley 2003; Remane et al. 2010). These interference effects appear to be compound dependent and quantitative analysis by MS should require an internal standard for each metabolite to be quantified accurately. Due to the high physical-chemical similarities between the labeled and the analyte, variations in instrumental response, and ion interference effects and matrix effect in LC-ESI-MS can be compensated (Bennett et al. 2008; Remane et al. 2010). Biosynthesis *in situ* of labeled metabolites in living organisms for MS techniques has been employed with low emission of pollutants and energy consumption. Metabolic labeling with ^{13}C, ^{2}D, ^{15}N and ^{34}S for preparation of labeled intracellular metabolites are possible for mycobacteria, yeast, algae, *E. coli* and plants (Patzelt et al. 1999; Mougous et al. 2002; Wu et al. 2005).

In the present study a hydroponic set-up was used for the *in situ* ^{34}S-metabolic labeling of sulfur-containing metabolites in *A. thaliana*. The ^{34}S metabolites biosynthesized in *A. thaliana* were extracted and used as internal standards to directly and simultaneously quantify sulfur-containing metabolites in *A. thaliana* and *O. banksiaefolia* fern spores by LC-ESI-MS/MS without prior derivatization.

Glutathione (GSH), oxidized glutathione (GSSG), perchloric acid (PCA) were purchased from Sigma (St. Louis, MO, USA), methionine from Acros Organics, $Na_2{}^{34}SO_4$ from ICON Isotope (Summit, NJ, USA). All chemicals used for Murashige and Skoog nutrient medium were obtained from Sigma-Aldrich, ACROS Organics, Merck (Darmstadt, Germany). HPLC-grade solvents were from J. K. Baker. *Arabidopsis thaliana* L. Columbia-0 (Col-0) seeds were sterilized and sown on a 50% Murashige and Skoog nutrient medium or soil. After sowing, media or soil were conserved 2 days at 4°C, after which they were placed in a temperature controlled room (22°C) and the photoperiod was 16 h at a photon fluence rate of 95 μmol $m^{-2}s^{-1}$ (PAR 400–700 nm). After 8 weeks growth shoot and root or flower stem were separated, weighed, and immediately frozen in liquid N_2 and stored at −80°C until analysis. For metabolite extraction, plant tissue was ground with mortar and pestle in liquid N_2. Fern spores were powdered in Eppendorf tube containing eight clean small stainless steel metal balls by a mixer-mill grinding. Tissue powder of *A. thaliana* and fern spores were transferred into a micro-centrifuge tube, mixed with 0.2% (v/v) PCA in a cool room, and thereafter boiled for 5 min in a waterbath. Subsequently 60 μl methanol was added and the sample was homogenized by a small plastic pestle. The homogenates were centrifuged at 18,000 g for 30 min at 4°C. Chromatographic separation were performed on a Thermo Accela LC system by using a Thermo Hypercarb column (5 μm, 2.1 mm × 150 mm) with a Hypercarb Drop-in Guards pk2 (5 μm, 10 mm × 2.1 mm) for separation. Separations were performed under gradient conditions at a flow rate of 0.25 ml min^{-1}. The mobile phases consisted of 0.1% formic acid (A) and 0.1% formic ac id in acetonitrile (B). Gradient conditions were as follows: 0–1 min 100% A, 1–6 min solvent A decreased linearly to 60% and solvent B increased linearly to 40%. Electrospray ionization – mass spectrometric (ESI-MS) analysis was performed on a LTQ Velos mass spectrometry

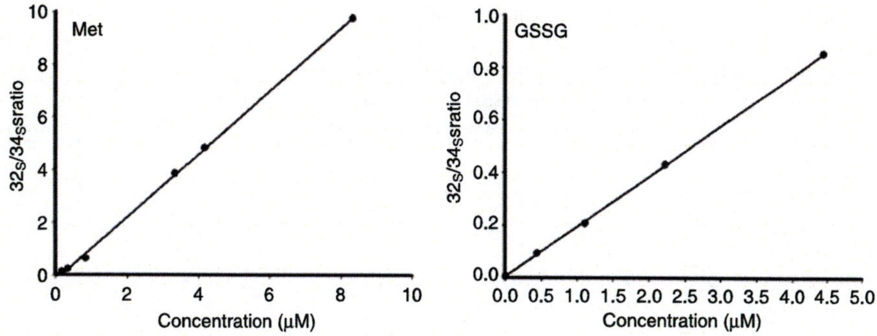

Fig. 1 Standard calibration curves for GSSG and methionine. GSSG: y = 0.1910x − 0.0035 (r^2 = 0.9998). Methionine: y = 1.1955x − 0.1939 (r^2 = 0.9993)

Table 1 Content of methionine, GSH and GSSG in rosette leaves and flower stem of *A. thaliana*

	Methionine (nmol g^{-1})	GSH (nmol g^{-1})	GSSG (nmol g^{-1})	GSH + GSSG (nmol g^{-1})	GSH/GSSG
Rosette leaves	2.8 ± 0.0	85.7 ± 2.5	7.1 ± 0.3	92.9 ± 3.2	12.0 ± 0.5
Flower stem	14.9 ± 0.8	99.1 ± 5.0	77.4 ± 0.5	177.5 ± 5.4	1.3 ± 0.1

Data represent the mean of three measurements with three plants in each (±SD)

(Thermo Fisher Scientific Inc., USA). Metabolites were detected in the positive ionization mode using the multiple reaction monitoring (MRM) scanning mode. The spray voltage was set to 3.5 kV, ion transfer capillary temperature to 280°C, sheath gas pressure to 50 (arbitrary units), auxiliary gas pressure to 15 (arbitrary units).

Calibration curves corrected by normalization by using ^{34}S-metabolic labeling of *A. thaliana* were analyzing solutions of standards (Fig. 1). For building up calibration curves, the ^{34}S-reference tissue extract was spiked with increasing concentrations of a mixture of ^{32}S-reference compounds in 0.1% PCA. Calibration curves were constructed by normalizing the signals related to ^{32}S-reference compounds to the signals corresponding to ^{34}S metabolites in tissue extracts. Consequently, absolute quantification can be achieved in plant samples by comparing the $^{32}S/^{34}S$ ratios obtained for the samples to the calibration curves.

The content of GSH, GSSG and methionine in 8-week-old *A. thaliana* were determined in rosette leaves and flower stem by LC/ESI-MS/MS. The results showed that the content of GSSG and methionine were higher in flower stem than in rosette leaves, while the ratio of GSH/GSSG was higher in rosette leaves than in flower stem (Table 1).

O. banksiaefolia spores are chlorophyll-containing (green spore) and unlike non-green spores, these spores contain relatively large amounts of H_2O and are known to be metabolically active and short-lived (Lloyd and Klekowski 1970).

Table 2 Comparison of the levels of endogenous GSH and GSSG, and viability of *O. banksiaefolia* fern spores stored at different temperature. The calibration curves corrected by using [^{34}S]GSH and [^{34}S]GSSG from ^{34}S-labeled *A. thaliana* as internal standards

	Germination rate	GSH (nmol mg^{-1})	GSSG (nmol mg^{-1})	GSH+GSSG (nmol mg^{-1})	GSH/GSSG
Dry storage at −80°C for 3 weeks	99%	7.81±1.14	2.31±0.18	10.11±1.32	3.39±0.24
Dry storage at 4°C for 3 weeks	97%	1.69±0.93	7.62±0.74	9.31±0.20	0.22±0.14
Dry storage at 25°C for 3 weeks	0%	0.67±0.65	6.46±0.69	7.14±1.35	0.10±0.09

Data represent the mean of three measurements (±SD)

O. banksiaefolia fern spores stored at different storage temperature showed different germination rates (Table 2). The GSH and GSSG contents of spores were determined in order to examine the relationship between the glutathione content, its redox state and viability of the spores. The total glutathione (GSSG+GSH) content was higher in fresh spores and the spores stored at 4°C than that of spores stored at 25°C, which had a low viability (Table 2). However, storage at both 4°C and 25°C resulted in substantial decrease in the GSH/GSSH ratio, showing no direct relation between the redox state of glutathione and the viability of the fern spores. Undoubtedly, the significance of glutathione in viability of spores needs further to be evaluated.

Acknowledgments The authors acknowledge Dr. Ding-Tzai Li for fruitful discussions and Mass Solution Technology Co., Ltd for excellent technical support.

References

Annesley TM (2003) Ion suppression in mass spectrometry. Clin Chem 49:1041–1044

Bajad S, Shulaev V (2007) Highly-parallel metabolomics approaches using LC-MS2 for pharmaceutical and environmental analysis. Trends Anal Chem 26:625–636

Bennett BD, Yuan J, Kimball EH, Rabinowitz JD (2008) Absolute quantitation of intracellular metabolite concentrations by an isotope ratio-based approach. Nat Protoc 3:1299–1311

Harwood DT, Kettle AJ, Brennan S, Winterbourn CC (2009) Simultaneous determination of reduced glutathione, glutathione disulphide and glutathione sulphonamide in cells and physiological fluids by isotope dilution liquid chromatography-tandem mass spectrometry. J Chromatogr B 877:3393–3399

Hawkesford MJ, De Kok LJ (2006) Managing sulphur metabolism in plants. Plant Cell Environ 29:382–395

Jacob C (2006) A scent of therapy: pharmacological implications of natural products containing redox-active sulfur atoms. Nat Prod Rep 23:851–863

Koralewska A, Buchner P, Stuiver CEE, Posthumus FS, Kopriva S, Hawkesford MJ, De Kok LJ (2009) Expression and activity of sulfate transporters and APS reductase in curly kale in response to sulfate deprivation and re-supply. J Plant Physiol 166:168–179

Lloyd RM, Klekowski EJ (1970) Spore germination and viability in pteridophyta: evolutionary significance of chlorophyllous spores. Biotropica 2:129–137

Lu W, Bennett BD, Rabinowitz JD (2008) Analytical strategies for LC-MS-based targeted metabolomics. J Chromatogr B 871:236–242

Mougous JD, Leavell MD, Senaratne RH, Leigh CD, Williams SJ, Riley LW, Leary JA, Bertozzi CR (2002) Discovery of sulfated metabolites in mycobacteria with a genetic and mass spectrometric approach. Proc Natl Acad Sci USA 99:17037–17042

Patzelt H, Keβler B, Oschkinat H, Oesterhelt D (1999) The entire metabolite spectrum of the green alga *Scenedesmus obliquus* in isotope-labelled form. Phytochemistry 50:215–217

Remane D, Wissenbach DK, Meyer MR, Maurer HH (2010) Systematic investigation of ion suppression and enhancement effects of fourteen stable-isotope-labeled internal standards by their native analogues using atmospheric-pressure chemical ionization and electrosprayionization and the relevance for multi-analyte liquid chromatographic/mass spectrometric procedures. Rapid Commun Mass Spectrom 24:859–867

Shahbaz M, Tseng MH, Stuiver CEE, Koralewska A, Posthumus FS, Venema JH, Parmar S, Schat H, Hawkesford MJ, De Kok LJ (2010) Copper exposure interferes with the regulation of the uptake, distribution and metabolism of sulfate in Chinese cabbage. J Plant Physiol 167:438–446

Wu L, Mashego MR, van Dam JC, Proell AM, Vinke JL, Ras C, van Winden WA, van Gulik WM, Heijnen JJ (2005) Quantitative analysis of the microbial metabolome by isotope dilution mass spectrometry using uniformly 13C-labeled cell extracts as internal standards. Anal Biochem 336:164–171

Plant Response to Mineral Ion Availability: Transcriptome Responses to Sulfate, Selenium and Iron

Mutsumi Watanabe, Hans-Michael Hubberten, and Rainer Hoefgen

Abstract The response of plants to sulfate starvation has been described at the level of agronomy, physiology, biochemistry, metabolomics and transcriptomics. The aim of these studies was the description of regulatory and control processes governing plant response to sulfate nutrient stress. However, these results are impaired by pleiotropic responses seemingly overlapping with the response to other mineral nutrient ion starvations or effects of soil borne ions taken up by the plant. This paper suggests making use of these often overlapping responses to various ions in order to distinguish between specific and pleiotropic effects. Here we compare sulfate starvation, iron depletion, and selenium exposure at the transcriptome level.

Introduction

Plant growth and vigour relies on CO_2, water, light and mineral nutrients together with physical factors such as temperature, pH, and soil parameters (Marschner 1995; Buchanan et al. 2007; Haensch and Mendel 2009; Amtmann and Blatt 2009; Amtmann and Armengaud 2009). Justus von Liebig, under the impression of severe famines in 1816 (the year without summer), started to study the interdependency between mineral availability and plant growth and yield, eventually resulting in the concept of mineral fertilisation termed as "Agrikulturchemie" (von Liebig 1840) which brought about tremendous improvements of agricultural yields. Mineral ions necessary for plant growth are generally distinguished into macronutrients

M. Watanabe • H.-M. Hubberten • R. Hoefgen (✉)
Max Planck Institute of Molecular Plant Physiology, Department of Molecular Physiology, Science Park Potsdam – Golm, Potsdam D-14424, Germany
e-mail: watanabe@mpimp-golm.mpg.de; Hubberten@mpimp-golm.mpg.de; Hoefgen@mpimp-golm.mpg.de

L.J. De Kok et al. (eds.), *Sulfur Metabolism in Plants*, Proceedings of the International Plant Sulfur Workshop 1, DOI 10.1007/978-94-007-4450-9_17,

(N, P, S, Mg, K, Ca) and micronutrients (Fe, Cu, Mn, Ni, Zn, Cl, B, Mo). Some of these elements are only necessary in trace amounts such as Mo and Ni (Marschner 1995; Buchanan et al. 2007). Silicon (Si) plays an important role in plant cell wall structure in gramineae or some algae, such as diatoms (Currie and Perry 2007). Some non-essential elements seem to improve plant performance and are sometimes termed beneficial (F, Br, I, Na, Co, and the group of rare earths) though it remains disputable whether these elements are indeed providing a relevant physiological response (Aller et al. 1990; Hu et al. 2004). Further elements might be taken up due to inaccuracies of the uptake systems (Al, Se, Cd, Pb, Hg, W, Cr, Ag, As, V, Cs, Rb, Sr). Often this leads to negative effects *e.g.* production of ROS (reactive oxygen species) or competition with essential nutrients. For example, selenium and arsenic both interfere with sulfur metabolism and arsenate with phosphate metabolism. Selenium is not essential for plants but taken up through the sulfur uptake system and metabolised as sulfur because of the close chemical properties. Selenocysteine and selenomethionine are incorporated into dysfunctional proteins (Sors et al. 2005; Pilon-Smits et al. 2009). Arsenic (As) is taken up by plants due to its similarity to phosphate and interferes with phosphate, iron, silicon and sulfur metabolism, as *e.g.* arsenite binds to thiol groups in proteins and disturbs functionality. Arsenate is taken up through the phosphate transporter system and arsenite via the silicon uptake mechanisms (Zhao et al. 2010).

A balanced supply of nutrients is essential. Deficiency as well as over-accumulation can lead to negative effects on plants (Hoefgen and Hesse 2008). For example, copper (Cu) is an essential element for plants but over-accumulation beyond 20–30 mg kg^{-1} DW leads to iron deficiency and oxidative damage (Puig et al. 2007). Likewise, normal iron levels are about 100 mg Fe kg^{-1} leaf DW, whereas levels of more than 500 mg Fe kg^{-1} leaf DW lead to the generation of reactive oxygen species and hydroxyl radicals through the electron transfer properties of $Fe^{2+} <-> Fe^{3+} + e^-$ and the electron being transferred to oxygen to yield a superoxide radical O_2^- (Staiger 2002; Pilon et al. 2009; Jeong and Guerinot 2009).

The interactions and competitions between minerals at the level of uptake have been described in various studies (Kopriva and Rennenberg 2004; Amtmann and Armengaud 2009; Watanabe et al. 2010). An interrelation between sulfate and selenium (Shinmachi et al. 2010; Zhu et al. 2009; Pilon-Smits et al. 2009), and sulfate and iron (Astolfi et al. 2010; Zuchi et al. 2009) has been described. Iron uptake in barley and tomato has been shown to be dependent on sulfate availability (Astolfi et al. 2010). Iron and sulfur directly co-operate in iron-sulfur-clusters being essential prosthetic groups and catalysts in various enzymes such as nitrite reductase, sulfite reductase, ferredoxins, superoxide dismutases, catalases, aconitase, xanthine oxidase (Buckhout et al. 2009; Pilon et al. 2009; Jeong and Guerinot 2009; Schwarz et al. 2009; Haensch and Mendel 2009). Interestingly, sulfate availability directly affects iron homoeostasis. Under sulfate depletion iron transporters and iron reductase are down-regulated, even in the presence of iron. Further, phytosiderphore biosynthesis and internal iron transport are impaired as the nicotianamine synthase is essentially switched off (Astolfi et al. 2010; Zuchi et al. 2009; Cassin et al. 2009; Klatte et al. 2009). In a recent study molecular features common to various mineral nutrient depletions or exposures (N, P, K, S, Se) could be deduced (Watanabe et al. 2010). Due to the obvious metabolic and regulatory interactions of minerals at the physiological and the molecular level we see a need for a comparative analysis of the overlapping effects of mineral nutrients on plant metabolism, which will aid in understanding specificity in comparison to combined or even downstream pleiotropic molecular and physiological responses (Marschner 1995; Watanabe et al. 2010).

Here we provide a case study of interdependencies of sulfate with iron and selenium. Selenium is being taken up by sulfate transporters and assimilated through the sulfur metabolic pathway. Selenate is reduced to selenide and the corresponding selenium amino acids, selenocysteine and selenemethionine, are formed (Pilon-Smits et al. 2009). Selenate is not essential for plants, but can be tolerated. High concentrations though are toxic. Some plant species have evolved mechanisms to detoxify selenium either through formation of less toxic derived compounds such as methyl-selenocysteine or volatilisation via dimethylselenide (Sors et al. 2005). In order to unravel mutual relationships we compared the transcriptome of roots of *Arabidopsis thaliana* exposed to selenium or deprived of iron to the respective responses of sulfate depletion.

The Root Transcriptome Response to Sulfur, Selenium and Iron Treatment

Roots take up nutrient ions from the soil and are the first plant organs to be exposed to soil nutrient availability. Thus, we compared the response of the *Arabidopsis* root transcriptome to sulfate starvation, iron starvation and selenate exposure. Root transcriptome data from *Arabidopsis thaliana* plants grown on 15 μM sulfate deficient medium for 10 days (Maruyama-Nakashita et al. 2006), plants grown with a non-lethal concentration of 40 μM selenate for 10 days (Van Hoewyk et al. 2008), and 7 days old plants shifted from 100 μM iron to 0 μM iron for 2 days (Dinneny et al. 2008) were compared (Affymetrix ATH1 chip). As visualized in Venn diagrams (Fig. 1), sets of genes overlapped between either pairwise comparisons and even between all three treatments indicate common response mechanisms. When applying a threshold of 2-fold for up-regulation and 0.5-fold for down-regulation of gene expression from the approximately 24,000 gene models on the chip 626 were down- and 1,514 up-regulated upon sulfate-depletion; upon selenium treatment 1,114 were down- and 1,625 up-regulated and iron starvation lead to 4,739 genes being down- and 2,224 up-regulated in root tissue (Fig. 1). When applying a threshold of factor 5, we observed 259 genes being upregulated and 103 down-regulated for sulfate depletion, 392 up- and 128 down-regulated for selenium treatment, and 447 up-regulated and 1,342 genes down-regulated upon iron starvation (Fig. 1). When looking at high stringency (factor 5) 25 genes were up-regulated both in sulfur starvation and selenium treatment, 10 genes were up-regulated both in sulfur and

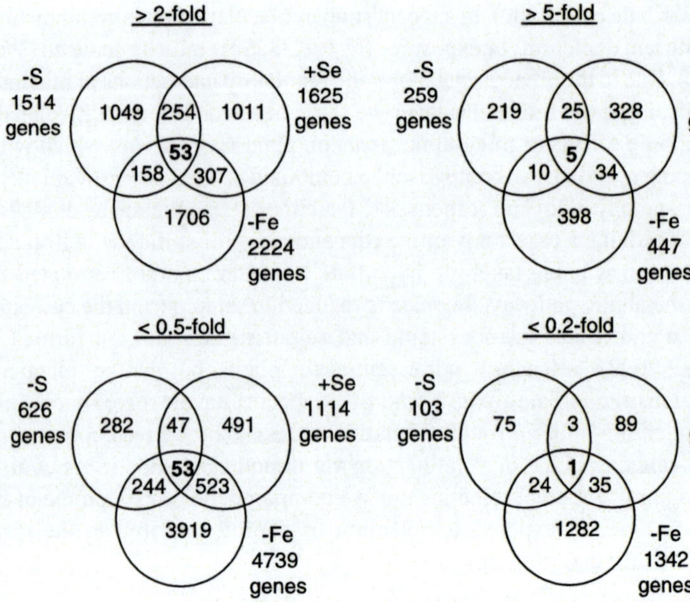

Fig. 1 Venn diagrams showing overlap between data sets, –S, +Se and –Fe. The data sets were obtained from the publications, Maruyama-Nakashita et al. (2006); –S, Van Hoewyk et al. (2008); +Se and Dinneny et al. (2008); –Fe. Thresholds of 2- or 0.5-fold and 5- or 0.2-fold, respectively, were applied

iron starvation, and 34 genes were up-regulated both under iron starvation and selenium exposure, with 5 genes up-regulated in all cases and, respectively, 3, 24, and 35 down-regulated in pairwise comparisons and 1 gene in all three.

Specific Features of Iron Starvation

While selenium exposure and sulfate starvation response showed common features the root transcriptome response to iron starvation hardly overlapped and rather displayed specific features typical for iron starvation. Nicotianamine synthase 1 and 4 (At5g04950; At1g56430) were induced 4- and 23-fold, respectively; as well as the iron reductase *FRO3* (At1g23020), fourfold, the iron transporter *OPT3* (At4g16370); *YSL1* (At4g24120) and *3* (At5g53550), about twofold, the tonoplast iron exporter *NRAMP4* (At5g67330), fivefold, and the citrate efflux transporter *FRD3* (At3g08040), fivefold. Among the induced genes were as well transcription factors such as *bHLH101* (At5g04150), tenfold, further *bHLH* genes (At3g56980; At3g47640), both fourfold, and a zinc finger transcription factor (At3g18290), sixfold.

With respect to interaction of iron starvation to other heavy metals which might substitute iron functionally (Puig et al. 2007; Haensch and Mendel 2009) an up-regulation of copper transporter is evident with *COPT1* and *2* (At5g59030; At3g46900) being induced three- and fivefold, respectively, and a zinc transporter, *ZIP9* (At4g33020) being up-regulated threefold. However, providing a complex response pattern, a set of other zinc transporters (*ZIP3*, *4*, *5*, *7* and *10*: At2g32270, At1g10970, At1g05300, At2g04032 and At1g31260, respectively) was down-regulated (lower than 0.5-fold change). Notably the molybdate transporter *SULTR5;2* (or *MOT1*) was down-regulated by a factor of 10 under iron starvation. *SULTR5;2* responded to all three conditions with down-regulation.

Expression Response of Sulfate Metabolism Related Genes

The expression response of the sulfate metabolism related genes were extracted for all three conditions (Table 1) and presented as fold changes to the respective unstressed controls. When applying a factor of 2 for the change in expression, for sulfate starvation 12 genes were altered with 9 up and 3 down, and 4 genes being specific, for selenium exposure 12 genes were altered with 8 up and 9 down with 1 gene responding exclusively, while for iron starvation 15 genes were altered, 7 up and 8 down, 7 specifically responding only to iron starvation. 5 genes showed a response in all three treatments: *SULTR2;1* and *4;2* being up-regulated, while *SULTR5;2*, the molybdate transporter, was in all cases down-regulated as well as *ATPS4*. Under sulfate starvation and selenate exposure the high affinity sulfate transporter *SULTR1;1* was up-regulated, while iron starvation lead to a down-regulation of its expression. Beyond this overlap sulfur and iron starvation shared no common effects on the expression of genes of the sulfate metabolism, while selenium exposure overlapped for additional three genes with *APR2* and *3* and *SERAT3;1* being over-expressed. Notable was also a down-regulation of *BSAS1;1* and *4;3* under iron starvation and an up-regulation of *BSAS5;1* together with a significant induction of the phytochelatine synthase, *PCS2*.

When applying a hierarchical cluster analysis of the root transcriptome data selenium and sulfur treatments clustered close together while iron was quite distant (data not shown). At a high threshold of tenfold for over-expression compared to controls, 16 genes could be identified for sulfate starvation (Table 2), 11 of these were also up-regulated under selenium treatment, while only 3 for iron starvation. Eight of 16 genes were of unknown identity and function, while *BGLU28*, *LSU*, *SULTR1;1*, *MS5*, *Cha-C-like*, a putative sugar transporter and carbonic anhydrase were known responders from sulfate starvation analyses (Nikiforova et al. 2003; Hirai et al. 2003; Maruyama-Nakashita et al. 2003). For the iron response, three genes were in contrast strongly down-regulated, *i.e. SULTR1:1* (see also Table 1) and two unknowns while *SRG2*, a senescence related beta glucosidase, and an unknown protein were up-regulated.

Table 1 Expression levels as x-fold changes of sulfur pathway relevant genes

Gene	AGI	–S	+Se	–Fe	Gene	AGI	–S	+Se	–Fe
SULTR1;1	At4g08620	91.29	10.33	0.01	SIR	At5g04590	0.85	1.26	0.29
SULTR1;2	At1g78000	4.45	1.72	0.59	SERAT1;1	At5g56760	1.16	0.89	0.74
SULTR1;3	At1g22150	0.77	0.63	1.86	SERAT2;1	At1g55920	1.28	1.21	1.14
SULTR2;1	At5g10180	6.91	6.43	2.56	SERAT2;2	At3g13110	1.10	1.01	0.72
SULTR2;2	At1g77990	1.58	1.07	0.61	SERAT3;1	At2g17640	2.58	2.13	0.70
SULTR3;1	At3g51900	ND	ND	ND	SERAT3;2	At4g35640	ND	ND	ND
SULTR3;2	At4g02700	1.36	1.11	0.75	BSAS1;1	At4g14880	1.05	0.93	0.36
SULTR3;3	At1g23090	1.66	0.55	0.53	BSAS1;2	At3g22460	1.00	0.55	1.94
SULTR3;4	At3g15990	2.88	1.51	1.44	BSAS2;1	At2g43750	1.34	1.49	0.65
SULTR3;5	At5g19600	0.98	0.91	0.66	BSAS2;2	At3g59760	0.98	0.95	0.74
SULTR4;1	At5g13550	1.92	1.69	1.72	BSAS3;1	At3g61440	1.05	0.93	0.90
SULTR4;2	At3g12520	6.19	2.96	3.03	BSAS4;1	At5g28020	1.63	1.72	0.90
SULTR5;1	At1g80310	0.80	0.47	1.40	BSAS4;2	At3g04940	1.16	0.85	0.62
SULTR5;2	At2g25680	0.44	0.20	0.11	BSAS4;3	At5g28030	0.64	3.61	0.05
ATPS1	At3g22890	0.89	1.65	1.27	BSAS5;1	At3g03630	0.76	1.11	2.26
ATPS2	At1g19920	0.71	0.55	0.69	GSH1	At4g23100	0.74	0.66	0.40
ATPS3	At4g14680	1.98	5.92	3.63	GSH2	At5g27380	1.14	1.11	1.09
ATPS4	At5g43780	0.21	0.25	0.50	AtPCS1	At5g44070	1.44	1.24	1.54
APK1	At2g14750	1.10	1.17	1.21	AtPCS2	At1g03980	1.80	0.93	3.96
APK2	At4g39940	0.32	1.25	1.24	CGS1	At3g01120	0.86	0.88	1.15
APK3	At3g03900	1.22	0.94	2.31	CGS2	At1g33320	1.32	0.71	1.98
APK4	At5g67520	1.01	1.04	0.19	CBL1	At1g64660	2.68	1.72	1.75
APR1	At4g04610	1.57	3.13	2.09	CBL2	At3g57050	0.81	0.95	1.15
APR2	At1g62180	2.57	2.57	1.33	MS2 c	At3g03780	1.11	1.01	0.94
APR3	At4g21990	4.50	5.12	0.87	MS3 p	At5g20980	1.26	1.17	1.11

Data sets were obtained from the same publications as Fig. 1. Fold changes relative to respective controls are shown. >2-fold and <0.5-fold changes are dark and light grey, respectively

ND not detected

Table 2 Selected highest responding genes (>10-fold) under either sulfur depletion or selenium treatment

Gene	AGI	–S	+Se	–Fe
BGLU28 (putative myrosinases)	At2g44460	250.35	239.19	2.46
Expressed protein	At1g12030	188.13	92.58	67.93
LSU1/Vikin/UP9	At3g49580	92.85	12.39	0.92
SULTR1;1	At4g08620	91.29	10.33	0.01
MS5/ATSDI1	At5g48850	88.83	33.51	2.04
ChaC-like family protein	At5g26220	86.17	95.20	3.38
BGLU30/DIN2/SRG2	At3g60140	81.53	30.76	15.11
Putative sugar transporter	At3g05400	41.48	12.92	5.02
BCA3 (beta carbonic anhydrase)	At1g23730	37.08	44.52	3.20
Expressed protein	At4g31330	15.85	8.63	2.61
GNAT family protein	At2g23060	13.35	16.37	5.17
Hypothetical protein	At5g57730	12.50	9.67	0.07
Transposable element gene	At2g07080	12.14	9.56	0.03
Expressed protein	At5g53820	10.82	7.26	2.20
Hypothetical protein	At1g20530	10.71	13.54	0.64
Transposable element gene	At1g37063	10.00	7.44	1.25

Data sets were obtained from the same publications as Fig. 1. Fold changes relative to the respective controls are shown. >10-fold changes are marked dark grey

Discussion

The effects of distinct nutrient starvations have been analyzed with respect to gene expression, metabolic and physiological effects (reviewed in Amtmann and Armengaud 2009). However, gross physiological responses such as chlorosis and growth retardation and the underlying molecular patterns such as changes of the transcriptome and the metabolome overlap between various nutrient alterations. A systematic analysis of these responses thus appears to be a helpful approach to identify either specific or common response patterns (Watanabe et al. 2010) to provide candidate genes for future investigations. Further, it has to be considered that the response patterns are not static but highly dynamic. Revealing these processes will help to understand plant sulfate metabolism (Hesse and Hoefgen 2003, 2008; Hoefgen and Hesse 2007) or effects of other nutrients on plant physiology.

In this study we compared root transcriptome data of plants depleted for iron, sulfate or exposed to excess selenium, because these soil ionic components have been shown to interact or influence one another (Maruyama-Nakashita et al. 2006; Van Hoewyk et al. 2008; Dinneny et al. 2008). High soil contents of selenium lead to a response of the transcriptome being highly congruent to sulfate depletion (Fig. 1; Table 1). Especially, the most highly expressed genes under sulfate deprivation were similar to a subset of genes being induced upon selenium treatment. In contrast, despite the known mutual influence of iron and sulfate, iron depletion hardly overlapped at the root transcriptome level (Table 2). The ratio of expressed genes compared to all genes is lower in roots (40%) than in rosette leaf tissues

(90%) displaying distinct differences representing functional specialization between green, floral and root tissues as displayed by cluster analysis (Ma et al. 2005). When applying a greater than fivefold threshold for relative gene expression changes sulfate starvation affected 4% of the gene set expressed in roots, selenium exposure 6%, iron starvation 19% (Fig. 1).

It was striking that most of the genes of sulfate uptake, assimilation and down-stream reactions such as GSH biosynthesis did not respond to sulfate starvation, iron starvation or selenium exposure in the root (Table 1). Of the sulfate uptake system only the high affinity sulfate transporter *SULTR1;1* and *1;2* were induced under S limitation, and, to a lesser extent as well under selenium exposure while iron starvation essentially down-regulated the high affinity sulfate uptake system. *SULTR2;1*, a low affinity sulfate transporter expressed in vascular tissue, probably involved in transport from root to shoot (Awazuhara et al. 2005), was induced. The vacuolar sulfate exporter *SULTR4;2* was induced upon sulfate and iron depletion and selenium exposure indicating that the plant sensed an internal sulfate depletion which corresponded to the induction of *APR2* and *APR3*. *BSAS* and *SERAT* genes did hardly or inconsistently respond to the treatments.

Iron depletion lead to rather specific responses corresponding to previous reports (Zuchi et al. 2009; Astolfi et al. 2010; Pilon et al. 2009). Iron transporter and iron reductase were induced. Among the sulfate metabolism related genes we observed an increase of the phytochelatine synthase *PC2*, though iron is not complexed by PCs (Table 1). This might indicate that the increased efforts of the plant to take up iron might be coupled to increased uptake of other, potentially toxic heavy metals. This seems to be supported by the fact that copper and zinc transporters were induced. Iron and copper (and zinc?) are coupled in a delicate balance *e.g.* for the function of superoxide dismutases (SOD) involved in ROS detoxification (Pilon et al. 2009). Interestingly, the molybdate transporter, *SULTR5;2,* was significantly down-regulated under all conditions, but especially under iron depletion. It might be worthwhile to investigate the reasons for the down-regulation of high-affinity sulfate transporters (*SULTR1:1* and *1;2*) under these conditions and whether they might be correlated to heavy metal uptake. However, in parallel a counter intuitive down-regulation of the GSH synthesizing genes *BSAS1;1* and *BSAS4;3* involved in cysteine biosynthesis, one of the GSH precursors, and of *GSH1*, the gamma-glutamyl-cysteine synthase, was observed. Detailed metabolomics and ionomic studies of plants under the same treatments would be helpful to resolve the physiological relevance of the observed expression patterns and whether these are converted into down-stream products, such as increased phytochelatin levels and increased copper, zinc, and manganese concentrations.

Among those genes highly induced in roots both under sulfate depletion or increased selenium exposure (Table 2) a set of "known" sulfate marker genes appeared: *LSU*, *SULTR1;1*, *MS5*, *Cha-C-like*, and a putative sugar transporter (Nikiforova et al. 2003; Hirai et al. 2003; Maruyama-Nakashita et al. 2003). However, the molecular functions and physiological consequences of these gene inductions are still unknown. When using the highest expressor of this gene set (Table 2; At2g44460) as query using the ATTED II software (http://atted.jp) a stable co-expression network over 1,388 available arrays was generated (Fig. 2). As this

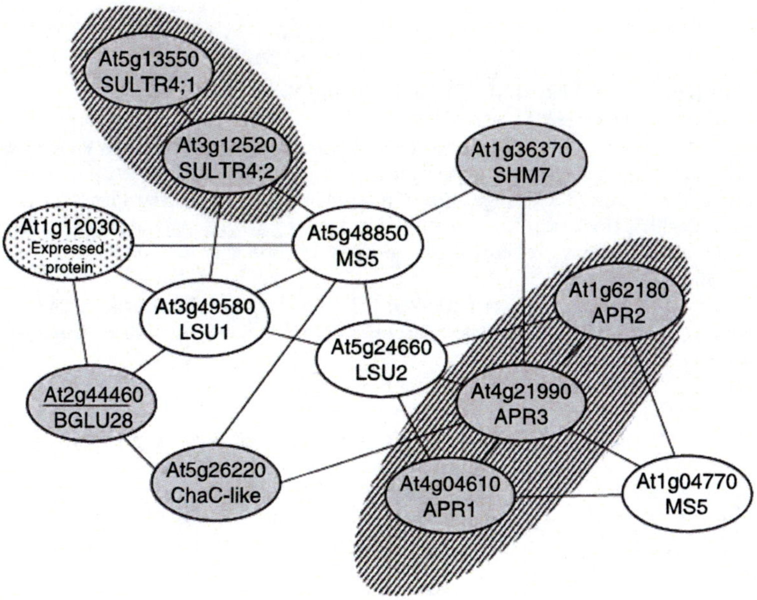

Fig. 2 Co-expressed gene network related to At2g44460 (*BGLU28*). At2g44460 was the highest responding gene under sulfur starvation in Table 1. ATTED II (http://atted.jp/) was used to search for correlated genes. *Grey circles* indicate pathway or structural genes. *White circles* indicate potential regulator genes. *Dotted circles* indicate genes of unknown function. Sulfur pathway genes are in hatched underlay

co-expression network is not based on the comparisons as executed in this analysis it indeed indicates a very robust expressional relation of these genes. Intriguing is the fact that structural genes seem to be tightly correlated to assumed regulatory genes as *LSU* and *MS5* (*SDI*) (Howarth et al. 2005, 2009; Lewandowska et al. 2010). This has to be included into the ongoing discussion of the regulatory processes controlling sulfate metabolism (Nikiforova et al. 2004; Maruyama-Nakashita et al. 2004a, b, c, 2005, 2006; Kasajima et al. 2007; Falkenberg et al. 2008; Gojon et al. 2009; Kawashima et al. 2009; Watanabe et al. 2010).

In summary, sulfate depletion and selenium exposure share definite parts of their molecular response patterns and selenium exposure mimics to a certain extent sulfate deprivation. Iron metabolism is dependent on sulfate availability and both directly interact in iron-sulfur-clusters. Despite these links the degree of mutually shared gene expression responses at the level of the root transcriptome is low rather indicating specific than mutually shared regulatory mechanisms and response patterns. We suggest here that comparative transcriptomics (and metabolomics) analyses of either single or multiple nutrient treatments will provide results to close the gap between the available agricultural and physiological knowledge of the effects of various mineral ions on plant growth and development and of the molecular and regulatory basis of these effects. Focussing on specific versus common responses will allow discerning targeted from downstream effects and specific from pleiotropic patterns.

References

Aller AJ, Bernal JL, del Nozal MJ, Deban L (1990) Effects of selected trace elements on plant growth. J Sci Food Agric 51:447–479

Amtmann A, Armengaud P (2009) Effects of N, P, K and S on metabolism: new knowledge gained from multi-level analysis. Curr Opin Plant Biol 12:275–283

Amtmann A, Blatt MR (2009) Regulation of macronutrient transport. New Phytol 181:35–52

Astolfi S, Zuchi S, Hubberten HM, Pinton R, Hoefgen R (2010) Supply of sulphur to S-deficient young barley seedlings restores their capability to cope with iron shortage. J Exp Bot 61: 799–806

Awazuhara M, Fujiwara T, Hayashi H, Watanabe-Takahashi A, Takahashi H, Saito K (2005) The function of SULTR2;1 sulfate transporter during seed development in *Arabidopsis thaliana*. Physiol Plant 125:95–105

Buchanan B, Gruissem W, Jones RL (2007) Biochemistry & molecular biology of plants. Wiley, New York

Buckhout TJ, Yang TJ, Schmidt W (2009) Early iron-deficiency-induced transcriptional changes in *Arabidopsis* roots as revealed by microarray analyses. BMC Genomics 10:147

Cassin G, Mari S, Curie C, Briat JF, Czernic P (2009) Increased sensitivity to iron deficiency in *Arabidopsis thaliana* overaccumulating nicotianamine. J Exp Bot 60:1249–1259

Currie HA, Perry CC (2007) Silica in plants: biological, biochemical and chemical studies. Ann Bot 100:1383–1389

Dinneny JR, Long TA, Wang JY, Jung JW, Mace D, Pointer S, Barron C, Brady SM, Schiefelbein J, Benfey PN (2008) Cell identity mediates the response of *Arabidopsis* roots to abiotic stress. Science 320:942–945

Falkenberg B, Witt I, Zanor MI, Steinhauser D, Mueller-Roeber B, Hesse H, Hoefgen R (2008) Transcription factors relevant to auxin signalling coordinate broad-spectrum metabolic shifts including sulphur metabolism. J Exp Bot 59:2831–2846

Gojon A, Nacry P, Davidian JC (2009) Root uptake regulation: a central process for NPS homeostasis in plants. Curr Opin Plant Biol 12:328–338

Haensch R, Mendel RR (2009) Physiological functions of mineral micronutrients (Cu, Zn, Mn, Fe, Ni, Mo, B, Cl). Curr Opin Plant Biol 12:259–266

Hesse H, Hoefgen R (2003) Molecular aspects of methionine biosynthesis. Trends Plant Sci 8: 259–262

Hesse H, Hoefgen R (2008) Sulfur and cysteine metabolism. In: Jez J (ed) Sulfur; a missing link between soils, crops, and nutrition. American Society of Agronomy, Madison, pp 83–104

Hirai MY, Fujiwara T, Awazuhara M, Kimura T, Masaaki N, Saito K (2003) Global expression profiling of sulfur-starved *Arabidopsis* by DNA macroarray reveals the role of *O*-acetyl-L-serine as a general regulator of gene expression in response to sulfur nutrition. Plant J 33: 651–663

Hoefgen R, Hesse H (2007) Sulfur in plants as part of a metabolic network. In: Hawkesford MJ, de Kok LJ (eds) Sulfur in plants: an ecological perspective. Springer, Dordrecht, pp 107–142

Hoefgen R, Hesse H (2008) Sulfur and cysteine metabolism. In: Jez J (ed) Sulfur: a missing link between soils, crops and nutrition. American Society of Agronomy, Madison, pp 83–104

Howarth JR, Barraclough PB, Hawkesford MJ (2005) A highly specific sulfate-deficiency induced gene (*sdi1*) from wheat. In: Saito K, De Kok LJ, Stulen I, Hawkesford MJ, Schnug E, Sirko A, Rennenberg H (eds) Sulfur transport and assimilation in plants in the post genomic era. Backhuys Publishers, Leiden, pp 161–164

Howarth JR, Parmar S, Barraclough PB, Hawkesford MJ (2009) A sulphur deficiency-induced gene, *sdi1*, involved in the utilization of stored sulphate pools under sulphur-limiting conditions has potential as a diagnostic indicator of sulphur nutritional status. Plant Biotechnol J 7:200–209

Hu Z, Richter H, Sparovek G, Schnug E (2004) Physiological and biochemical effects of rare earth elements on plants and their agricultural significance: a review. J Plant Nutr 27:183–220

Jeong J, Guerinot ML (2009) Homing in on iron homeostasis in plants. Trends Plant Sci 14: 280–285

Kasajima I, Ohkama-Ohtsu N, Ide Y, Hayashi H, Yoneyama T, Suzuki Y, Naito S, Fujiwara T (2007) The BIG gene is involved in regulation of sulfur deficiency-responsive genes in *Arabidopsis thaliana*. Physiol Plant 129:351–363

Kawashima CG, Yoshimoto N, Maruyama-Nakashita A, Tsuchiya YN, Saito K, Takahashi H, Dalmay T (2009) Sulfur starvation induces the expression of microRNA-395 and one of its target genes but in different cell types. Plant J 57:313–321

Klatte M, Schuler M, Wirtz M, Fink-Straube C, Hell R, Bauer P (2009) The analysis of *Arabidopsis* nicotianamine synthase mutants reveals functions for nicotianamine in seed iron loading and iron deficiency responses. Plant Physiol 150:257–271

Kopriva S, Rennenberg H (2004) Control of sulphate assimilation and glutathione synthesis: interaction with N and C metabolism. J Exp Bot 55:1831–1842

Lewandowska M, Wawrzynska A, Moniuszko G, Lukomska J, Zientara K, Piecho M, Hodurek P, Zhukov I, Liszewska F, Nikiforova V, Sirko A (2010) A contribution to identification of novel regulators of plant response to sulfur deficiency: characteristics of a tobacco gene UP9C, its protein product and the effects of UP9C silencing. Mol Plant 3:347–360

Ma L, Sun N, Liu X, Jiao Y, Zhao H, Deng XW (2005) Organ-specific expression of *Arabidopsis* genome during development. Plant Physiol 138:80–91

Marschner H (1995) Mineral nutrition of higher plants, 2nd edn. Academic, London/San Diego

Maruyama-Nakashita A, Inoue E, Watanabe-Takahashi A, Yamaya T, Takahashi H (2003) Transcriptome profiling of sulphur-responsive genes in *Arabidopsis* reveals global effects of sulfur nutrition on multiple metabolic pathways. Plant Physiol 132:597–605

Maruyama-Nakashita A, Nakamura Y, Watanabe-Takahashi A, Yamaya T, Takahashi H (2004a) Induction of SULTR1;1 sulfate transporter in *Arabidopsis* roots involves protein phosphorylation/dephosphorylation circuit for transcriptional regulation. Plant Cell Physiol 45:340–345

Maruyama-Nakashita A, Nakamura Y, Yamaya T, Takahashi H (2004b) A novel regulatory pathway of sulfate uptake in *Arabidopsis* roots: implication of CRE1/WOL/AHK4-mediated cytokinin-dependent regulation. Plant J 38:779–789

Maruyama-Nakashita A, Nakamura Y, Yamaya T, Takahashi H (2004c) Regulation of high-affinity sulphate transporters in plants: towards systematic analysis of sulphur signalling and regulation. J Exp Bot 55:1843–1849

Maruyama-Nakashita A, Nakamura Y, Watanabe-Takahashi A, Inoue E, Yamaya T, Takahashi H (2005) Identification of a novel cis-acting element conferring sulfur deficiency response in *Arabidopsis* roots. Plant J 42:305–314

Maruyama-Nakashita A, Nakamura Y, Tohge T, Saito K, Takahashi H (2006) *Arabidopsis* SLIM1 is a central transcriptional regulator of plant sulfur response and metabolism. Plant Cell 18:3235–3251

Nikiforova V, Freitag J, Kempa S, Adamik M, Hesse H, Hoefgen R (2003) Transcriptome analysis of sulfur depletion in *Arabidopsis thaliana*: interlacing of biosynthetic pathways provides response specificity. Plant J 33:633–650

Nikiforova VJ, Gakière B, Kempa S, Adamik M, Willmitzer L, Hesse H, Hoefgen R (2004) Towards dissecting nutrient metabolism in plants: a systems biology case study on sulphur metabolism. J Exp Bot 55:1861–1870

Pilon M, Cohu CM, Ravet K, Abdel-Ghany SE, Gaymard F (2009) Essential transition metal homeostasis in plants. Curr Opin Plant Biol 12:347–357

Pilon-Smits EA, Quinn CF, Tapken W, Malagoli M, Schiavon M (2009) Physiological functions of beneficial elements. Curr Opin Plant Biol 12:267–274

Puig S, Andres-Colas N, Garcia-Molina A, Penarrubia L (2007) Copper and iron homeostasis in *Arabidopsis*: responses to metal deficiencies, interactions and biotechnological applications. Plant Cell Environ 3:271–290

Schwarz G, Mendel RR, Ribbe MW (2009) Molybdenum cofactors, enzymes and pathways. Nature 460:839–847

Shinmachi F, Buchner P, Stroud JL, Parmar S, Zhao FJ, McGrath SP, Hawkesford MJ (2010) Influence of sulfur deficiency on the expression of specific sulfate transporters and the distribution of sulfur, selenium, and molybdenum in wheat. Plant Physiol 153:327–336

Sors TG, Ellis DR, Salt DE (2005) Selenium uptake, translocation, assimilation and metabolic fate in plants. Photosynth Res 86:373–389

Staiger D (2002) Chemical strategies for iron acquisition in plants. Angew Chem Int Ed Engl 41:2259–2264

Van Hoewyk D, Takahashi H, Inoue E, Hess A, Tamaoki M, Pilon-Smits EA (2008) Transcriptome analyses give insights into selenium-stress responses and selenium tolerance mechanisms in *Arabidopsis*. Physiol Plant 132:236–253

Von Liebig J (1840) Die organische Chemie in ihrer Anwendung auf Agrikultur und Physiologie. Vieweg, Braunschweig

Watanabe M, Hubberten H-M, Saito K, Hoefgen R (2010) General regulatory patterns of plant mineral nutrient depletion as revealed by *serat* quadruple mutants disturbed in cysteine synthesis. Mol Plant 3:438–466

Zhao FJ, McGrath SP, Meharg AA (2010) Arsenic as a food chain contaminant: mechanisms of plant uptake and metabolism and mitigation strategies. Annu Rev Plant Biol 61:535–559

Zhu YG, Pilon-Smits EA, Zhao FJ, Williams PN, Meharg AA (2009) Selenium in higher plants: understanding mechanisms for biofortification and phytoremediation. Trends Plant Sci 14: 436–442

Zuchi S, Cesco S, Varanini Z, Pinton R, Astolfi S (2009) Sulfur deprivation limits Fe-deficiency responses in tomato plants. Planta 230:85–94

Regulation of Iron-Sulfur Cluster Formation in Plastids

Karl Ravet, Douglas Van Hoewyk, and Marinus Pilon

Abstract Iron-sulfur (Fe-S) clusters have pivotal roles in chloroplast processes including photosynthetic electron transport, pigment synthesis, nitrogen and sulfur assimilation and carbon metabolism. There are five cluster types found in plant plastids, which harbor their own Fe-S cluster assembly system. The cysteine desulfurase of the chloroplast, CpNifS, is an essential protein that requires a SufE protein for activation of its activity. SufE proteins serve to accept sulfane sulfur from the cysteine desulfurase and presumably donate S to downstream scaffolds on which cluster assembly takes place. There are three SufE proteins in plastids and SufE1 is the major activator of CpNifS. SufE1 consists of a SufE domain linked to a BolA domain. It is likely that the BolA domain of SufE1 has a regulatory role perhaps in providing feedback to the Fe-S assembly system. Theoretical predictions suggest that the SufE1-BolA region interacts with a glutaredoxin (GRX) domain protein. There are several GRX domain proteins in plastids and some of these can bind iron-sulfur clusters. Whereas, there are several potential Fe-S cluster assembly scaffold proteins identified in the chloroplast the exact role of each one in the assembly pathway still needs to be revealed. We discuss possible models for this regulation and explore areas for future research.

K. Ravet (✉) • M. Pilon
Biology Department and Program in Molecular Plant Biology, Colorado State University, 1878 Campus delivery, Fort Collins, CO 80523-1878, USA
e-mail: kravet@lamar.colostate.edu

D. Van Hoewyk
Biology Department, Coastal Carolina University, Conway, SC 29526, USA

L.J. De Kok et al. (eds.), *Sulfur Metabolism in Plants*, Proceedings of the International Plant Sulfur Workshop 1, DOI 10.1007/978-94-007-4450-9_18,

Introduction

Iron-sulfur (Fe-S) clusters are among the oldest and most versatile cofactors found in living organisms. They participate in electron transfer, catalysis, and regulatory processes (Beinert et al. 1997). In plants five different cluster types exist: the ferredoxin type 2Fe-2S, the Rieske type 2Fe-2S; the 4Fe-4S cluster, 3Fe-4S cluster and the 4Fe-4S-siroheme cluster (Ye et al. 2006a). Differences in the ligands and in the protein environment of the Fe–S clusters allow a wide range of redox potentials (Beinert et al. 1997). Fe–S proteins are therefore pivotal components of the electron transfer chains in the respiratory complexes of mitochondria and in the photosynthetic apparatus in the chloroplasts. In addition, Fe-S proteins also have catalytic roles in plastids, mitochondria and in the cytosol and nucleus (Balk and Pilon 2011). Photosynthesis is the process that determines plant productivity and ultimately drives all life on earth. Its importance for biology, agriculture and the environment cannot be overstated and its efficiency is directly dependent on Fe-S clusters. Furthermore, plant iron status is of importance for human health, since billions of people worldwide are suffering from iron deficiency, and most of the dietary iron in developing countries is derived from vegetarian food sources (White and Broadley 2009). Therefore a thorough understanding of how these Fe-S clusters are formed in plant chloroplasts and insight into factors that may affect cluster assembly, such as S and Fe deficiency and oxidative stress, is of utmost interest and importance.

General Principles of Biological Fe-S Assembly

Iron–sulfur (Fe–S) clusters are chemically simple and can be formed *in vitro* from sulfide and an iron salt and inserted "spontaneously" into certain apo-proteins such as ferredoxin if it has free reduced thiols (cysteines) to receive the cluster. However, most apo-proteins are inefficiently or not at all reconstituted in a test tube: *in vivo* all Fe-S proteins need protein cofactors to mature (Lill 2009). The pioneering work of Dean and coworkers in nitrogen fixing bacteria and in *Escherichia coli* revealed a general framework for the assembly of Fe-S clusters in living cells. *In vivo* three steps can be distinguished: (1) mobilization of sulfur from cysteine and of Fe from as yet mostly undetermined sources, (2) assembly of a cluster on a scaffold and (3) transfer of the cluster from the scaffold to an apoprotein (Lill 2009). Nitrogen fixing bacteria have a gene cluster (*Nif* operon) that encodes a specialized machinery dedicated to the synthesis of the Fe-S cofactors of nitrogen reductase. A second gene cluster termed the *Isc* (for Iron Sulfur Cluster assembly) operon mediates Fe-S assembly for other "housekeeping" Fe-S proteins in these bacteria and is also found in *E. coli*. In *E. coli* and several other bacteria another major gene cluster termed the *Suf* (for Sulfur limitation) operon encodes for a machinery for Fe-S assembly under sulfur limitation, iron starvation and oxidative stress conditions. All these bacterial gene clusters have in common that they include a gene encoding a cysteine desulfurase (involved in the release of S from cysteine) as well as genes encoding possible scaffolds on which clusters may be pre-formed before transfer to target apo-proteins (Lill 2009). The need for different Fe-S cluster types probably necessitates the existence of multiple scaffolds or carriers of Fe-S intermediates and factors that can modulate the structure and redox state of clusters (Beinert et al. 1997; Lill 2009).

To form their Fe-S clusters, plants have three assembly machineries, which are localized in the plastids, mitochondria and cytoplasm, respectively (Ye et al. 2006a; Balk and Pilon 2011). The mitochondrial Fe-S assembly machinery in plants and other eukaryotes resembles the bacterial *Isc* (type I) system, while the plastidial machinery shares several components with the bacterial *Suf* (type II) systems Balk and Pilon 2011). The Fe-S machinery in the cytosol of eukaryotes, called the CIA (for Cytosolic Iron-sulfur Assembly) system, comprises a set of almost ubiquitously conserved scaffolds and transfer proteins. This machinery seems to depend on the cysteine desulfurase activity in the mitochondria and on the export of a still unknown mitochondrial compound into the cytosol, at least in yeast and in plants (Balk and Pilon 2011).

Fe–S Protein Biogenesis in the Plastid: The Central Role of CpNifS

Nuclear-encoded chloroplast proteins such as ferredoxin acquire their cofactors after import into the organelle. Components that may be involved in Fe transport into plastids include a ferric reductase (Jeong et al. 2008) and a conserved putative Fe transporter (Duy et al. 2007). It is not known what the direct iron donor is for Fe-S assembly but it is at least clear that ferritins do not have this function (Ravet et al. 2009). Fe-S cluster assembly in Fd was observed in isolated chloroplasts with cysteine as the sulfur donor, a reaction that further requires light or ATP and NADPH (Takahashi et al. 1986). Therefore, the plastids of plants harbor a complete Fe-S cluster assembly machinery which is probably optimized to function in an oxygen-saturated environment and to facilitate the biogenesis of numerous Fe-S proteins that function in photosynthetic electron transfer, nitrogen and sulfur assimilation, as well as chlorophyll metabolism and protein import in plastids (Balk and Pilon 2011).

The known proteins that function in plastid Fe-S assembly are nuclear-encoded and imported from the cytosol (see Fig. 1 for a summary). Many of the components are identified and an understanding of how the chloroplastic Fe-S pathway operates has started to emerge. At the core of the system is the chloroplastic NifS-like cysteine desulfurase CpNifS/NFS2 (Leon et al. 2002; Pilon-Smits et al. 2002). CpNifS is in the stroma of *Arabidopsis* chloroplasts and is expressed in all major plant tissues at about equal levels. The purified enzyme had both cysteine desulfurase and selenocysteine lyase activity, the former being ~300-fold lower than the latter (Pilon-Smits et al. 2002). CpNifS was shown to be active by itself in stimulating holo-fd formation (Ye et al. 2005). Not only can CpNifS mediate Fe-S formation for ferredoxin in "*in vitro*" assays, it is required, as shown by biochemical depletion

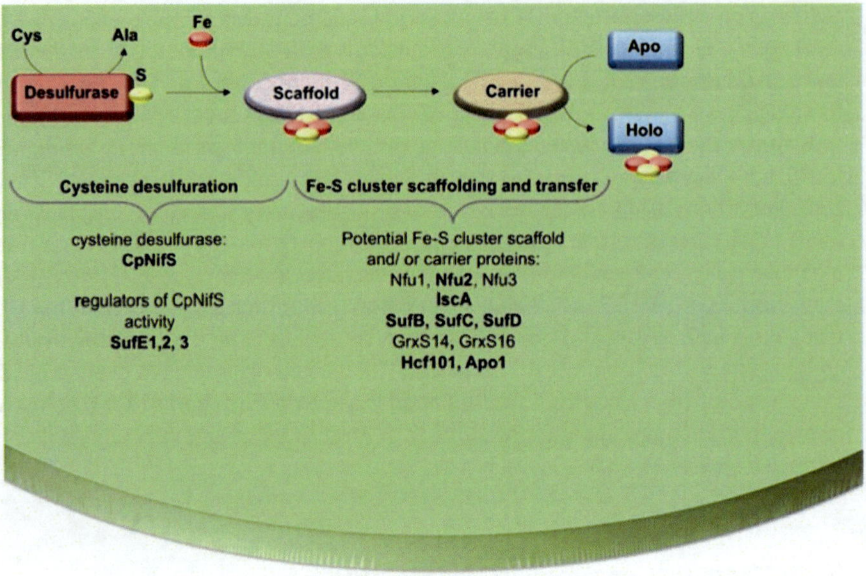

Fig. 1 Working model for plastid Fe-S cluster biogenesis. All the proteins predicted to be involved in the process are listed. In *bold* are the proteins shown to be involved both *in vitro* and *in planta*. See text for details

experiments. The apparent reconstitution activity of chloroplast stroma was 50–80 times higher than that of pure CpNifS protein (Ye et al. 2005). Thus, stromal components activate CpNifS. The effects of CpNifS overexpression and its downregulation have been studied. In the over-expression studies S metabolism was investigated together with the effects on selenium (Se) metabolism, because NifS-like proteins have selenocysteine lyase activity (Pilon et al. 2003). Overexpression led to enhanced S and Se levels, as well as enhanced Se tolerance and prevention of Se incorporation into protein (Van Hoewyk et al. 2005). Inducible RNAi knockdown in transgenics led to a reduction or complete disappearance of plastidic Fe-S proteins, chlorosis, thylakoid degradation and a severe impairment of photosynthesis, and eventually death (Van Hoewyk et al. 2007). An ethanol-inducible system was used to test the direct phenotypic effects of CpNifS depletion (Van Hoewyk et al. 2007). This inducible system has great advantages compared to constitutive systems, because it permits to follow the effect on Fe-S assembly activity soon after induction of the depletion, avoiding pleiotropic defects. It is known that several knock-out mutants are lethal and we can learn very little from genotypes that cause embryonic lethality.

Type-II cysteine desulfurases such as CpNifS need a SufE-like protein for full activity. There are three SufE proteins in plastids: SufE1 is the main activator in most tissues, SufE2 is pollen specific, and SufE3 is composed of a SufE domain linked to quinolinate synthase (Ye et al. 2006b; Murthy et al. 2007). The SufE region of SufE3 is dedicated to the formation of a 4Fe-4S cluster in its quinolinate synthase domain (Murthy et al. 2007). The biochemistry of the CpNifS/SufE1 complex is well described; the two proteins form a dynamic complex where CpNifS transfers the S from the substrate cysteine to a cys residue in SufE1 (Ye et al. 2005, 2006b). Both proteins are essential (Xu and Moller 2006; Ye et al. 2006b; Van Hoewyk et al. 2007). It was proposed that SufE1 activates CpNifS activity by accepting sulfur from CpNifS (Ye et al. 2006b). Thus, SufE1 (or the complex CpNifS-SufE1) should donate S to downstream scaffolds for Fe-S assembly.

Assembly Scaffolds

In the bacterial and mitochondrial Isc systems, the cysteine desulfurase IscS binds to the scaffold IscU (Lill 2009). However, there is no IscU in plastids. Among the possible targets for interaction with the CpNifS-SufE1 complex are the following potential scaffolds: the NifU-like proteins Nfu1-3 (Touraine et al. 2004; Yabe et al. 2004), the ubiquitous IscA/SufA and the SufB/C/D complex (Abdel-Ghany et al. 2005; Xu and Moller 2004; Xu et al. 2005), the P-loop ATPase Hcf101 (Schwenkert et al. 2009), and the monothiol glutaredoxins GrxS14/GrxS16 (Bandyopadhyay et al. 2008). It must be noted that such interactions have not yet been demonstrated in plants.

The 3 NFU genes encode chloroplast proteins that are differentially expressed but closely related in sequence to each other and similar to cyanobacterial Nfu and the C-terminus of NifU and proposed to function as Fe-S assembly scaffolds (Leon et al. 2003; Yabe et al. 2004). In leaves, both Nfu2 and Nfu3 are expressed. Nfu2 is the most abundant isoform and forms a transient Fe-S cluster (Leon et al. 2003) that can be passed on to apo-ferredoxin *in vitro*. *nfu2* insertion mutants have a dwarf phenotype and are deficient in some but not all plastid Fe-S proteins (Yabe et al. 2004; Touraine et al. 2004). However, direct links between CpNifS and Nfu proteins have not been reported so far.

Initial functional studies of the potential scaffold proteins have focused on IscA/SufA (Abdel-Ghany et al. 2005). Upon incubation with CpNifS *in vitro*, purified IscA acquires a transient Fe-S cluster, which it can subsequently transfer to apo-Fd to form holo-Fd. However, it should be noted that the role of the conserved and almost ubiquitous IscA/SufA proteins as a scaffold is controversial in the Fe-S field. Loss of function mutants for *iscA*/*sufA* in microbes show feeble phenotypes – if any – which argues against both an essential Fe carrier or scaffold role (see also Balk and Pilon 2011). In *Arabidopsis* Yabe and Nakai (2006) found that the *iscA* T-DNA insertion lines show no appreciable phenotypes. In cyanobacteria the deletion of IscA or SufA did not give a growth phenotype, yet deregulation (overexpression) of the Suf operon was seen in a *sufA* deletion strain under conditions of Fe-deficiency and these observations suggest a regulatory role for SufA/IscA (Balasubramanian et al. 2006). Interestingly, Yabe and Nakai (2006) reported that *Arabidopsis* IscA accumulation is affected by deficiency of Nfu2, and thus IscA may act downstream. Furthermore IscA expression was affected by mutation of PIC1 (Permease In Chloroplasts 1), a possible chloroplast Fe transporter (Duy et al. 2007). When stromal proteins were fractionated by gel filtration, it was observed that IscA almost exclusively eluted in a high molecular weight fraction (~600 kDa), and about 10% of both CpNifS and SufE were found in this same fraction (Ye et al. 2005, 2006b; Abdel-Ghany et al. 2005). Taken together, the available data point to a role of IscA as either a protein with a regulatory function or as a buffer of Fe-S intermediates.

Monothiol glutaredoxin proteins such as GrxS14 and GrxS16 could serve as Fe-S scaffolds based on *in vitro* studies, however the *in planta* support for such a function is lacking (Bandyopadhyay et al. 2008). Additional components that are linked to plastid Fe-S assembly but for which a molecular role in Fe-S assembly is not fully clear yet are the SufB, C and D proteins, and Hcf101 (for discussion, see Balk and Pilon 2011).

The Need for Regulation

Plants grow in fluctuating environments that require constant remodeling of metabolism in order to adjust their physiology. Fe and S are both crucial elements for plant physiology. Starvation for S or Fe dramatically affects plant photosynthesis and growth. In this regard it must be noted that APS reductase (APR; adenosine 5'-phosphosulfate reductase) and sulfite reductase, two key enzymes in the S assimilation pathway in plastids, are themselves Fe-S enzymes (Ye et al. 2006a). The chloroplast is also a major site of reactive oxygen species production in plant cells particularly under environmental stresses or high light intensities. Oxidative stress is thought to be deleterious for photosynthetic activity, mainly through its effect on Fe-S protein stability. Moreover, ROS scavenging systems in the chloroplast are highly dependent on Fe co-factors and glutathione. It is therefore to be expected that Fe-S cluster assembly will be regulated in coordination with the physiology of the plant, particularly in the chloroplast. CpNifS, which is required for Fe-S formation, also affects Se tolerance (Van Hoewyk et al. 2005). Interestingly, APR2 may also function in Se tolerance because loss of APR2 function severely decreases plant tolerance to selenate. In contrast, selenite tolerance was not affected in APR2 mutants (Grant et al. 2011); this latter result is expected, because selenite enters the sulfur/selenium reduction pathway downstream of APR2.

It is also possible that the activity of key components of the Fe-S assembly pathway is regulated in response to the activity of other components of the pathway. For instance, we propose that Fe and S are incorporated as a Fe-S cluster onto a scaffold, then transferred to carrier proteins that donate the Fe-S cluster to the appropriate target proteins. In such a system, sensing steps are required in order to provide feedback to the early steps of the process, *i.e.* the entry of Fe and S in the pathway (Fig. 2). Such a hypothesis implies that alterations in the pathway should affect the regulation of the system.

An intellectually appealing regulatory model is that the Fe-S pathway could be auto-regulated, by sensing the requirement for *de novo* Fe-S cluster synthesis in response to physiological needs (Fig. 2). Most of the proteins suspected to be involved

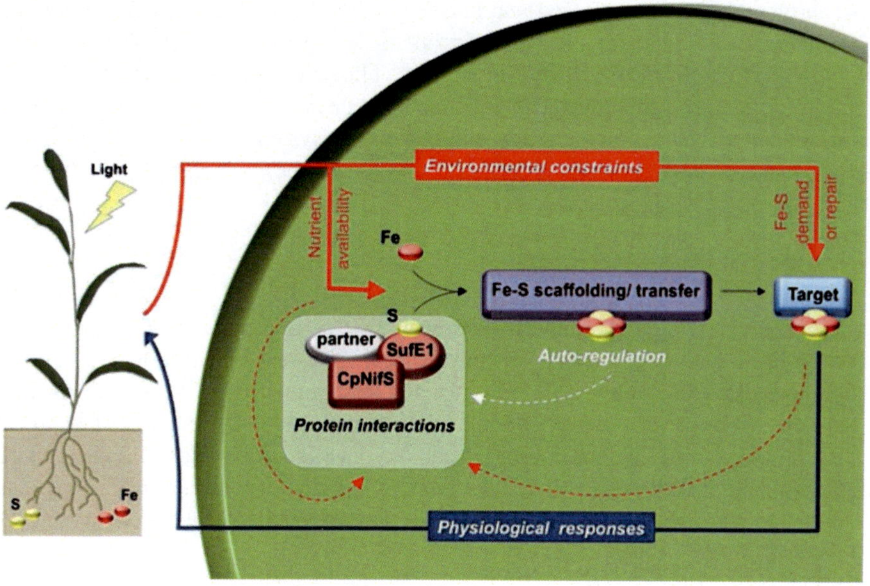

Fig. 2 Fe-S assembly, from the protein to the phenotype. Overview of possible regulatory processes involving the chloroplast Fe-S assembly machinery

in Fe-S cluster scaffolding and/or transfer in plants were identified by their sequence similarity with analogous proteins in other organisms (Balk and Pilon 2011). For some of them, their ability to accommodate a cluster and to transfer it to a target apo-protein has been shown *in vitro*, but *in planta* evidence for such a scaffolding function is still far from clear. The *Nfu2* knock-out line is the only mutant exhibiting a macroscopic phenotype associated with altered growth, chlorosis, and a decrease in the abundance of (2Fe-2S) and (4Fe-4S) proteins (Touraine et al. 2004; Yabe et al. 2004), which is consistent with a direct role in the Fe-S cluster building or transfer. For the other potential scaffold proteins, no obvious phenotype has been associated with the absence of the protein. Perhaps some of the components of the assembly pathway act as sensors of the Fe-S cluster status, the modification of the abundance of these proteins should provide feed-back into the early steps of the Fe-S cluster building, possibly through the modulation of CpNifS activity. SufE1 has a c-terminal domain with similarity to BolA. This region is not required to activate CpNifS *in vitro*. We consider it likely that the BolA-domain of SufE1 regulates cysteine desulfurase activity, perhaps in response to demand for Fe-S assembly (Ye et al. 2006b). Bioinformatic analysis suggests that glutaredoxins (Grx) could be involved in such regulation (Huynen et al. 2005). Indeed, in yeast, the Bol-A like protein FRA2 interacts with a Grx and this interaction is important for the proper regulation of the Fe-responsive transcription factor AFT1 (Kumanovics et al. 2008). This signaling pathway which involves FRA2 somehow "reads" the activity of the mitochondrial Fe-S assembly system because AFT is activated under iron starvation and when mitochondrial NSF1 (the cysteine desulfurase in yeast) is inactivated.

The recombinantly expressed and purified yeast FRA2 (BolA) -GRX3/4 complex contains an Fe-S cluster with unique properties (Li et al. 2009). These observations provide strong incentive to investigate if in chloroplasts an interaction between the BolA domain of SufE1 and a GRX- domain protein occurs and to study the biological significance of such interaction within this unique system.

Conclusions

While much progress has been made several key questions remain unanswered regarding plastid Fe-S cluster synthesis. How do S and Fe reach the scaffolds *in planta*? How is the activity of this core machinery regulated? How does the plastidic system of Fe-S assembly respond to environmental stresses? We believe that protein interaction studies in plants and in isolated organelles and that exploit the CpNifS-SufE1 complex as a molecular entry point into the pathway might reveal new candidates for the regulation of the assembly pathway and provide direct insight into the scaffolding process. Photosynthesis is pivotal to biomass production and it is important that we understand its limiting factors in view of the pressures imposed by environmental impacts and global change.

Acknowledgement Work in the authors' laboratory is supported by the US National Science Foundation (grant # IOS-0847442 and MCB 0950726). KR is supported by an European-FP7-International Outgoing Fellowship (Marie Curie, IntegrRegulFeSPlast; PIOF-GA-2010-273586).

References

Abdel-Ghany SE, Ye H, Garifullina GF, Zhang L, Pilon-Smits EAH, Pilon M (2005) Iron-sulfur cluster biogenesis in chloroplasts: involvement of the scaffold protein CpIscA. Plant Physiol 138:161–172

Balasubramanian R, Shen G, Bryant DA, Golbeck JH (2006) Regulatory roles for IscA and SufA in iron homeostasis and redox stress responses in the Cyanobacterium Synechococcus sp. strain PCC 7002. J Bacteriol 188:3182–3191

Balk J, Pilon M (2011) Ancient and essential: the assembly of iron-sulfur clusters in plants. Trends Plant Sci 16:218–226

Bandyopadhyay S, Gama F, Molina-Navarro MM, Gualberto JM, Claxton R, Naik SG, Huynh BH, Herrero E, Jacquot JP, Johnson MK, Rouhier N (2008) Chloroplast monothiol glutaredoxins as scaffold proteins for the assembly and delivery of (2Fe-2S) clusters. EMBO J 27:1122–1133

Beinert H, Holm RH, Münck E (1997) Iron-sulfur clusters: nature's modular, multipurpose structures. Science 277:653–659

Duy D, Wanner G, Meda AR, von Wirén N, Soll J, Philippar K (2007) PIC1, an ancient permease in *Arabidopsis* chloroplasts, mediates iron transport. Plant Cell 19:986–1006

Grant K, Carey NM, Mendoza M, Schulze J, Pilon M, Pilon-Smits EA, Van Hoewyk D (2011) Adenosine 5'-phosphosulfate reductase (APR2) mutation in *Arabidopsis* implicates glutathione depletion in selenate toxicity. Biochem J 438:325–335

Huynen MA, Spronk CA, Gabaldon T, Snel B (2005) Combining data from genomes, Y2H and 3D structure indicates that BolA is a reductase interacting with a glutaredoxin. FEBS Lett 579:591–596

Jeong J, Cohu C, Kerkeb L, Pilon M, Connolly EL, Guerinot ML (2008) Chloroplast Fe(III) chelate reductase activity is essential for seedling viability under iron limiting conditions. Proc Natl Acad Sci USA 105:10619–10624

Kumanovics A, Chen O, Li L, Bagley D, Adkins E, Lin H, Dingra NN, Outten CE, Keller G, Winge D, Ward DM, Kaplan J (2008) Identification of FRA1 and FRA2 as genes involved in regulating the yeast iron regulon in response to decreased mitochondrial iron-sulfur cluster synthesis. J Biol Chem 283:10276–10286

Leon S, Touraine B, Briat JF, Lobreaux S (2002) The AtNFS2 gene from *Arabidopsis thaliana* encodes a NifS-like plastidial cysteine desulphurase. Biochem J 366:557–564

Léon S, Touraine B, Ribot C, Briat JF, Lobréaux S (2003) Iron-sulphur cluster assembly in plants: distinct NFU proteins in mitochondria and plastids from *Arabidopsis thaliana*. Biochem J 371:823–830

Li H, Mapolelo DT, Dingra NN, Naik SG, Lees NS, Hoffman BM, Riggs-Gelasco PJ, Huynh BH, Johnson MK, Outten CE (2009) The yeast iron regulatory proteins Grx3/4 and Fra2 form heterodimeric complexes containing a (2Fe-2S) cluster with cysteinyl and histidyl ligation. Biochemistry 48:9569–9581

Lill R (2009) Function and biogenesis of iron-sulphur proteins. Nature 460:831–838

Murthy NUM, Ollagnier-de-Choudens S, Sanakis Y, Abdel-Ghany SE, Rousset C, Ye H, Fontecave M, Pilon-Smits EAH, Pilon M (2007) Characterization of *Arabidopsis thaliana* SufE2 and SufE3: functions in chloroplast iron-sulfur cluster assembly and NAD synthesis. J Biol Chem 282:18254–18264

Pilon M, Owen JD, Garifullina GF, Kurihara T, Mihara H, Esaki N, Pilon-Smits EAH (2003) Enhanced selenium tolerance and accumulation in transgenic *Arabidopsis thaliana* expressing a mouse selenocysteine lyase. Plant Physiol 131:1250–1257

Pilon-Smits EAH, Garifullina G, Abdel-Ghany S, Kato S, Mihara H, Hale K, Burkhead J, Esaki N, Kurihara T, Pilon M (2002) Characterization of a NifS-like chloroplast protein from *Arabidopsis*: implications for its role in sulfur and selenium metabolism. Plant Physiol 130:1309–1318

Ravet K, Touraine B, Boucherez J, Briat JF, Gaymard F, Cellier F (2009) Ferritins control interaction between iron homeostasis and oxidative stress in *Arabidopsis*. Plant J 57:400–412

Schwenkert S, Netz DJ, Frazzon J, Pierik AJ, Bill E, Gross J, Lill R, Meurer J (2009) Chloroplast HCF101 is a scaffold protein for [4Fe-4S] cluster assembly. Biochem J 425:207–214

Takahashi Y, Mitsui A, Hase T, Matsubara H (1986) Formation of the iron sulfur cluster of ferredoxin in isolated chloroplasts. Proc Natl Acad Sci USA 83:2434–2437

Touraine B, Boutin JP, Marion-Poll A, Briat JF, Peltier G, Lobréaux S (2004) Nfu2: a scaffold protein required for [4Fe-4S] and ferredoxin iron-sulphur cluster assembly in *Arabidopsis* chloroplasts. Plant J 40:101–111

Van Hoewyk D, Garifullina GF, Ackley AR, Abdel-Ghany SE, Marcus MA, Fakra S, Ishiyama K, Ioue E, Pilon M, Takahashi H, Pilon-Smits EAH (2005) Overexpression of AtCpNifS enhances selenium tolerance and accumulation in *Arabidopsis*. Plant Physiol 139:1518–1528

Van Hoewyk D, Abdel-Ghany SE, Cohu C, Herbert S, Kugrens P, Pilon M, Pilon-Smits EAH (2007) The *Arabidopsis* cysteine desulfurase CpNifS is essential for maturation of iron-sulfur cluster proteins, photosynthesis, and chloroplast development. Proc Natl Acad Sci USA 104: 5686–5691

White PJ, Broadley MR (2009) Biofortification of crops with seven mineral elements often lacking in human diets–iron, zinc, copper, calcium, magnesium, selenium and iodine. New Phytol 182:49–84

Xu XM, Møller SG (2004) AtNAP7 is a plastidic SufC-like ATP-binding cassette/ATPase essential for *Arabidopsis* embryogenesis. Proc Natl Acad Sci USA 101:9143–9148

Xu XM, Møller SG (2006) AtSufE is an essential activator of plastidic and mitochondrial desulfurases in *Arabidopsis*. EMBO J 25:900–909

Xu XM, Adams S, Chua NH, Møller SG (2005) AtNAP1 represents an atypical SufB protein in *Arabidopsis* plastids. J Biol Chem 280:6648–6654

Yabe T, Nakai M (2006) *Arabidopsis* AtIscA-I is affected by deficiency of Fe-S cluster biosynthetic scaffold AtCnfU-V. Biochem Biophys Res Commun 340:1047–1052

Yabe T, Morimoto K, Kikuchi S, Nishio K, Terashima I, Nakai M (2004) The *Arabidopsis* chloroplastic NifU-like protein CnfU, which can act as an iron-sulfur cluster scaffold protein, is required for biogenesis of ferredoxin and photosystem I. Plant Cell 16:993–1007

Ye H, Garifullina G, Abdel-Ghany S, Zhang L, Pilon-Smits EAH, Pilon M (2005) The Chloroplast NifS-like protein of *Arabidopsis thaliana* is required for iron-sulfur cluster formation in ferredoxin. Planta 220:602–608

Ye H, Pilon M, Pilon-Smits EAH (2006a) Iron sulfur cluster biogenesis in chloroplasts. New Phytol 171:285–292

Ye H, Abdel-Ghany SE, Anderson TD, Pilon-Smits EAH, Pilon M (2006b) CpSufE activates the cysteine desulfurase CpNifS for chloroplastic Fe-S cluster formation. J Biol Chem 273: 13264–13272

Plant Accumulation of Sulfur's Sister Element Selenium – Potential Applications and Ecological Implications

Elizabeth A.H. Pilon-Smits

Abstract Selenium (Se) is an essential nutrient for many organisms including humans, but is also toxic at higher levels. Both Se deficiency and Se toxicity are problems worldwide. Although Se has not been shown to be essential for higher plants, it is considered a beneficial element, providing enhanced antioxidant activity. Selenium is chemically similar to sulfur (S) and readily taken up and assimilated via sulfur (S) transporters and enzymes. Thus, Se can replace S in many S compounds, including volatile forms. Some plants native to Se-rich soils can hyperaccumulate Se to levels around 1% of plant dry weight. They grow poorly without Se and thus appear to profit from Se physiologically. Selenium can also serve ecological functions as an elemental defense against pathogens and herbivores, and in elemental allelopathy. The ability of plants to (hyper)accumulate and volatilize Se may be used for phytoremediation of polluted soils or waters, and also to produce nutritionally enhanced crops. These applications will benefit from better insight into the mechanisms that control Se tolerance and accumulation in plants, and the potential ecological implications. This review gives an overview of our current knowledge of plant Se metabolism, including Se tolerance and hyperaccumulation mechanisms. It also summarizes what is known about ecological implications of plant Se (hyper)accumulation.

Introduction

Because of the chemical similarity between the elements selenium (Se) and sulfur (S), organisms mistakenly take up and metabolize Se via S transporters and biochemical pathways (Terry et al. 2000). This can cause toxicity, due to oxidative stress and replacement of S by Se in proteins and other S compounds, which disrupts their

E.A.H. Pilon-Smits (✉)
Biology Department, Colorado State University,
Fort Collins, CO 80523, USA
e-mail: epsmits@lamar.colostate.edu

L.J. De Kok et al. (eds.), *Sulfur Metabolism in Plants*, Proceedings of the International Plant Sulfur Workshop 1, DOI 10.1007/978-94-007-4450-9_19,

function. While toxic at higher levels, Se is also an essential trace element for many organisms such as mammals, some bacteria, and certain green algae. For higher plants, Se has not been shown to be essential but it is considered a beneficial nutrient (Pilon-Smits et al. 2009). The basis for the essentiality of Se for some organisms is their reliance on a small set of selenoproteins that contain selenocysteine (SeCys) in their active site; to date, no selenoproteins have been confirmed to exist in higher plants. Since selenoproteins have redox functions, including the scavenging of free radicals, Se deficiency often enhances the probability of developing cancers or viral infections; problems associated with Se deficiency include Keshan disease and male infertility in humans, and white muscle disease in livestock. There is only a narrow window between the amount of Se that is required as a nutrient and the amount that is toxic, which is why both Se deficiency and toxicity are common worldwide. Extremely high Se levels (1,000–10,000 mg kg^{-1} DW) occur in so-called hyperaccumulator plant species that are endemic on seleniferous soils in the Western USA and parts of China. In these seleniferous areas chronic or acute Se poisoning are serious problems for humans, livestock and wildlife (Terry et al. 2000).

Plants can take up inorganic Se from their environment using S transporters, and incorporate this selenate and selenite into organic compounds via S assimilation enzymes (Anderson 1993; Fig. 1 shows an overview of Se metabolism in plants). First selenate undergoes two sequential reduction steps via selenite to selenide, which can be combined with *O*-acetylserine (OAS) to form selenocysteine (SeCys). The non-specific incorporation of this seleno-aminoacid into proteins leads to toxicity. To prevent this toxicity, SeCys can be broken down to elemental Se and alanine. Elemental Se is insoluble and relatively innocuous. SeCys can also be methylated to form methyl-SeCys, which can be safely accumulated since it is not incorporated into proteins. Methyl-SeCys can also act as a precursor for the production of a form of volatile Se, dimethyldiselenide (DMDSe) (Terry et al. 2000; Sors et al. 2005). This is the main volatile form of Se emitted by hyperaccumulator species. Finally, SeCys can be converted to selenomethionine (SeMet), which can be converted to another volatile compound, dimethylselenide (DMSe) (Terry et al. 2000).

Plant accumulation and volatilization of Se are potentially useful processes to alleviate both Se deficiency and toxicity. In Se phytoremediation, plants may be used to clean up Se-polluted soil or water. They may in part release the Se into the atmosphere as relatively non-toxic volatile compounds, and in part accumulate it in their harvestable tissues. This may produce a Se-fortified crop with enhanced nutritional quality that can be used to prevent Se deficiency in other areas. Even in Se-replete individuals Se can have beneficial effects. Some selenocompounds (particularly methyl-SeCys) have potent anticarcinogenic properties; these are found in Se hyperaccumulator species as well as in some members of the S-rich *Brassica* and *Allium* genera (Lyi et al. 2005). These "sulfur-loving" species, can accumulate Se to fairly high levels (up to 0.1% of dry weight (DW), or 1,000 mg Se kg^{-1} DW) and have been called Se accumulator species (not to be confused with hyperaccumulators). True Se hyperaccumulators are found in the genera *Stanleya* (Brassicaceae), *Astragalus* (Fabaceae) and *Xylorhiza* and *Oonopsis* (Asteraceae), and accumulate Se to 1,000–15,000 mg Se kg^{-1} DW (0.1–1.5%) (Ellis et al. 2004;

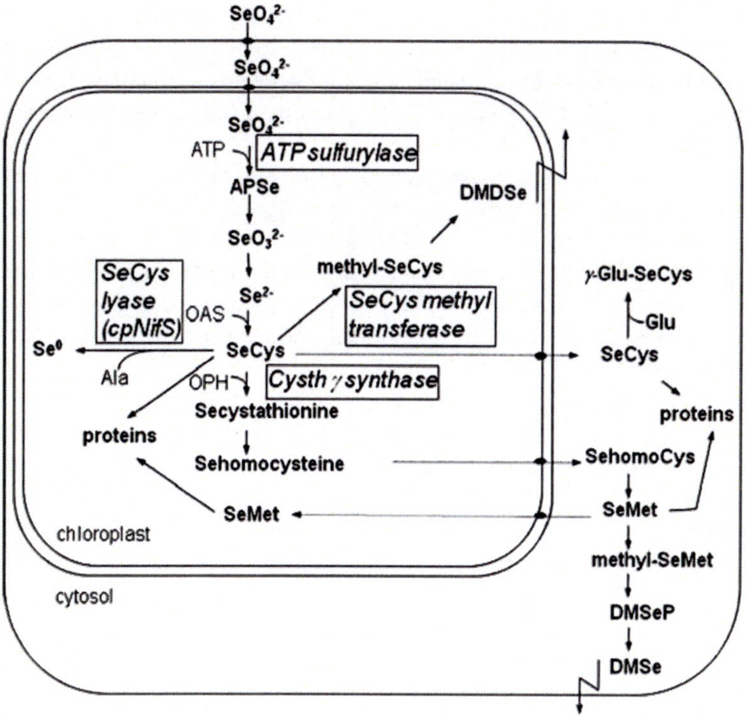

Fig. 1 Overview of plant Se metabolism and genetic engineering approaches that have been used successfully to enhance plant Se tolerance, accumulation and/or volatilization. Overexpressed enzymes are shown in *boxes*

LeDuc et al. 2004). Hyperaccumulators preferentially take up Se over S and show different patterns for these two elements in terms of seasonal fluctuations and tissue distribution; they may be able to distinguish between S and Se (White et al. 2007). Since hyperaccumulators are endemic to seleniferous soils, they may physiologically or ecologically require Se.

Genetic Engineering of Plant Se Metabolism, and Potential Applications

Through up-regulation of key genes involved in sulfur/selenium assimilation and volatilization, it has been possible to enhance Se accumulation, tolerance, and volatilization by plants (Fig. 1, see Pilon-Smits and LeDuc 2009 for further references). In a first approach, overexpression of ATP sulfurylase (APS), involved in selenate-to-selenite conversion in *Brassica juncea* (Indian mustard) led to enhanced selenate reduction. The transgenic APS plants accumulated an organic form of Se when supplied with selenate, while wildtype controls accumulated selenate. The APS transgenics accumulated two- to threefold more Se than the wild type, and tolerated the accumulated Se better than wild type, perhaps because of the organic form of Se accumulated. They also contained 1.5-fold more S than the wild type.

In a second approach, overexpression in *B. juncea* of the first enzyme in the conversion of SeCys to SeMet, cystathionine gamma synthase (CgS), enhanced volatilization rates two- to threefold compared to wildtype plants. The CgS transgenics accumulated 40% less Se in their tissues than wild type, probably because of their higher volatilization rates. The CgS transgenics were also more Se tolerant than wildtype plants, perhaps owing to their lower tissue Se levels.

In a third transgenic approach, a mouse selenocysteine lyase (SL) was expressed in *A. thaliana* and *B. juncea*. This enzyme breaks down SeCys into alanine and elemental Se. The SL transgenics showed reduced Se incorporation into proteins and up to twofold enhanced Se accumulation, compared to wildtype plants. When an *A. thaliana* homologue of the mouse SL (called CpNifS) was overexpressed, similar results were obtained: the CpNifS transgenics showed less Se incorporation in proteins, twofold enhanced Se accumulation, as well as enhanced Se tolerance. Incidentally, the main function of CpNifS is to provide elemental S for iron-sulfur cluster formation (Van Hoewyk et al. 2007).

In another approach, SeCys methyltransferase (SMT) from hyperaccumulator *A. bisulcatus* was overexpressed in *A. thaliana* and *B. juncea* (Ellis et al. 2004; LeDuc et al. 2004). The SMT transgenics showed enhanced Se accumulation in the form of methyl-SeCys, and enhanced Se tolerance and volatilization, with more Se volatilized in the form of DMDSe. The APS and SMT *B. juncea* transgenics were crossed to create double-transgenic plants. These APS × SMT double transgenics contained up to nine times higher Se levels than wild type, and most of the Se in the double transgenics was in the form of methyl-SeCys (LeDuc et al. 2006). The APS × SMT plants accumulated up to eightfold more methyl-SeCys than wild type and nearly twice as much as the SMT single transgenics. Selenium tolerance was similar in the single and double transgenics.

These studies show convincingly that the sulfate assimilation and volatilization pathway is responsible for selenate assimilation and volatilization. The level of APS appears to be rate-limiting for the assimilation of selenate to organic Se, and CgS for DMSe volatilization. Elevated APS expression seems to trigger selenate uptake and Se and S accumulation, perhaps via upregulation of sulfate transporter expression. The SL and CpNifS transgenics have revealed that breakdown of SeCys can reduce non-specific incorporation of Se into proteins, enhancing Se tolerance. Also, the finding that overexpression of SL or CpNifS led to enhanced Se accumulation suggests that introduction of this new sink for Se upregulates Se and S uptake. The SMT transgenics have shown that SMT is a key enzyme for Se hyperaccumulation. However, for improved selenate assimilation and detoxification, APS needs to be overexpressed together with SMT. While APS × SMT double transgenics show significantly enhanced Se tolerance and accumulation, they still do not approach the performance of true Se hyperaccumulator species. Further research is needed to identify additional Se tolerance and accumulation genes in these specialized plant species.

The different transgenics showed up to ninefold higher Se accumulation, up to threefold faster Se volatilization, and enhanced Se tolerance, all under laboratory conditions. These are useful properties for phytoremediation and biofortification. To test the potential of these transgenics under more realistic field conditions, they were analyzed for their capacity to accumulate Se from seleniferous soil and from Se-contaminated sediment. When grown on naturally seleniferous soil in a greenhouse pot experiment the APS transgenics accumulated Se to threefold higher levels than wildtype *B. juncea*, and the CgS transgenics contained 40% lower Se levels than wild type, all in agreement with the laboratory results. Plant biomass was the same for all plant types. Furthermore, two field experiments were carried out by Gary Bañuelos and coworkers on Se-contaminated sediment in the San Joaquin Valley (CA, USA) (Bañuelos et al. 2005, 2007). The APS transgenics accumulated Se to fourfold higher levels than wildtype *B. juncea*, and cpSL and SMT transgenics showed twofold higher Se accumulation than wildtype *B. juncea*, all in agreement with earlier laboratory experiments. Biomass production was comparable for the different plant types. Thus, in the field as well as the lab the various transgenics showed enhanced Se accumulation, volatilization and/or tolerance, promising traits for use in phytoremediation or as Se-fortified foods.

New Insights into Se Tolerance and Accumulation Mechanisms

Different varieties of the model plant species *A. thaliana* and the related hyperaccumulator *S. pinnata* were compared at the molecular and physiological level, with the goal to provide new insight into key genes for Se uptake, hyperaccumulation, and volatilization. In a first approach, a Se tolerance study was performed using recombinant inbred lines (RIL) of model species *A. thaliana*, which identified two QTLs for selenate tolerance (Zhang et al. 2006a). Tolerance to selenate and selenite appear to be controlled by different loci. In another study, tolerance to and accumulation of Se were found to not be correlated in a study comparing 19 different ecotypes of *Arabidopsis* with variable tolerance to Se (Zhang et al. 2006b). In another approach, the transcriptomes were compared between *A. thaliana* plants grown with or without selenate or selenite (Tamaoki et al. 2008; Van Hoewyk et al. 2008). The hormones ethylene and jasmonic acid appear to be important for Se tolerance. Genes involved in ethylene and jasmonic acid biosynthesis were upregulated by Se, mutants with a defect in ethylene synthesis, ethylene signaling, and jasmonic acid signaling showed reduced tolerance to Se, and overexpression of a protein involved in ethylene signaling increased Se resistance. The resistance mechanism appears to involve enhanced sulfate uptake and reduction; this may serve to prevent Se from replacing S in proteins and other S compounds.

In *Stanleya*, similar Se tolerance mechanisms were found to those described above for *Arabidopsis* (Freeman et al. 2010). The plant hormones JA and ethylene, as well as the hormone salicylic acid appear to play a role in regulating Se stress response in *Stanleya*. Probably as a response to the elevated levels of these hormones, hyperaccumulators have constitutively upregulated expression of sulfur transporters and assimilatory enzymes, and hence higher levels of total S, reduced S compounds (including the antioxidant glutathione), and higher levels of total Se. The *Stanleya* hyperaccumulator also showed an interesting Se sequestration pattern not observed in non-hyperaccumulators: the Se accumulated as methyl-SeCys in vacuoles of leaf epidermal cells (Freeman et al. 2006a, 2010).

Ecological Aspects of Se Phytoremediation

High Se levels in plants may have important ecological effects (Fig. 2). Hyperaccumulated Se has long been known to be toxic and even lethal for mammalian herbivores (Terry et al. 2000). Thus, hyperaccumulators may sequester Se as a defense against herbivory. In support of this elemental defense hypothesis (Boyd 2010), laboratory and field studies showed that Se accumulation may protect plants from a range of herbivores and pathogens, from prairie dogs to a variety of arthropods and fungi (Hanson et al. 2003, 2004; Freeman et al. 2006a, 2007, 2009; Quinn et al. 2008, 2010). Selenium protected the plants both through deterrence of herbivores and toxicity, and in some cases even low levels (10 mg kg^{-1} DW) already protected plants. In further support of the elemental defense hypothesis, Se hyperaccumulating species harbored fewer arthropod species and individuals in their natural habitat than comparable Se non-accumulators (Galeas et al. 2008). While Se hyperaccumulation is an effective plant defense mechanism against herbivory, some herbivores have evolved tolerance. A population of diamondback moth living in a seleniferous area on hyperaccumulator *S. pinnata* was shown to be completely Se tolerant (Freeman et al. 2006b).

Selenium in plants may also affect other ecological interactions, such as those with associated microbes, neighboring plants, and pollinators. Rhizospheric and saprophytic fungi associated with Se-rich plants appear to have evolved enhanced Se tolerance (Wangeline et al. 2011; Quinn et al. 2011a). Hyperaccumulators also appear to affect their neighboring plants by enriching their surrounding soil and neighbors with Se. Depending on whether the neighbors are Se tolerant or sensitive, this may have a positive or negative effect (El Mehdawi et al. 2011). Flowers can accumulate substantial Se levels, and this may affect pollen germination, depending on plant Se tolerance (Prins et al. 2011). Visitation by pollinators was not affected by flower Se concentration (Quinn et al. 2011b), and therefore the health effects of pollinator-ingested Se warrants further study. Since Se is a micronutrient for insects, these may be positive or negative.

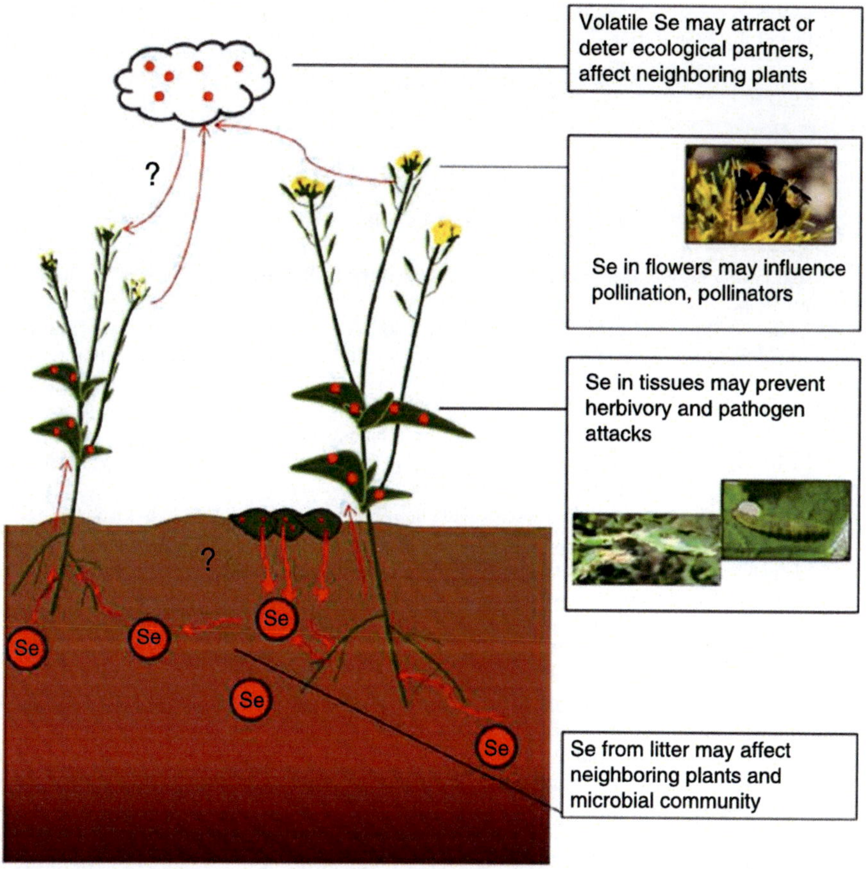

Fig. 2 Overview of potential ecological implications of Se accumulation in plants

Conclusions

Plants metabolize Se via S transporters and enzymes, via non-specific mechanisms. As a result, plants can accumulate Se to substantial levels, and also volatilize it. Different species vary with respect to the level and form of Se accumulated, as well as its tissue localization. Via genetic engineering of S enzymes, it has been possible to enhance Se tolerance, accumulation and volatilization up to ~10 fold, which has potential applications in phytoremediation and biofortification. If plants, transgenic or otherwise, are to be used on a large scale, it is important to have insight into potential ecological implications. Studies so far show that plant-accumulated Se has profound effects on plant-herbivore interactions, deterring a wide variety of herbivores and causing toxicity when ingested. High-Se plants may also affect associated microbes, as well as surrounding vegetation. Further studies of these ecological interactions will help optimize phytoremediation and biofortification practices while preventing ecological problems.

Acknowledgements National Science Foundation grant # IOS-0817748 to EAHPS supported the writing of this manuscript.

References

Anderson JW (1993) Selenium interactions in sulfur metabolism. In: De Kok LJ, Stulen I, Rennenberg H, Brunold C, Rauser WE (eds) Sulfur nutrition and assimilation in higher plants – regulatory, agricultural and environmental aspects. SPB Academic Publishing, The Hague, pp 49–60

Bañuelos G, Terry N, LeDuc DL, Pilon-Smits EAH, Mackey B (2005) Field trial of transgenic Indian mustard plants shows enhanced phytoremediation of selenium contaminated sediment. Environ Sci Technol 39:1771–1777

Bañuelosn G, LeDuc DL, Pilon-Smits EAH, Tagmount A, Terry N (2007) Transgenic Indian mustard overexpressing selenocysteine lyase, selenocysteine methyltransferase, or methionine methyltransferase exhibit enhanced potential for selenium phytoremediation under field conditions. Environ Sci Technol 41:599–605

Boyd RS (2010) Heavy metal pollutants and chemical biology. Exploring new frontiers. J Chem Ecol 36:46–58

El Mehdawi AF, Quinn CF, Pilon-Smits EAH (2011) Effects of selenium hyperaccumulation on plant-plant interactions: evidence for elemental allelopathy? New Phytol 191:120–131

Ellis DR, Sors TG, Brunk DG, Albrecht C, Orser C, Lahner B, Wood KV, Harris HH, Pickering IJ, Salt DE (2004) Production of Se-methylselenocysteine in transgenic plants expressing selenocysteine methyltransferase. BMC Plant Biol 4:1–11

Freeman JL, Quinn CF, Marcus MA, Fakra S, Pilon-Smits EAH (2006a) Selenium tolerant diamondback moth disarms hyperaccumulator plant defense. Curr Biol 16:2181–2192

Freeman JL, Zhang LH, Marcus MA, Fakra S, McGrath SP, Pilon-Smits EAH (2006b) Spatial imaging, speciation and quantification of selenium in the hyperaccumulator plants *Astragalus bisulcatus* and *Stanleya pinnata*. Plant Physiol 142:124–134

Freeman JL, Lindblom SD, Quinn CF, Fakra S, Marcus MA, Pilon-Smits EAH (2007) Selenium accumulation protects plants from herbivory by orthoptera due to toxicity and deterrence. New Phytol 175:490–500

Freeman JL, Quinn CF, Lindblom SD, Klamper EM, Pilon-Smits EAH (2009) Selenium protects the hyperaccumulator *Stanleya pinnata* against black-tailed prairie dog herbivory in native seleniferous habitats. Am J Bot 96:1075–1085

Freeman JL, Tamaoki M, Stushnoff C, Quinn CF, Cappa JJ, Devonshire J, Fakra S, Marcus MA, McGrath S, Van Hoewyk D, Pilon-Smits EAH (2010) Molecular mechanisms of selenium tolerance and hyperaccumulation in *Stanleya pinnata*. Plant Physiol 153:1630–1652

Galeas ML, Klamper EM, Bennett LE, Freeman JL, Kondratieff BC, Pilon-Smits EAH (2008) Selenium hyperaccumulation affects plant arthropod load in the field. New Phytol 177:715–724

Hanson BR, Garifullina GF, Lindblom SD, Wangeline A, Ackley A, Kramer K, Norton AP, Lawrence CB, Pilon-Smits EAH (2003) Selenium accumulation protects *Brassica juncea* from invertebrate herbivory and fungal infection. New Phytol 159:461–469

Hanson BR, Lindblom SD, Loeffler ML, Pilon-Smits EAH (2004) Selenium protects plants from phloem-feeding aphids due to both deterrence and toxicity. New Phytol 162:655–662

LeDuc DL, AbdelSamie M, Montes-Bayon M, Wu CP, Reisinger SJ, Terry N (2006) Overexpressing both ATP sulfurylase and selenocysteine methyltransferase enchances selenium phytoremediation traits in Indian mustard. Environ Poll 144:70–76

LeDuc DL, Tarun AS, Montes-Bayon M, Meija J, Malit MF, Wu CP, Abdel Samie M, Chiang C-Y, Tagmount A, de Souza MP, Neuhierl B, Bock A, Caruso JA, Terry N (2004) Overexpression of selenocysteine methyltransferase in *Arabidopsis* and Indian mustard increases selenium tolerance and accumulation. Plant Physiol 135:377–383

Lyi SM, Heller LI, Rutzke M, Welch RM, Kochian LV, Li L (2005) Molecular and biochemical characterization of the selenocysteine Se-methyltransferase gene and Se-methylselenocysteine synthesis in broccoli. Plant Physiol 138:409–420
Pilon-Smits EAH, LeDuc DL (2009) Phytoremediation of selenium using transgenic plants. Curr Opin Biotechnol 20:207–212
Pilon-Smits EAH, Quinn CF, Tapken W, Malagoli M, Schiavon M (2009) Physiological functions of beneficial elements. Curr Opin Plant Biol 12:267–274
Prins CN, Hantzis LJ, Quinn CF, Pilon-Smits EAH (2011) Effects of selenium accumulation on reproductive functions in *Brassica juncea* and *Stanleya pinnata*. J Exp Bot 62:5633–5640
Quinn CF, Freeman JF, Galeas ML, Klamper EM, Pilon-Smits EAH (2008) Selenium protects plants from prairie dog herbivory – implications for the functional significance and evolution of Se hyperaccumulation. Oecologia 155:267–275
Quinn CF, Freeman JL, Reynolds RJB, Lindblom SD, Cappa JJ, Marcus MA, Fakra SF, Pilon-Smits EAH (2010) Selenium hyperaccumulation protects plants from cell disruptor herbivores. BMC Ecol 10:19
Quinn CF, Wyant K, Wangeline AL, Shulman J, Galeas ML, Valdez JR, Paschke MW, Pilon-Smits EAH (2011a) Selenium hyperaccumulation increases leaf decomposition rate in a seleniferous habitat. Plant Soil 341:51–61
Quinn CF, Prins CN, Gross AM, Hantzis L, Reynolds RJB, Freeman JL, Yang SI, Covy PA, Bañuelos GS, Pickering IJ, Fakra SF, Marcus MA, Arathi HS, Pilon-Smits EAH (2011b) Selenium accumulation in flowers and its effects on pollination. New Phytol 192:727–737
Sors TG, Ellis DR, Salt DE (2005) Selenium uptake, translocation, assimilation and metabolic fate in plants. Photosynth Res 86:373–389
Tamaoki M, Freeman JL, Pilon-Smits EAH (2008) Cooperative ethylene and jasmonic acid signaling regulates selenite resistance in *Arabidopsis thaliana*. Plant Physiol 146:1219–1230
Terry N, Zayed AM, de Souza MP, Tarun AS (2000) Selenium in higher plants. Annu Rev Plant Physiol Plant Mol Biol 51:401–432
Van Hoewyk D, Abdel-Ghany SE, Cohu C, Herbert S, Kugrens P, Pilon M, Pilon-Smits EAH (2007) The *Arabidopsis* cysteine desulfurase CpNifS is essential for maturation of iron-sulfur cluster proteins, photosynthesis, and chloroplast development. Proc Natl Acad Sci USA 104:5686–5691
Van Hoewyk D, Takahashi H, Hess A, Tamaoki M, Pilon-Smits EAH (2008) Transcriptome and biochemical analyses give insights into selenium-stress responses and selenium tolerance mechanisms in *Arabidopsis*. Physiol Plant 132:236–253
Wangeline AL, Valdez JR, Lindblom SD, Bowling KL, Reeves FB, Pilon-Smits EAH (2011) Selenium tolerance in rhizosphere fungi from Se hyperaccumulator and non-hyperaccumulator plants. Am J Bot 98:1139–1147
White PJ, Bowen HC, Marshall B, Broadley MR (2007) Extraordinarily high leaf selenium to sulfur ratios define 'Se-accumulator' plants. Ann Bot 100:111–118
Zhang L, Byrne PF, Pilon-Smits EAH (2006a) Mapping quantitative trait loci associated with selenate tolerance in *Arabidopsis thaliana*. New Phytol 170:33–42
Zhang L, Ackley AR, Pilon-Smits EAH (2006b) Variation in selenium tolerance and accumulation among nineteen *Arabidopsis* ecotypes. J Plant Physiol 164:327–336

Identification and Sequence Analysis of Sulfate/Selenate Transporters in Selenium Hyper- and Non-accumulating *Astragalus* Plant Species

Emmanuelle Cabannes, Peter Buchner, and Malcolm J. Hawkesford

Abstract Selenium is an essential element for human health, consumed mainly from plant sources. In plants, there is good evidence that selenate is transported via sulfate transporters, and that this anion will interact competitively with sulfate. Some of selenium (Se)-hyper-accumulating plant species belonging to the genus *Astragalus* are known for their capacity to accumulate up to 0.6% of their foliar dry weight as Se. Comparative analyses of sulfate transporter genes of contrasting Se hyper-accumulator and non-accumulator species will be helpful for a better understanding of sulfate/selenium uptake processes and provide insights on the substrate binding site for the transported anion. The gene family of sulfate transporters is subdivided into five groups with distinct and/or functional characteristics. cDNAs for sulfate transporters belonging to the Group 1, 2, 3 and 4 of the sulfate transporter families were isolated from both Se hyper-accumulating (*Astragalus racemosus, Astragalus bisulfcatus, Astragalus crotalariae*) and the closely related non-accumulating species (*Astragalus glycophyllos, Astragalus drummondii*). Sequence analysis was performed to identify sequence variations which may contribute to selectivity of sulfate and selenate uptake and distribution.

In trace amounts, Se is an essential micronutrient for humans and animals as part of the selenoenzymes, thioredoxin reductase and glutathione peroxidase. However, excess Se may also be toxic (Läuchli 1993; Schromburg et al. 2004).

Selenium has no confirmed physiological role in higher plants. In most plants, excess selenate uptake effects protein synthesis, causing symptoms including chlorosis and stunting that mimic sulfate starvation, as well as withering and drying

E. Cabannes • P. Buchner • M.J. Hawkesford (✉)
Plant Science Department, Rothamsted Research, Harpenden AL5 2JQ, UK
e-mail: malcolm.hawkesford@rothamsted.ac.uk

L.J. De Kok et al. (eds.), *Sulfur Metabolism in Plants*, Proceedings of the International Plant Sulfur Workshop 1, DOI 10.1007/978-94-007-4450-9_20,

of leaves and premature death (Terry et al. 2000). However some resistant plants growing on seleniferous soils, including some species of *Astragalus* and Stanleya, are able to accumulate Se to very high concentrations (Feist and Parker 2001; Pickering et al. 2003; Cabannes et al. 2011).

As Se is consumed in the diet mainly from plant sources, biofortification of crops through Se fertilization is a feasible strategy to enhance human Se intake (Broadley et al. 2006). Se is taken up by plants in the forms of selenate, selenite and organic Se. Selenate uptake occurs through sulfate transporters in the plasma membrane of plant roots and is transported to the shoot where it can be redistributed within the plant or accumulated in older leaves as in Se hyper-accumulating plants (Pickering et al. 2003). The assimilation of selenate follows the sulfate pathway (Kopsell and Kopsell 2007; Sors et al. 2005) as the selenate molecule has similar size and charge to sulfate.

It is widely known that the uptake and assimilation of sulfate is regulated by the nutrient status of the plant (Smith et al. 1995). Several studies have shown that a decrease in root sulfate availability results in a several fold enhanced expression of sulfate transporter genes, which enhances the capacity for sulfate uptake and effectively enhances the concentration of Se (Hawkesford 2000, 2003; Shinmachi et al. 2010). Studies with the *Arabidopsis* selenate-resistant high-affinity sulfate transporter (SULTR1;2) mutant indicated that selenate uptake occurs by sulfate transporters (Shibagaki et al. 2002). The plant sulfate transporter gene family is composed of five subgroups. Detailed expression studies in wheat have shown that in addition to initial uptake, Group 1 high affinity sulfate transporters, as well as the Group 2 low affinity and Group 4 vacuole efflux sulfate transporters are involved in tissue/organ and subcellular distribution of selenate within the plant (Shinmachi et al. 2010). In contrast to wheat, in which increased Se accumulation is only found under sulfur limited conditions, some *Astragalus* subspecies are able to accumulate high amounts of Se independent of the plant S-nutritional status (Sors et al. 2005).

Plants differ in their ability to accumulate Se; White et al. (2004) suggested that the transporters responsible for the uptake and distribution of sulfate and selenate are selective either for sulfate (in non-accumulating plants) or selenate (in hyper-accumulating plants).

Discrimination mechanisms between sulfate and selenate transport may be involved at various transport steps, from the uptake level, the root to shoot translocation to the vacuole transport. Sequence differences may be important for influencing the selectivity towards sulfate or selenate. To investigate the role of sulfate transporters in selenate transport, cDNAs for sulfate transporters belonging to different groups of the sulfate transporter gene family were cloned from both Se hyper- and closely related non-accumulating *Astragalus* species (Table 1) by PCR techniques.

Plants differ in their ability to accumulate Se; White et al. (2004) suggested that the transporters responsible for the uptake and distribution of sulfate and selenate are selective either for sulfate (in non-accumulating plants) or selenate (in hyper-accumulating plants).

Discrimination mechanisms between sulfate and selenate transport may be involved at various transport steps, from the uptake level, the root to shoot translocation

Table 1 EMBL Nucleotide Sequence Database accessions of putative *Astragalus* sulfate transporter genes belonging to the Groups 1–4 of the sulfate transporter gene family

	Gene (mRNA)	Accession
Se hyper-accumulating *Astragalus* species		
A. racemosus	Sultr1a	FN689510
A. racemosus	Sultr1b	FN689511
A. racemosus	Sultr1c	FN689512
A. racemosus	Sultr2	FN689513
A. racemosus	Sultr3;4	FN689514
A. racemosus	Sultr4	FR819655
A. bisulcatus	Sultr1b	FN689517
A. bisulcatus	Sultr2	FN689518
A. bisulcatus	Sultr3;4	FN689519
Se non-accumulating *Astragalus* species		
A. glycyphyllos	Sultr1a	FN689515
A. glycyphyllos	Sultr1b	FN689516
A. drummondii	Sultr1b	FN689520
A. drummondii	Sultr2	FN689521
A. drummondii	Sultr3;4	FN689522
A. crotalariae	Sultr1b	FN689523

to the vacuole transport. Sequence differences may be important for influencing the selectivity towards sulfate or selenate. To investigate the role of sulfate transporters in selenate transport, cDNAs for sulfate transporters belonging to different groups of the sulfate transporter gene family were cloned from both Se hyper- and closely related non-accumulating *Astragalus* species (Table 1) by PCR techniques.

Closely related high-affinity Group 1 sulfate transporter genes/cDNA could be isolated from all hyper- as well as non-accumulating *Astragalus* species. The different cDNAs encoded proteins of 658–662 amino acids. Furthermore Group 2 (Sultr2) and Group 3 (Sultr3;4) sulfate transporter isoforms could be identified from *Astragalus racemosus*, *Astragalus drummondii* and *Astragalus bisulcatus*. One Group 4 sulfate transporter isoform (Sultr4) and a further Group 3 (Sultr3;1) were isolated from *Astragalus racemosus*. In the phylogenetic analysis, the *Astragalus* Group 1 clade is distinct from both the *Arabidopsis* and rice equivalent isoforms, and is closer to the legume *Stylosanthes hamata* Group 1 sulfate transporter (Smith et al. 1995; Fig. 1). A similar pattern is seen for the *Astragalus* Group 2 sulfate transporter isoform, which is closely related to the *Stylosanthus* SHST3 isoform but distinct to the *Arabidopsis* and rice Group 2 sulfate transporters (Fig. 1).

The Group 3 *Astragalus* isoforms, Sultr3;1 and Sultr3;4, are closely related to the *Arabidopsis* homologues, Sultr3;1 and Sultr3;4. The *Astragalus* Group 4 type identified was distinct from *Arabidopsis* 4;1 and 4;2 homologues and was thus named Sultr4. It has been shown that the number of homologues per group differs depending on the species (Fig. 1). So far the RT-PCR technique using degenerated primer identified only one Group 4 sulfate transporter isoform.

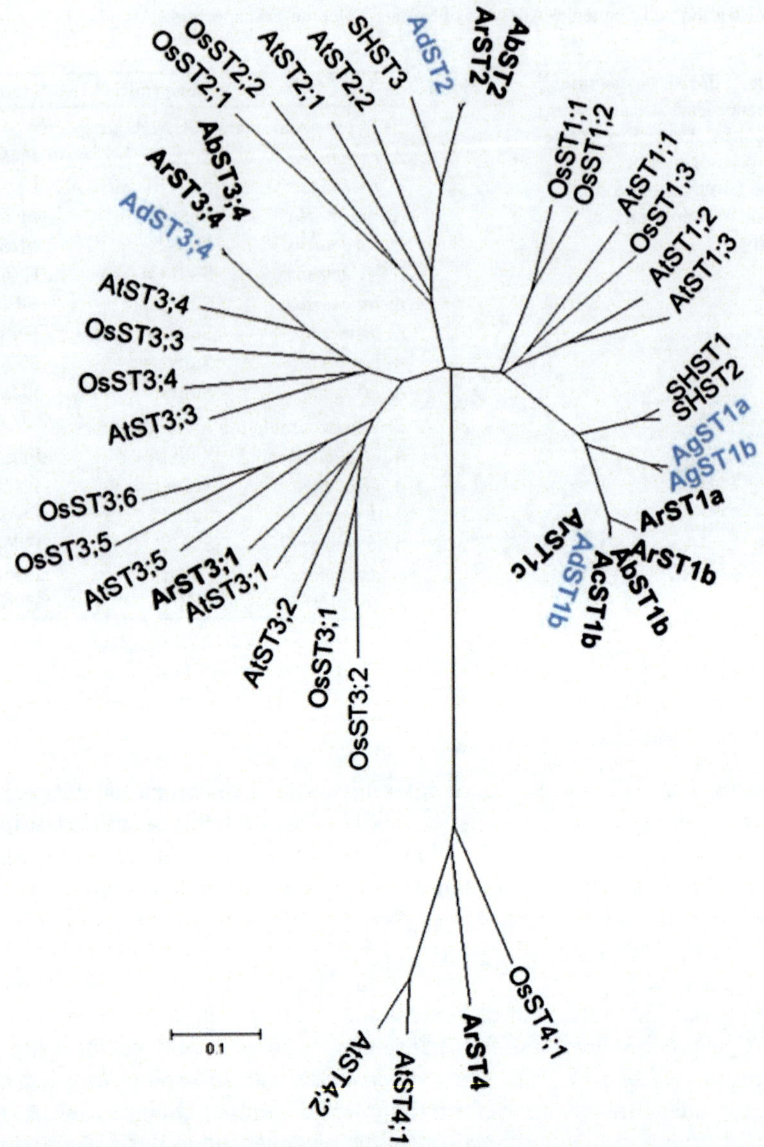

Fig. 1 Phylogenetic tree of the coding cDNAs of the *Arabidopsis thaliana*, *Oryza sativa*, *Stylosanthes hamata* and *Astragalus* sulfate transporter family. *A. thaliana* (At) Group 1–4 sulfate transporter are: AtST1.1, At4g08620; AtST1.2, At1g78000; AtST1.3, At1g22150; AtST2.1, At5g10180, AtST2.2, At1g77990; AtST3.1, At3g51895, AtST3.2, At4g02700; AtST3.3, At1g23090; AtST3.4, At3g15990; AtST3.5, At5g19600; AtST4.1, At5g13550; AtST4.2, At3g12520. *O. sativa* (Os) Group 1–4 sulfate transporters are from genomic sequences, Buchner et al. (2004). *S. hamata* Group 1 sequences are: SHST1, X82255; SHST2, X82256; and Group 2:SHST3, X82454. Full length Group 1 type, Group 2 type and Group 3;4 type and incomplete Group 3;1 and Group 4 type *Astragalus* transporter genes isolated in this study are: (in *black bold*) Se-hyperaccumulator (Ar), (Ab) and (Ac) *A. racemosus*, *A. bisulcatus*, and *A. crotalariae*, (Ad) and (Ag), (in *blue bold*) Se-non-accumulator *A. drummondii* and *A. glycyphyllos*. ClustalW (Thompson et al. 2002) was used to perform the multiple alignment (all positions containing gaps and missing data were eliminated from the dataset). Phylogenetic analyses were conducted in MEGA4 (Tamura et al. 2007) and the Neighbour-Joining tree was drawn to scale

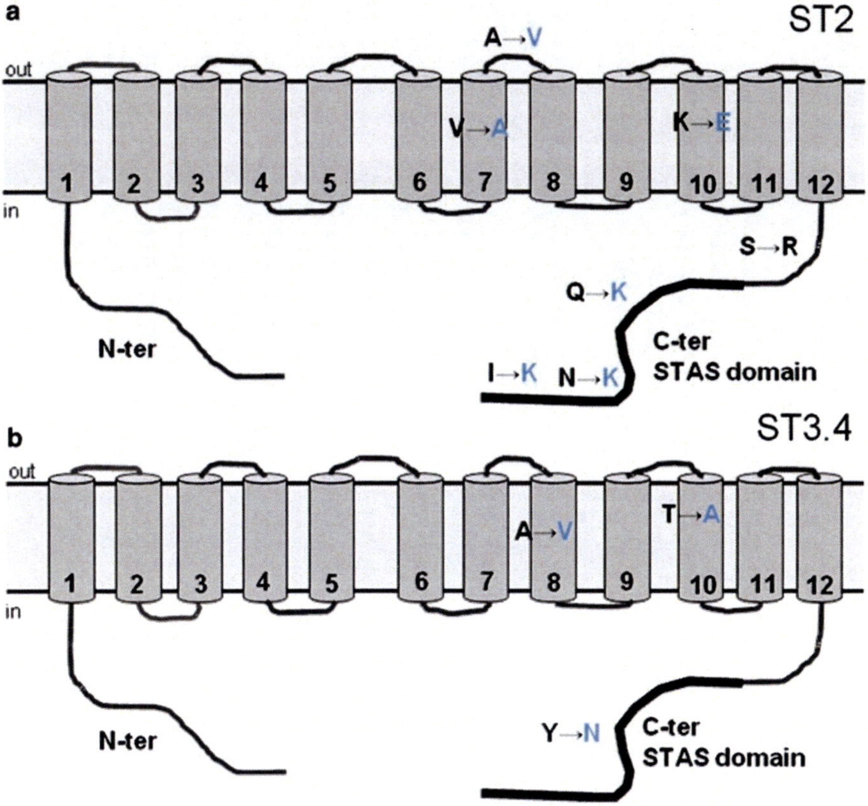

Fig. 2 Predictive model highlighting amino acid variation between putative sulfate/selenate transporters sequences from the Se-hyper-accumulator and the related non accumulator *Astragalus* species. Amino acid conserved within the sequences from the Se-hyperaccumulator, *A. racemosus* and *A. bisulcatus* (*black bold*), but differing from the sequence of the related non accumulators, *A. drummondii*, and from the *Arabidopsis* sequence (*blue bold*) are indicated. (**a**) Group 2 type sequences. (**b**) Group 3;4 sequences

Conservation of sequences across different members of the sulfate transporter family may indicate important residues with structural or functional roles. It is hypothesized that sequence variation changing charge, size or polarity of selected residues between the equivalent isoforms from hyper- and non-accumulators would have the potential to modify the ratio of selectivity of sulfate and selenate transport and as a consequence may contribute to the accumulation phenotype. Identification of conserved residues amongst the sulfate transporter family showing variation between *Astragalus* types would enable targeting of experimental research to regions most likely to be involved in selenate transport specificity. Several amino acid changes were apparent when comparing the sequences of Group 2 or Group 3;4 type putative sulfate/selenate transporters from the Se-hyperaccumulator, *Astragalus racemosus* and *A. bisulcatus*, and the related non-accumulator, *A. drummondii* (Fig. 2). Comparison of Group 1 *Astragalus* sequences with other SulP sequences

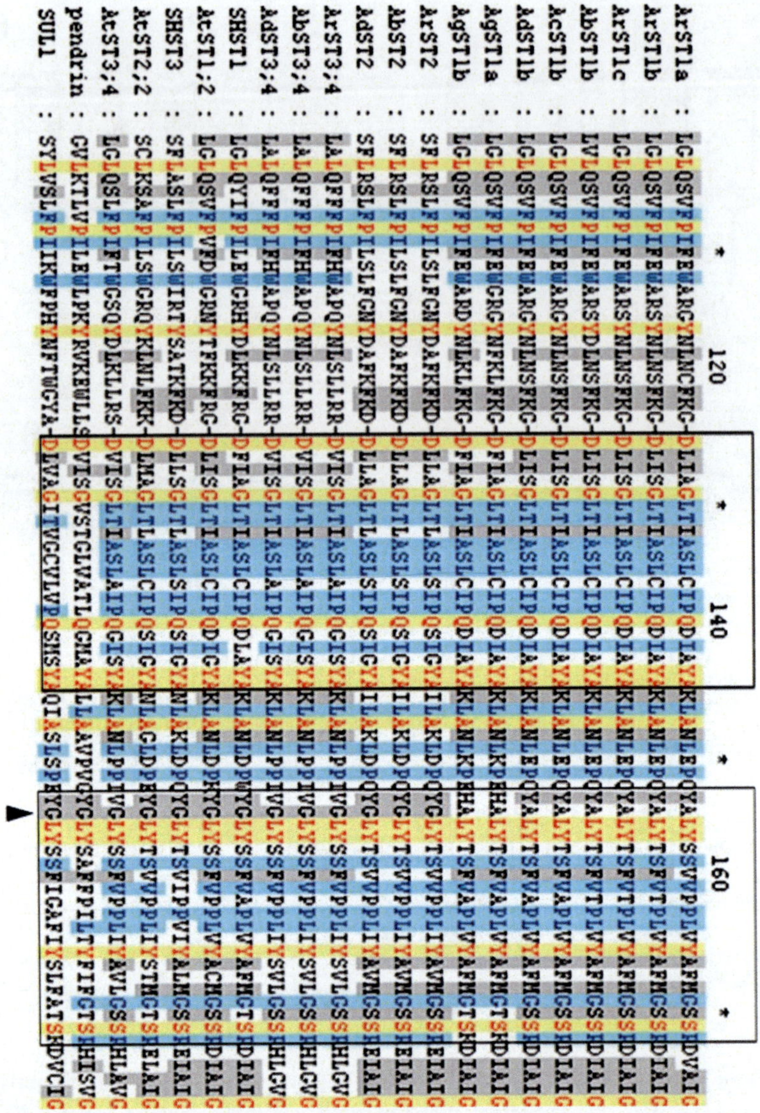

Fig. 3 Alignment of selected members of the SulP family showing the first two predicted transmembrane helices and highlighting an alanine residue conserved between Group 1 *Astragalus* sequences, otherwise conserved as a glycine residue within sequences from members of the SulP family. The alignment was performed using ClustalW (Thompson et al. 2002). The sequences shown, with gene identifier numbers in parentheses when available are: *Astragalus racemosus* ArST1A (FN689510), ArST1B (FN689511), ArST1C (FN689512) and ArST3.4 (FN689514); *A.s bisulcatus* AbST1B (FN689517); *A. glycyphyllos* AgST1A (FN689515) and AgST1B (FN689516); *Arabidopsis thaliana* SULTR1.1 (At4g08620); *Stylosanthes hamata* SHST1 (CAA57710) and SHST3 (CAA57831) high and low affinity sulfate transporters, respectively; *Saccharomyces cerevisiae* SUL1 (P38359); and *Homo sapiens* pendrin (AAC51873)

from various organisms highlighted an amino acid, which is an alanine for all the cloned Group 1 type *Astragalus* sequences, whereas it is a conserved glycine for all other identified transporters belonging to the eukaryotic sulfate permease (SulP) family, including yeast and human transporter proteins (Fig. 3). This amino acid is located in the second position of the second trans-membrane α-helice which has been shown to be important for transport function and that might determine the ion-binding size and confer selectivity of the filter (Shelden et al. 2003; Leves et al. 2008). As this Gly to Ala transition is found in Sultr1 sulfate transporter genes of the all *Astragalus* species, it is unlikely to be responsible for the increased Se accumulation in the hyper-accumulating *Astragalus* species.

On the other hand, single amino acid alterations in an ion binding site or in the transport channel have been shown with other transporters to be sufficient to modify transport selectivity (Rogers et al. 2000; Yang et al. 2009). Mutation studies in a heterologous system will be necessary to examine the roles of *Astragalus* candidate amino acids in conferring S versus Se selectivity.

Acknowledgements Rothamsted Research receives grand-aided support from the Biotechnology and Biological Sciences Research Council (BBSRC) of the U.K. E.C. was supported by the Lawes Agricultural Trust.

References

Broadley MR, White PJ, Bryson RJ, Meacham MC, Bowen HC, Johnson SE, Hawkesford MJ, McGrath SP, Zhao FJ, Breward N, Harriman M, Tucker M (2006) Biofortification of UK crops with selenium. Proc Nutr Soc 65:169–181

Buchner P, Takahashi H, Hawkesford MJ (2004) Plant sulphate transporters: co-ordination of uptake, intracellular and long-distance transport. J Exp Bot 55:1785–1798

Cabannes E, Buchner P, Broadley MR, Hawkesford MJ (2011) A comparison of sulfate and selenium accumulation in relation to expression of sulfate transporter genes in *Astragalus* species. Plant Physiol. doi:10.1104/pp. 111.183897

Feist LJ, Parker DR (2001) Ecotypic variation in selenium accumulation among populations of *Stanleya pinnata*. New Phytol 149:61–69

Hawkesford MJ (2000) Plant responses to sulfur deficiency and the genetic manipulation of sulfate transporters to improve S-utilization efficiency. J Exp Bot 51:131–138

Hawkesford MJ (2003) Transporter gene families in plants: the sulphate transporter gene family – redundancy or specialization? Physiol Plant 117:155–165

Kopsell DA, Kopsell DE (2007) Selenium. In: Barker AE, Pilbeam DJ (eds) Handbook of plant nutrition. CRC Press Book, New York, pp 515–549

Läuchli A (1993) Selenium in plants: uptake, functions, and environmental toxicity. Bot Acta 106:455–468

Leves FP, Tierney ML, Howitt SM (2008) Polar residues in a conserved motif spanning helices 1 and 2 are functionally important in the SulP transporter family. Int J Biochem Cell Biol 40:2596–2605

Pickering IJ, Wright C, Bubner B, Ellis D, Persans MW, Yu EY, George GN, Prince RC, Salt DE (2003) Chemical form and distribution of selenium and sulfur in the selenium hyperaccumulator *Astragalus bisulcatus*. Plant Physiol 131:1460–1467

Rogers EE, Eide DJ, Guerinot ML (2000) Altered selectivity in an Arabidopsis metal transporter. Proc Nat Acad Sci USA 97:12356–12360

Schromburg L, Schweizer U, Köhrle J (2004) Selenium and selenoproteins in mammals: extraordinary, essential, enigmatic. Cell Mol Life Sci 61:1988–1995

Shelden MC, Loughlin P, Tierney ML, Howitt SM (2003) Interactions between charged amino acid residues within transmembrane helices in the sulfate transporter SHST1. Biochemistry 42:12941–12949

Shibagaki N, Rose A, McDermott JP, Fujiwara T, Hayashi H, Yoneyama T, Davies JP (2002) Selenate-resistant mutants of *Arabidopsis thaliana* identify Sultr1;2, a sulfate transporter required for efficient transport of sulfate into roots. Plant J 29:475–486

Shinmachi F, Buchner P, Stroud JL, Parmar S, Zhao F-J, McGrath SP, Hawkesford MJ (2010) Influence of sulfur deficiency on the expression of specific sulfate transporters and the distribution of sulfur, selenium and molybdenum in wheat. Plant Physiol 153:327–336

Smith FW, Ealing PM, Hawkesford MJ, Clarkson DT (1995) Plant members of a family of sulfate transporters reveal functional subtypes. Proc Natl Acad Sci USA 92:9373–9377

Sors TG, Ellis DR, Na GN, Lahner B, Lee S, Leustek T, Pickering IJ, Salt DE (2005) Analysis of sulfur and selenium assimilation in *Astragalus* plants with varying capacities to accumulate selenium. Plant J 42:785–797

Tamura K, Dudley J, Nei M, Kumar S (2007) MEGA4: Molecular Evolutionary Genetics Analysis (MEGA) software version 4.0. Mol Biol Evol 24:1596–1609

Terry N, Zayed AM, De Souza MP, Tarun AS (2000) Selenium in higher plants. Annu Rev Plant Physiol Plant Mol Biol 51:401–432

Thompson JD, Gibson TJ, Higgins DG (2002) Multiple sequence alignment using ClustalW and ClustalX. Curr Protoc Bioinform 2, Unit 2.3

White PJ, Bowen HC, Parmaguru P, Fritz M, Spracklen WP, Spiby RE, Meacham MC, Mead A, Harriman M, Trueman LJ, Smith BM, Thomas B, Broadley MR (2004) Interactions between selenium and sulphur nutrition in *Arabidopsis thaliana*. J Exp Bot 55:1927–1937

Yang HS, Kim E, Lee S, Park HJ, Cooper DS, Rajbhandari I, Choi I (2009) Mutation of aspartate 555 of the sodium/bicarbonate transporter SLC4A4/NBCe1 induces chloride transport. J Biol Chem 284:15970–15979

Expression of Ferritins *ZmFer1* and *ZmFer2* Genes in Shoots and Roots of S-Deprived Young Maize Plants

Styliani N. Chorianopoulou, Maria Dimou, Chariclea Kazila, Philippa Maniou, Panagiotis Katinakis, and Dimitris L. Bouranis

Abstract A perturbation of the iron levels is observed in maize, as an early effect of sulfur deprivation. Ferritins (ferroxidase, EC 1.16.3.1), readily interact with Fe(II) inducing its oxidation and its deposition as an insoluble mineral iron core into a central enzyme cavity. Fourteen-day-old maize seedlings were subject to sulfur (−S) deprivation for 6 days, in a hydroponic culture, and the expression of *ZmFer1* and *ZmFer2* ferritin genes, in shoots and roots, was monitored at the indicated time intervals by means of real time RT-PCR. The response of the individual expression levels of the ferritin genes differed under sulfur deprivation. In control plants, the range of the relative expression ratio of *ZmFer1* gene was higher in the shoots than that of the roots. The same was observed for the expression of the *ZmFer2* gene. The relative expression of *ZmFer1* was higher compared with that of *ZmFer2* gene in both shoots and roots. S deprivation caused a decrease in *ZmFer1* gene expression and an increase in *ZmFer2*. The response of both ferritin genes takes place first in the root. *ZmFer1* gene expression apparently decreased because there was not an iron redundancy for storage. *ZmFer2* gene expression increased in response to sulfur stress and/or in coordination to antioxidant action, which takes place under the aforementioned experimental conditions.

S.N. Chorianopoulou (✉) • C. Kazila • P. Maniou • D.L. Bouranis
Plant Physiology Laboratory, Department of Plant Biology,
Faculty of Agricultural Biotechnology, Agricultural University of Athens,
75 Iera Odos, Athens 11855, Greece
e-mail: s.chorianopoulou@aua.gr

M. Dimou • P. Katinakis
Laboratory of General and Agricultural Microbiology, Department of Biochemistry, Enzyme Technology, Microbiology and Molecular Biology, Faculty of Agricultural Biotechnology, Agricultural University of Athens, 75 Iera Odos, Athens 11855, Greece

L.J. De Kok et al. (eds.), *Sulfur Metabolism in Plants*, Proceedings of the International Plant Sulfur Workshop 1, DOI 10.1007/978-94-007-4450-9_21,

In graminaceous species, iron acquisition depends upon phytosiderophore production, which in turn depends upon sulfur assimilation and specifically upon methionine biosynthesis. Therefore, an early consequence of sulfur deficiency is a perturbation in iron content (Astolfi et al. 2003; Bouranis et al. 2003).

The iron homeostasis mechanism includes ferritins, which are proteins characterized by highly conserved three-dimensional structures similar to spherical shells, designed to accommodate large amounts of iron in a safe, soluble and bioavailable form. All ferritins readily interact with Fe(II), in order to induce its oxidation and its deposition as an insoluble mineral iron core into a central enzyme cavity, a reaction catalyzed by a ferroxidase center. This is an anti-oxidant activity, which consumes Fe(II) and peroxides, the reagents that produce toxic free radicals in the Fenton reaction. The mechanism of ferritin iron incorporation has been characterized in detail, whilst that of iron release and recycling has been studied less thoroughly. Generally, ferritin expression is regulated by iron and by oxidative damage, and plays a central role in the control of cellular iron homeostasis (Arosio et al. 2009).

Ferritin synthesis is developmentally and environmentally controlled, partly through the differential expression of the various members of a small gene family. Ferritin gene regulation appears to be consisting of a complex interplay of transcriptional and posttranscriptional mechanisms, involving cellular relays such as plant hormones, oxidative steps and Ser: Thr phosphatase (Briat et al. 1999). In maize two subclasses of ferritin cDNAs and genes (*ZmFer1* and *ZmFer2*) have been characterised (Lobreaux et al. 1992; Fobis-Loisy et al. 1995). Expression of *ZmFer1* and *ZmFer2* is activated by iron. In addition, expression of the *ZmFer1* gene can be activated by hydrogen peroxide (Briat et al. 1999). Abscisic acid is required for the iron regulated expression of the *ZmFer2* gene but not for the iron-regulated expression of the *ZmFer1* gene, although an abscisic acid response element is present upstream of the promoter region of both genes. Moreover, two metal-regulatory elements and a drought-responsive element were observed in the *ZmFer2* promoter but not in the corresponding *ZmFer1* promoter, suggesting that *ZmFer2* could play a role in stress response (Fobis-Loisy et al. 1995; Proudhon et al. 1996).

The aim of the present study was to examine the response of *ZmFer1* and *ZmFer2* gene expression, in sulfur deprived maize plants grown in a hydroponic culture. For this purpose, 7-day-old maize seedlings were placed in full nutrient solution and 7 days later, half of them were subject to sulfur deprivation, while the rest of them remained in full nutrient solution for 6 more days (Fig. 1). The expression of *ZmFer1* and *ZmFer2* ferritin genes, in the shoots and roots of these seedlings, was monitored by means of real time RT-PCR. The oligonucleotide primers used for RT-qPCR are listed in Table 1. The primers were designed using Primer-BLAST (http://blast.ncbi.nlm.nih.gov/Blast.cgi) and optimised to an equal annealing temperature of 60°C. Each primer pair was further assessed for specificity with melting curve analysis and gel electrophoresis of amplification products. The efficiency of each RT-qPCR reaction was calculated using the LinRegPCR software (Ruijter et al. 2009). Relative expression of the target gene was calculated using the equation described by Pfaffl (2001). *ZmUBQ* was used as internal control. Statistical significance was obtained by two-tailed t-test at $p \leq 0.05$.

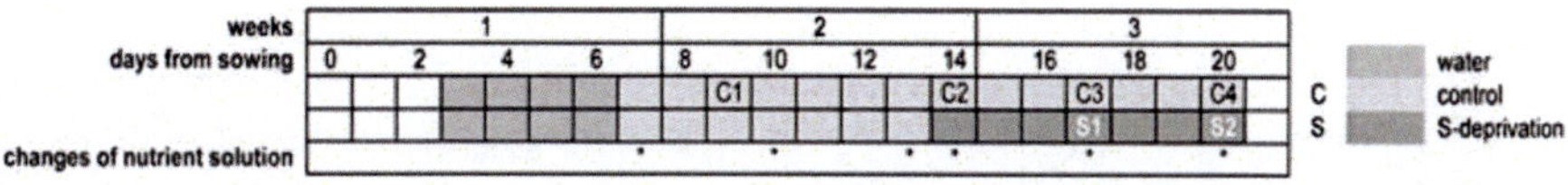

Fig. 1 Chart of the experimental schedule. *White cells*: days of imbibition process and seedling development. Afterwards, seedlings were transferred to the hydroponic system where the root system was in water for 4 days and then in full nutrient solution. Four samplings took place in the control treatment and two in the sulfur deprivation treatment at selected days. An *asterisk* indicates the change of the nutrient solution

Table 1 The primers used in the real-time RT-PCR

Gene	TC target	Forward primer	Reverse primer
ZmFer1	TC548925	5′-GTTGATCAGGCGG AGGAATA-3′	5′-AATCCCAACGAG CATAGCAC-3′
ZmFer 2	TC462167	5′-TGAGCTCTGGGTG GTGTATTC-3′	5′-CCTCCACGTAACA TCCATCA-3′
ZmUBQ	TC549976	5′-AAGCAGCTGGAGG ATGGCCGTA-3′	5′-ACGAAGATCTGCA TGCCACCCCT-3′

Under normal growth conditions, 2 days after full nutrient solution (and iron) supply, both ferritin genes were expressed in shoots as well as in roots although with differing accumulation patterns. This is in agreement with data described by Fobis-Loisy et al. (1995). Relative expression levels of both genes were much higher in the shoots than in the roots. Relative expression levels of *ZmFer1* were higher than that of *ZmFer2* in both shoots and roots of control seedlings (Fig. 2). Trendlines for *ZmFer1* gene expression revealed trend of increased expression in the shoots (Fig. 2a) and trend of decreased expression in the roots (Fig. 2), whilst *ZmFer2* showed only tendency of increased expression in both shoots and roots (Fig. 2b, d). The relative expression of both *ZmFer1* and *ZmFer2* genes presented oscillations around the trendline showing virtually the same pattern. The oscillation range was robust in the case of *ZmFer1* and rather marginal for *ZmFer2*.

In sulfur-deprived seedlings, at day 17 (3 days under sulfur deprivation), *ZmFer1* gene expression increased in the shoots (106%) while decreased in the roots (41%), compared to control plants (inserted table in Fig. 2). At the same day, *ZmFer2* gene expression increased in shoots (135%), as well as in roots (136%). Three days later (and 6 days under sulfur deprivation), the gene expression patterns were different. A decrease in *ZmFer1* gene expression was observed in both shoots and roots (75% and 40%, respectively), while *ZmFer2* gene although it showed a decreased expression in roots (37%), it also showed a strong increase in shoots (446%). The corresponding trendlines decrease in the case of *ZmFer1* gene expression and increase in the case of *ZmFer2*, indicating differential rates of accumulation. Interestingly, the oscillation pattern of *ZmFer2* gene expression, in sulfur-deprived roots has been reversed.

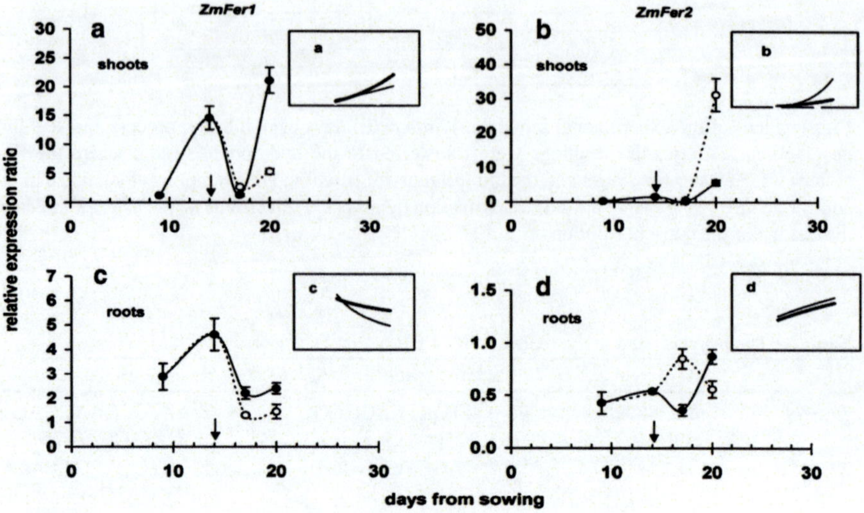

	ZmFer1	ZmFer2	ZmFer1	ZmFer2
	statistically significant % change			
	d3	d3	d6	d6
shoots	106.2	134.6	-74.6	446.3
roots	-40.9	136.1	-39.6	-35.6

Fig. 2 Expression of *ZmFer1* and *ZmFer2* genes, in shoots and roots of young maize seedlings, relative to the expression of ubiquitin gene (*ZmUBQ*). Plants in full nutrient solution, *closed circle* and *full line*; plants in sulfur depleted nutrient solution, *open circle* and *dashed line*. Error bars represent standard error (n=9). Sulfur deprivation treatment started at day 14 (indicated by *arrow*). Regression analysis has been performed based on the power function as the model and the resulted trendline has been applied in all data sets. These trendlines are shown in the inserted figures. *Bold trendline* indicates control treatment. Inserted table provides the statistically significant percent changes in *ZmFer1* and *ZmFer2* gene expression in shoots and roots of sulfur-deprived plants by means of a two-tailed t-test at $p \leq 0.05$

S deprivation increased the root/shoot molar ratio of Fe at day 15 (Fig. 3c) due to the perturbation of Fe levels in both shoots and roots (Fig. 3b). Considering the combined functions of ferritins as storage and antioxidant molecules, the decrease in sulfur content of the shoots at day 17 (Fig. 3e) should cause a redundancy of iron (Fig. 3b), which cannot be incorporated into Fe-S clusters. This iron redundancy could be handled by the simultaneous increased expression of both ferritin genes. In the root, the combination of the decreased *ZmFer1* expression levels with the increased *ZmFer2* expression levels, indicates a probable precedency of *ZmFer2,* possibly due to its role in stress responses. The strong increase of *ZmFer2* gene expression in the shoot at day 20 suggests a stronger need for antioxidant action during iron handling. Expression of *ZmFer2* in the root decreased with the same rate as *ZmFer1*. It should be stressed that the overall observed trend was a decline of *ZmFer1* gene expression accompanied by an increase of the corresponding *ZmFer2*.

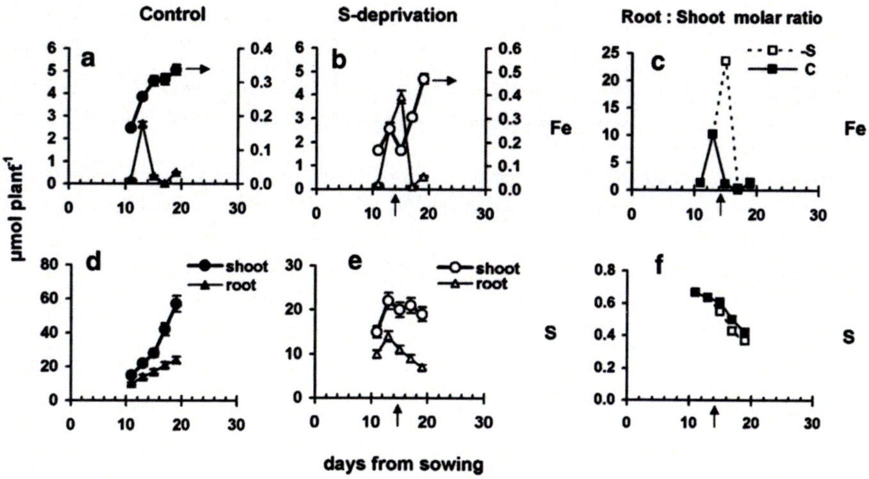

Fig. 3 Iron and S contents in the shoots and the roots of plants in full nutrient solution (**a, d**), along with the corresponding effects of S deprivation on Fe and S (**b, e**) contents and the root/shoot molar ratio respectively (**c, f**). Error bars represent standard error (n=9). Sulfur deprivation treatment started at day 14 (indicated by *vertical arrow*). *Horizontal arrow* indicates which curve corresponds to the *right* axis. Fresh weight per sample (shoot, root) was recorded, samples were oven-dried at 80°C, dry weight was recorded and the samples were ground to pass a 40-mesh screen using an analytical mill (IKA, model A10) prior to chemical analysis (Mills and Jones 1996). Fe and S were determined following a wet acid digestion procedure based on the combination of HNO_3 and 30% H_2O_2 for Fe analysis and dry ashing for S analysis (Mills and Jones 1996). The concentration of Fe was determined in the digests by atomic absorption spectrophotometry using a GBC Avanta spectrophotometer. S was extracted from the ash with 2% acetic acid and the concentration of S was determined by using a turbidimetric method (Sörbo 1987)

References

Arosio P, Ingrassia R, Cavadini P (2009) Ferritins: a family of molecules for iron storage, antioxidation and more. Biochim Biophys Acta 1790:589–599

Astolfi S, Zuchi S, Passera C, Cesco S (2003) Does the sulfur assimilation pathway play a role in the response to Fe deficiency in maize (*Zea mays* L.) plants? J Plant Nutr 26:2111–2121

Bouranis DL, Chorianopoulou SN, Protonotarios V, Siyiannis V, Hopkins L, Hawkesford MJ (2003) Leaf responses of young iron-inefficient maize plants to sulfur deprivation. J Plant Nutr 26:1189–1202

Briat JF, Lobrèaux S, Grignon N, Vansuyt G (1999) Regulation of plant ferritin synthesis: how and why. Cell Mol Life Sci 56:155–166

Fobis-Loisy I, Loridon K, Lobreaux S, Lebrun M, Briat JF (1995) Structure and differential expression in response to iron and abscisic acid of two maize ferritin genes. Eur J Biochem 231:609–619

Lobreaux S, Massenet O, Briat JF (1992) Iron induces ferritin synthesis in maize plantlets. Plant Mol Biol 19:563–575

Mills HA, Jones JB Jr (1996) Plant analysis handbook II. MicroMacro Publishing Inc, Athens

Pfaffl MW (2001) A new mathematical model for relative quantification in real-time RT-PCR. Nucleic Acids Res 29:e45

Proudhon D, Wei J, Briat JF, Theil EC (1996) Ferritin gene organization: differences between plants and animals suggest possible kingdom-specific selective constraints. J Mol Evol 42:325–336

Ruijter JM, Ramakers C, Hoogaars WM, Karlen Y, Bakker O, van den Hoff MJ, Moorman AF (2009) Amplification efficiency: linking baseline and bias in the analysis of quantitative PCR data. Nucleic Acids Res 37:45

Sörbo B (1987) Sulfate: turbidimetric and nephelometric methods. In: Jakoby WB, Griffith OW (eds) Methods in enzymology: sulfur and sulfur amino acids, vol 143. Academic, New York, pp 3–6

Molybdate Affects Sulfate Acquisition in *Brassica juncea* Plants

Mario Malagoli, Michela Schiavon, Markus Wirtz, Ruediger Hell, Elizabeth A.H. Pilon-Smits, Sara Trevisan, and Silvia Quaggiotti

Abstract In the last years *Brassica juncea* plants have been studied and utilized for heavy metal phytoremediation. In the present study plants of *Brassica juncea* were exposed for 24 h to 200 μM molybdate, and the interaction between molybdate and sulfate was evaluated. The presence of Mo in the medium affected the biomass of the plants. In particular, the effect was evident in roots, which appeared to be the primary target of molybdenum toxicity. In the plants supplied with Mo, the S content was reduced consistently with the decline of the sulfate uptake rates. However, the transcript level of the sulfate transporter *BjSultr2;1* in roots was unchanged, while Mo treatment clearly enhanced the *MOT1* transcript accumulation. The content of glutathione (GSH) decreased in roots and leaves of plants exposed to Mo, likely due to the competition of molybdate with sulfate for the access to the sulfur metabolic pathway.

Molybdenum (Mo) is an essential micronutrient for plants and is used for the biosynthesis of a pterin-based Mo cofactor (Moco), a component of some enzymes involved in sulfite detoxification, purine catabolism, nitrate assimilation, and abscisic acid biosynthesis (Schwarz et al. 2009; Kruse et al. 2010). It is known that plants take up molybdenum mainly in the dissolved form of molybdate (Kaiser et al. 2005). Mo concentration in soils can increase as a result of industrial, mining and agricultural activities. Excess Mo in soil can induce toxicity symptoms, including stunted

M. Malagoli (✉) • M. Schiavon • S. Trevisan • S. Quaggiotti
DAFNAE, University of Padova, Agripolis, Legnaro PD 35020, Italy
e-mail: mario.malagoli@unipd.it

M. Wirtz • R. Hell
Centre for Organismal Studies, University of Heidelberg, 69120 Heidelberg, Germany

E.A.H. Pilon-Smits
Biology Department, Colorado State University, Fort Collins, CO 80523, USA

L.J. De Kok et al. (eds.), *Sulfur Metabolism in Plants*, Proceedings of the International Plant Sulfur Workshop 1, DOI 10.1007/978-94-007-4450-9_22,

plant growth and low productivity (Nautiyal and Chatterjee 2004). The mechanisms of Mo transport, distribution and accumulation inside cells have been only partially elucidated to date. A molybdenum transporter (MOT1) was isolated in *A. thaliana* and proved to function as a regulator of total plant Mo accumulation and homeostasis (Tomatsu et al. 2007; Baxter et al. 2008; Tejada-Jiménez et al. 2009). MOT1 is localized to mitochondria and, in part, to plasma membranes and to vesicles (Tomatsu et al. 2007; Baxter et al. 2008). It is primarily expressed in roots, particularly in endodermis and stele cells, but its expression can also be detected in the shoot. Molybdenum is known to interfere with S transport (Fitzpatrick et al. 2008) and assimilation (Wangeline et al. 2004). On account of this, the relationship between S and Mo would attain significant attention in phytoremediation technologies as the managing of S application can influence Mo accumulation and distribution in plant tissues (MacLeod et al. 1997; Shinmachi et al. 2010). In the present study the interaction between molybdate and sulfate was investigated in *Brassica juncea*, a plant species known to accumulate molybdenum (Hale et al. 2001).

Seeds of *B. juncea* (L.) Czern. (Cv. PI 426314) were allowed to germinate and grow for 8 days in agar MS-medium inside a chamber with a 13 h light/11 h dark cycle, air T of 20/15°C, relative humidity of 70/85% and at a PFD of 280 mol m^{-2} s^{-1}. Seedlings were then transferred to 3 l plastic pots containing a thoroughly aerated nutrient solution with the following composition: 40 μM KH_2PO_4, 200 μM $Ca(NO_3)_2$, 200 μM KNO_3, 200 μM $MgSO_4$, 10 μM FeNaEDTA, 4.6 μM B, 1.1 μM Cl, 0.9 μM Mn, 0.09 μM Zn, 0.01 μM Mo. The nutrient solution was renewed every 2 days. After 4 days, plants were split in two groups: one group was kept in the same nutrient conditions and the second group was supplied with the nutrient solution added with molybdenum (200 μM) in the form of ammonium molybdate ($(NH_4)_6Mo_7O_{24}$). After 24 h, individual plants were harvested and their fresh weight was measured to evaluate the degree of tolerance to molybdate. S and Mo content in plant tissues were determined via ICP-AES. The content of glutathione (GSH) was determined according to Wirtz et al. (2004) and sulfate uptake experiments were performed as described by Quaggiotti et al. (2003). The analysis of *BjSultr2;1* and *MOT1* gene expression was performed via semi-quantitative RT-PCR according to Schiavon et al. (2008), and *PP2A* (phosphatase 2A), was used as an internal control. Measurements were made on three replicates for each treatment. Analysis of variance was performed followed by pair-wise post-hoc analyses to determine which means differed significantly. Statistically significant differences ($P<0.05$) are reported in the text and shown in the figures.

A 24 h exposure of plants to 200 μM molybdate negatively affected plant biomass production; it decreased with 13%. Mo mainly affected the fresh weight of roots, which hence appeared to be the primary target of molybdenum toxicity. This was probably because the majority of Mo accumulated in the root, although high levels of the metal were also measured in the leaves (Table 1). As a result of Mo addition to plants, the decrease of sulfur content was observed, both in roots and leaves (Table 1). The reduction of S content was consistent with the decline of the sulfate uptake rates (−30%) (Fig. 1), but the sulfate transporter *BjSultr2;1* transcript level in roots was unchanged (Fig. 2). This result indicates that Mo competed with S for

Table 1 The effect of molybdate on molybdenum and sulfur content of roots and leaves of *B. juncea*

	Control		200 μM molybdate	
	Roots	Leaves	Roots	Leaves
Molybdenum (mg kg^{-1} DW)	0.06±0.01c	n.d.	868±813a	487±56b
Sulfur (g kg^{-1} DW)	8.04±0.74b	10.11±1.29a	6.75±0.29c	6.49±0.83c

Plants were exposed to 200 μM molybdate for 24 h. Data represent the mean of five measurements with ten plants in each (±SD)
a, b, c indicate significant differences (P < 0.05)
n.d. not detected

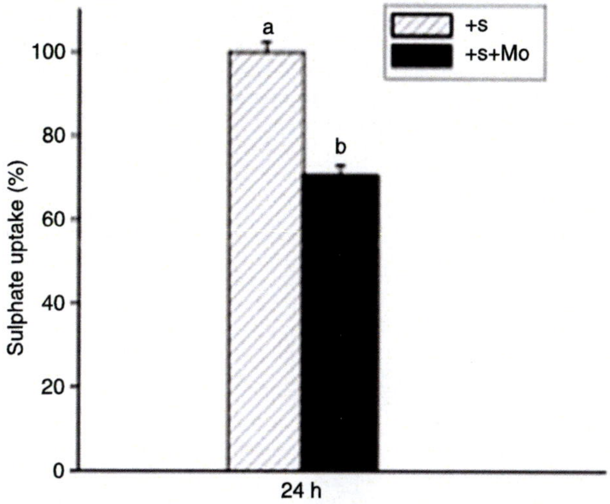

Fig. 1 Effect of molybdate treatment on sulfate uptake in *B. juncea* plants. Sulfate uptake was evaluated by measuring the rate of $^{35}SO_4^{2-}$ absorption over a 10 min pulse period. Mean sulfate uptake rate in control plants was 754±66 nmol SO_4^{2-} g^{-1} FW. Data are the means of three replicates (±SD) with eight plants in each per treatment. Letters above *bars* indicate significant differences ($P<0.05$)

the transport through sulfate permeases, but did not trigger the signals responsible of the up-regulation of *BjSultr2;1*, usually observed in S-deficiency conditions (Schiavon et al. 2008). However, the high accumulation of Mo in roots and leaves could not be only ascribed to the activity of sulfate transporters. Indeed, other transporters could be implied in molybdate transport, like MOT1. The increase of *MOT1* transcript accumulation we observed could determine the enhancement of the molybdenum transport system, and is consistent with previous studies, where the availability of Mo was shown to up-regulate the expression of MOT1 in *Arabidopsis thaliana* plants (Tomatsu et al. 2007; Baxter et al. 2008).

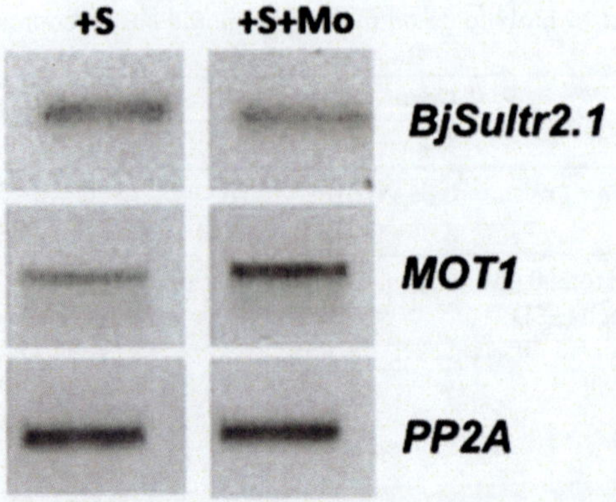

Fig. 2 Gene expression of root transcript accumulation of the low affinity sulfate transporter, *BjSultr2;1*, and molybdenum transporter, *MOT1*, in *B. juncea* plants grown with or without Mo. The constitutively expressed phosphatase 2A (PP2A) gene was used as an internal control. Between 14 and 30 cycles were tested to determine the optimal number of cycles, corresponding to the exponential phase in the amplification for each gene

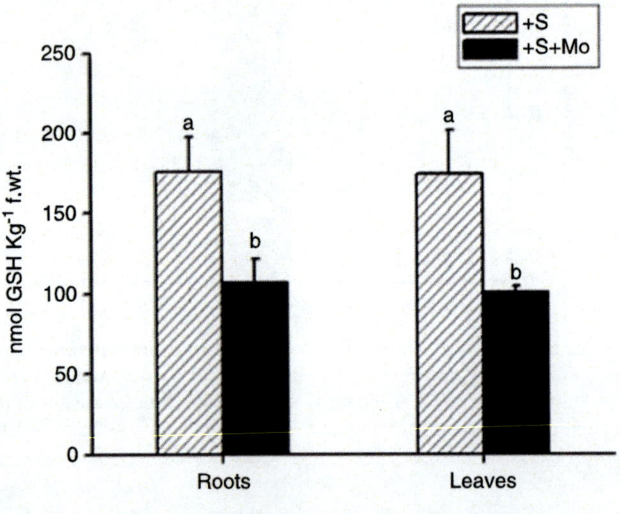

Fig. 3 The effect of molybdate on gluthathione content of roots and leaves of *B. juncea* plants. Data are the means of three replicates (±SD) with 30 plants in each per treatment. Letters above *bars* indicate significant differences ($P < 0.05$)

The level of glutathione (GSH) was comparable between roots and leaves, regardless of Mo treatment (Fig. 3). When plants were exposed to molybdenum, the content of GSH decreased to a similar degree in roots and leaves of roughly –40%. The opposite effect on GSH accumulation was observed in roots of plants exposed to the analogous oxyanion, chromate, and was supposed to be a consequence of oxidative stress (Schiavon et al. 2008). In the present study the drop of GSH level was likely due to the competition of molybdate with sulfate for the access to the sulfur metabolic pathway, since both anions function as substrates for ATP sulfurylase (Wangeline et al. 2004).

The presence of Mo in the plant substrate clearly interfere with S acquisition, confirming that the interaction between molybdate and sulfate is a key factor that should be taken into consideration when *B. juncea* plants are going to be used for Mo phytoremediation.

References

Baxter I, Muthukumar B, Park HC, Buchner P, Lahner B, Danku J, Zhao K, Lee J, Hawkesford MJ, Guerinot ML, Salt DE (2008) Variation in molybdenum content across broadly distributed populations of *Arabidopsis thaliana* is controlled by a mitochondrial molybdenum transporter (MOT1). PLoS Genet 4:e1000004

Fitzpatrick KL, Tyerman SD, Kaiser BN (2008) Molybdate transport through the plant sulfate transporter SHST1. FEBS Lett 582:1508–1513

Hale KL, McGrath S, Lombi E, Stack S, Terry N, Pickering IJ, George GN, Pilon-Smits EAH (2001) Molybdenum sequestration in *Brassica*: a role for anthocyanins? Plant Physiol 126:1391–1402

Kaiser BN, Gridley KL, Ngaire BJ, Phillips T, Tyerman SD (2005) The role of molybdenum in agricultural plant production. Ann Bot 96:745–754

Kruse T, Gehl C, Geisler M, Lehrke M, Ringel P, Hallier S, Hansch R, Mendel RR (2010) Identification and biochemical characterization of molybdenum cofactor-binding proteins from *Arabidopsis thaliana*. J Biol Chem 285:6623–6635

MacLeod JA, Gupta UC, Stanfield B (1997) Molybdenum and sulfur relationships in plants. In: Gupta UC (ed) Molybdenum in agriculture. Press Syndicate of the University of Cambridge, USA 229–244

Nautiyal N, Chatterjee C (2004) Molybdenum stress-induced changes in growth and yield of chickpea. J Plant Nutr 27:173–181

Quaggiotti S, Abrahamshon C, Malagoli M, Ferrari G (2003) Physiological and molecular aspects of sulphate uptake in two maize hybrids in response to S-deprivation. J Plant Physiol 160:167–173

Schiavon M, Pilon-Smits EAH, Wirtz M, Hell R, Malagoli M (2008) Interaction between chromium and sulfur nutrition in *Brassica juncea*. J Environ Qual 37:1536–1545

Schwarz G, Mendel RR, Ribbe MW (2009) Molybdenum cofactors, enzymes and pathways. Nature 460:839–847

Shinmachi F, Buchner P, Stroud JL, Parmar S, Zhao F-J, McGrath SP, Hawkesford MJ (2010) Influence of sulfur deficiency on the expression of specific sulfate transporters and the distribution of sulfur, selenium, and molybdenum in wheat. Plant Physiol 153:327–336

Tejada-Jimenez M, Galvan A, Fernandez E, Llamas A (2009) Homeostasis of the micronutrients Ni, Mo and Cl with specific biochemical functions. Curr Opin Plant Biol 12:358–363

Tomatsu H, Takano J, Takahashi H, Watanabe-Takahashi A, Shibagaki N, Fujiwara T (2007) An *Arabidopsis thaliana* high-affinity molybdate transporter required for efficient uptake of molybdate from soil. Proc Natl Acad Sci USA 104:18807–18812

Wangeline AL, Burkhead JL, Hale KL, Lindblom S-D, Terry N, Pilon M, Pilon-Smits EAH (2004) Overexpression of ATP sulfurylase in *Brassica juncea*: effects on tolerance and accumulation of twelve metals. J Environ Qual 33:54–60

Wirtz M, Droux M, Hell R (2004) *O*-acetylserine (thiol) lyase: an enigmatic enzyme of plant cysteine biosynthesis revisited in *Arabidopsis thaliana*. J Exp Bot 55:1785–1798

Impact of Enhanced Copper Levels and Sulfate Deprivation on the Uptake and Metabolism of Sulfate in Chinese Cabbage (*Brassica pekinensis*)

Muhammad Shahbaz, C. Elisabeth E. Stuiver, Freek S. Posthumus, and Luit J. De Kok

Abstract Exposure of Chinese cabbage (*Brassica pekinensis*) to an enhanced Cu^{2+} level (4 μM) resulted in a reduced plant biomass production and an increased shoot to root ratio at both sulfate-sufficient and sulfate-deprived conditions. However, sulfate deprivation had a more rapid negative effect on plant biomass production than an enhanced Cu^{2+} level, The expression and activity of the sulfate transporters and the expression of APS reductase in Chinese cabbage were rapidly up-regulated (already after 1 or 2 days) upon exposure to 4 μM Cu^{2+}, sulfate deprivation and their combination. Though the impact of sulfate deprivation on the expression and activity of the sulfate transporters was hardly further affected by Cu^{2+}.

The uptake, distribution and assimilation of sulfur are modulated by the plant sulfur status and the sulfur demand for growth (Hawkesford and De Kok 2006). Sulfate deprivation of Chinese cabbage resulted in a rapidly induced expression of Sultr1;1 and an enhanced expression of the constitutively expressed Sultr1;2 in the root, accompanied with an increased sulfate uptake capacity (Koralewska et al. 2008, 2010; Stuiver et al. 2009). A one week exposure of Chinese cabbage (*Brassica pekinensis*) to enhanced Cu^{2+} levels (≤5 μM) resulted in a decreased plant biomass production, an altered mineral composition, an enhanced expression of the Group 1 high affinity sulfate transporters and an enhanced sulfate uptake activity (Shahbaz et al. 2010a, b). The up-regulation of the sulfate transporters in Chinese cabbage upon Cu^{2+} exposure was likely not only due to a higher sulfur demand necessary for the synthesis of metal-binding compounds (*viz.* phytochelatins), but might also be the consequence of a direct interference/reaction of Cu with the signal transduction

M. Shahbaz
Institute of Agricultural Sciences, University of the Punjab, Lahore 54590, Pakistan

C.E.E. Stuiver • F.S. Posthumus • L.J. De Kok (✉)
Laboratory of Plant Physiology, University of Groningen,
P.O. Box 11103, 9700 CC Groningen, The Netherlands
e-mail: l.j.de.kok@rug.nl

L.J. De Kok et al. (eds.), *Sulfur Metabolism in Plants*, Proceedings of the International Plant Sulfur Workshop 1, DOI 10.1007/978-94-007-4450-9_23,

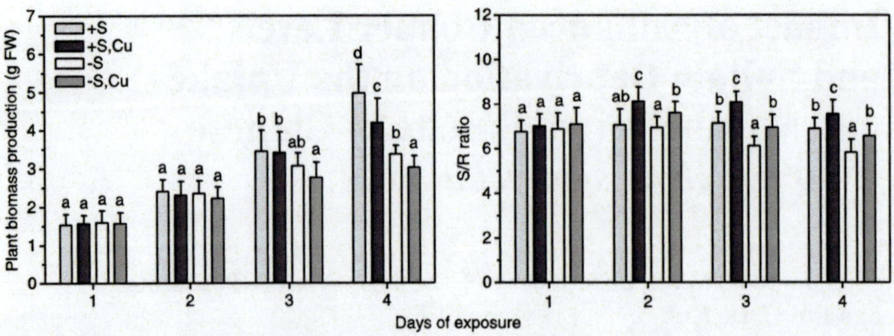

Fig. 1 Impact of Cu^{2+} and sulfate deprivation on plant biomass production (g FW) and shoot to root ratio of Chinese cabbage. Seedlings were grown on 25% Hoagland nutrient solution containing 0.5 mM sulfate in a climate-controlled room for 7 days, transferred to a fresh nutrient solution at 0.5 mM sulfate (+S) or 0 mM sulfate (–S) and containing supplemental 0 or 4 μM Cu^{2+} for 1, 2, 3 and 4 days. See for methods Koralewska et al. (2009) and Shahbaz et al. (2010). Data represent the mean of 15 measurements with 3 plants in each (±SD). Different letters indicate significant differences between treatments within the day of exposure ($p<0.01$, Student's t-test)

pathway involved in the regulation of the sulfate transporters (Shahbaz et al. 2010a). In the present study the interaction between an enhanced Cu^{2+} level and sulfur nutrition was further studied in Chinese cabbage.

Ten-day old seedlings of Chinese cabbage (*Brassica pekinensis* (Lour.) Rupr. cv. Kasumi F1 (Nickerson Zwaan, Made, The Netherlands) were grown on a 25% Hoagland nutrient solution (see for composition Shahbaz et al. 2010a) containing 0.5 mM sulfate in 30 l containers (20 sets of plants per container, 3 plants per set) in a climate-controlled room for 7 days. Subsequently, plants were transferred to a fresh nutrient solution containing 0.5 mM sulfate (+S) or 0 mM sulfate (–S) at 0 and/or 4 μM $CuCl_2$ and grown for 1, 2, 3 and 4 days. Biomass production, metabolite content, sulfate uptake capacity, extraction of RNA and determination of expression of sulfate transporters and APS reductase (APR) were carried out as described previously (Koralewska et al. 2009; Shahbaz et al. 2010a).

Exposure of Chinese cabbage to 4 μM Cu^{2+} resulted in a decreased plant biomass production, an increased dry matter content (data not shown) and an increased shoot to root ratio, as has been observed previously (Fig. 1; Shahbaz et al. 2010a). Root growth was more rapidly affected than shoot growth, resulting in an increased shoot to root ratio already after 2 days (Fig. 1). Similarly to previous observations (Koralewska et al. 2008; Stuiver et al. 2009), sulfate deprivation of Chinese cabbage resulted in a decreased plant biomass production and in a change in shoot to root biomass partitioning in favor of that of the root, as illustrated by a decrease in shoot to root ratio (Fig. 1). A simultaneous exposure of plants to sulfate deprivation and 4 μM Cu^{2+} resulted in an almost similar decrease of plant biomass production as observed upon Cu^{2+} exposure of sulfate-sufficient plants. The shoot to root biomass partitioning, however, was quite similar to that observed for Cu^{2+}-exposed sulfate-sufficient plants (Fig. 1). There was an increase in dry matter content of the root of

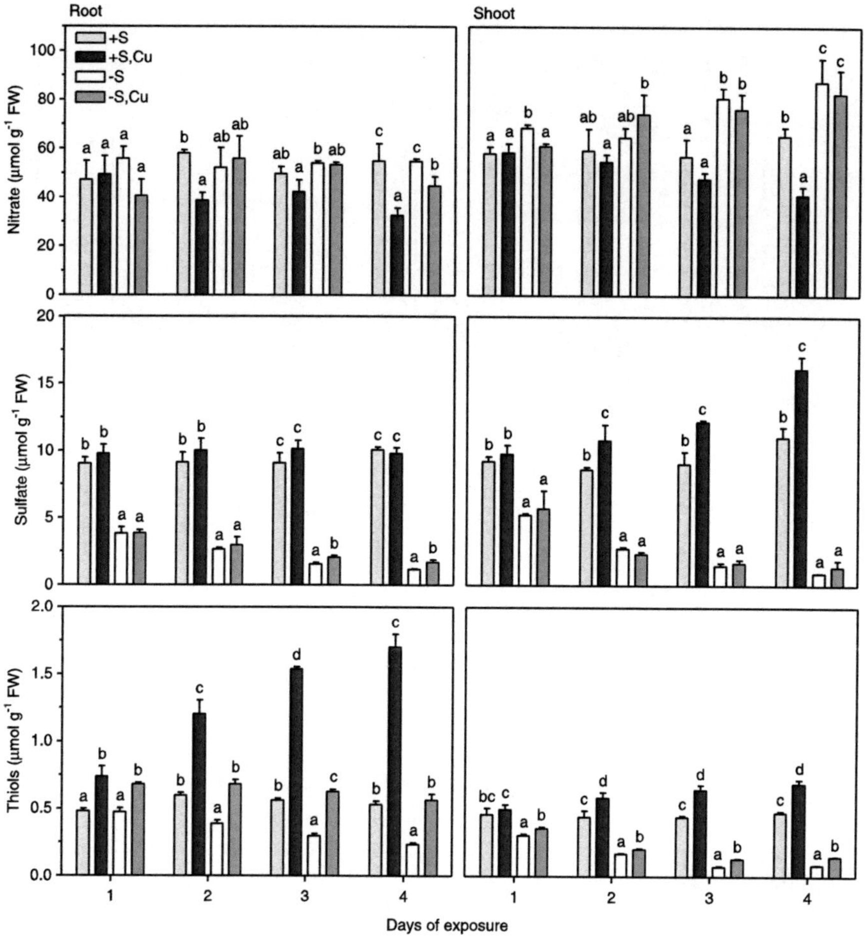

Fig. 2 Impact of Cu^{2+} and sulfate deprivation on nitrate, sulfate and thiol content (μmol g^{-1} FW) of Chinese cabbage. For experimental details see legend Fig. 1. Data represent the mean of three measurements with three plants in each (±SD). Different letters indicate significant differences between treatments ($p<0.01$, Student's t-test)

the Cu^{2+}-exposed sulfate-deprived plants after 3 days, whereas that of the shoot remained unaffected (data not shown). Evidently, upon a 4 day-exposure, the impact of sulfate deprivation on plant biomass production was more pronounced (more rapidly) than that at enhanced Cu^{2+} levels.

At sulfate-sufficient conditions, Cu^{2+} exposure resulted in a rapid increase in water-soluble non-protein thiol content in both root and shoot. However, the thiol accumulation was much more pronounced in the root than in the shoot; after 4 days its content had increased 4-fold and 1.5-fold in root and shoot, respectively (Fig. 2). It was evident that in Chinese cabbage only a small proportion of the increase in

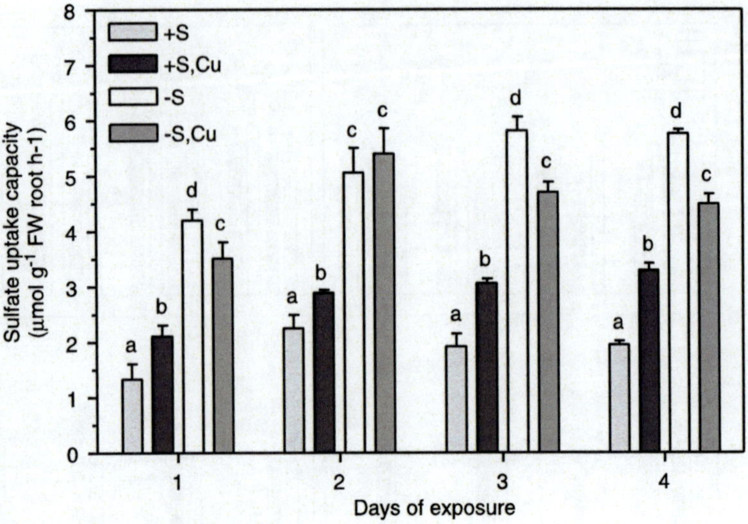

Fig. 3 Impact of Cu^{2+} and sulfate deprivation on sulfate uptake capacity (μmol g^{-1} FW root h^{-1}) of Chinese cabbage. For experimental details see legend Fig. 1. Data represent the mean of three measurements with three plants in each (±SD). Different letters indicates significant differences between treatments ($p<0.01$, Student's t-test)

thiol content could be ascribed to a Cu^{2+}-induced synthesis of phytochelatins (Shahbaz et al. 2010a).

The sulfate content of the root was hardly affected upon Cu^{2+} exposure, but its content in the shoot started to increase after 2 days up to 1.5-fold after 4 days of exposure (Fig. 2). In contrast, the nitrate content decreased in both root and shoot upon Cu^{2+} exposure (Fig. 2).

Upon sulfate deprivation the thiol and sulfate content were strongly decreased in both root and shoot. However, the nitrate content of the shoot was increased, whereas that of the root was not affected upon sulfate deprivation (Fig. 2; Koralewska et al. 2008; Stuiver et al. 2009). The thiol content in the root of plants simultaneously exposed to sulfate deprivation and 4 μM Cu^{2+} was increased after 1 day, thereafter it remained unaltered (Fig. 2). In the shoot, however, the thiol content was increased after 1 day and remained higher than that of the sulfate-deprived plants up to 4 days of exposure (Fig. 2). The strong decrease in thiol content in both root and shoot upon sulfate deprivation was most likely due to growth dilution and/or metabolism of thiol compounds to support the synthesis of other essential organic sulfur-containing compounds (*e.g.* proteins). The thiol content of plants simultaneously exposed to sulfate deprivation and 4 μM Cu^{2+} remained unaltered/higher in the root and shoot than in sulfate-deprived plants. This might indicate that part of the accumulated thiols upon Cu^{2+} exposure, presumably the phytochelatins fraction could not be re-metabolized. The nitrate content of Chinese cabbage simultaneously exposed to sulfate deprivation and Cu^{2+} was not substantially affected in the both root and shoot (Fig. 2).

Exposure of Chinese cabbage to 4 μM Cu^{2+} resulted in an enhanced expression and activity of the sulfate transporters in the root (Figs. 3 and 4). The expression of

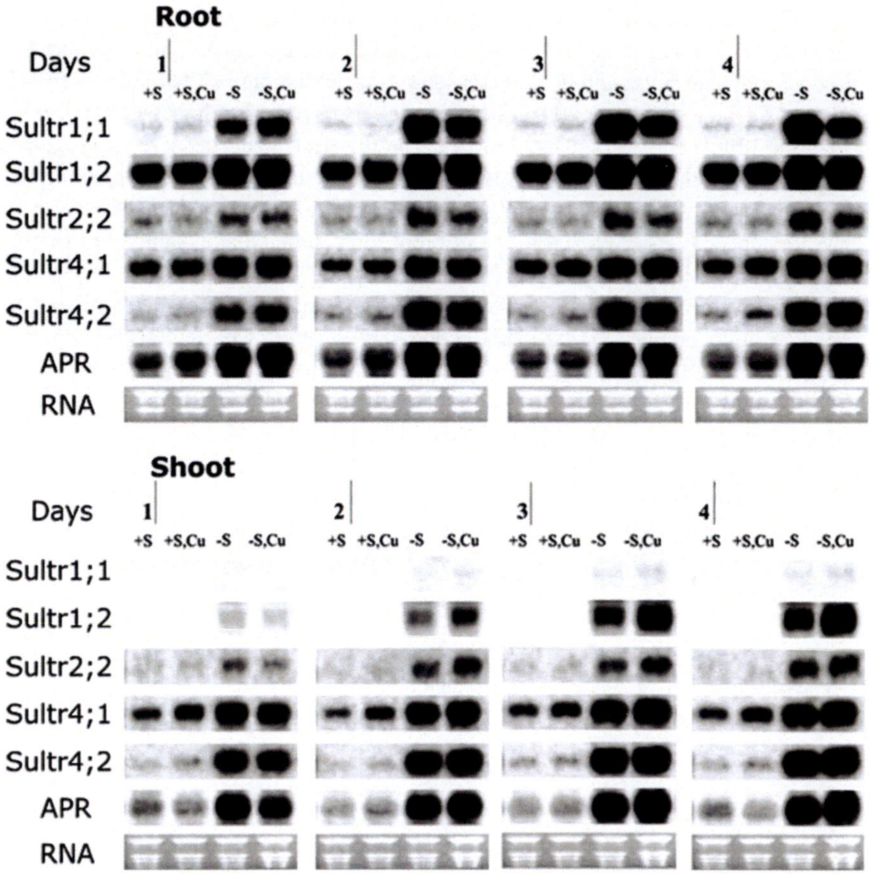

Fig. 4 Impact of Cu^{2+} and sulfate deprivation on mRNA abundance of sulfate transporters (*Sultr*) and APS reductase (*APR*) in the root and shoot of Chinese cabbage. For experimental details see legend Fig. 1

the constitutively abundant sulfate transporter Sultr1;2 and the sulfate uptake capacity were already up-regulated after 1 day of exposure (Figs. 3 and 4). In both root and shoot, also the Group 4 sulfate transporter Sultr4;1 was already up-regulated upon 1 day of Cu^{2+} exposure, whereas in the root also Sultr4;2 was slightly up-regulated upon 4 day of exposure (Fig. 3). The expression of APS reductase, the key regulating enzyme in sulfate reduction pathway, was first up-regulated upon 1 day of Cu^{2+} exposure in the root, but after 2 days it was down-regulated again to a level similar to that of non-exposed control plants (Fig. 4). The expression of APS reductase in the shoot, however, was not affected upon Cu^{2+} exposure (Fig. 4). The transient up-regulation of APR in the root may illustrate a temporary up-regulation of the sulfate reduction pathway possibly needed for a Cu-induced synthesis of thiols (presumably glutathione and for a lesser part phytochelatins; Shahbaz et al. 2010), in response to excess of Cu taken up by the root (Figs. 3 and 4).

Sulfate deprivation also resulted in an enhanced expression of Sultr1;1 and Sultr1;2 in the root, an enhanced expression of Sultr2;2, Sultr4;1, Sultr4;2 and APR in both root and shoot, and in an increased sulfate uptake capacity of the root (Figs. 3 and 4; Koralewska et al. 2008; Stuiver et al. 2009). A simultaneous exposure of plants to sulfate deprivation and 4 μM Cu^{2+} hardly affected the observed up-regulated expression of the sulfate transporters and APR, but after 2 days it resulted in a slightly less up-regulation of Sultr1;1 in the root along with the decrease in the sulfate deprivation-induced up-regulation of the sulfate uptake capacity of the root (Figs. 3 and 4).

From the current results it was evident that sulfate deprivation had a more rapid negative affect on plant biomass production than enhanced Cu^{2+} levels. Cu^{2+} exposure disturbed the sulfate metabolism very rapidly and already after 1 day, the expression and the activity of the sulfate transporters were up-regulated. It is unlikely that the up-regulation of the sulfate uptake could solely be attributed to a higher sulfur need upon Cu^{2+} exposure, since Cu^{2+} exposure also resulted in enhanced sulfate levels in the shoot. The consequences of sulfate deprivation on plant growth and the expression and activity of the sulfate transporters, *e.g.* the induction and up-regulation of sulfate transporters and APS reductase, was hardly further affected by an enhanced Cu^{2+} level, indicating that sulfate deprivation more or less surpassed the development of Cu toxicity.

Acknowledgements M. Shahbaz received a Ph.D. grant from the HEC-NUFFIC program by the Government of Pakistan.

References

Hawkesford MJ, De Kok LJ (2006) Managing sulfur metabolism in plants. Plant Cell Environ 29:382–395

Koralewska A, Stuiver CEE, Posthumus FS, Kopriva S, Hawkesford MJ, De Kok LJ (2008) Regulation of sulfate uptake, expression of the sulfate transporters Sultr1;1 and Sultr1;2, and APS reductase in Chinese cabbage (*Brassica pekinensis*) as affected by atmospheric H_2S nutrition and sulfate deprivation. Funct Plant Biol 35:318–327

Koralewska A, Buchner P, Stuiver CEE, Posthumus FS, Kopriva S, Hawkesford MJ, De Kok LJ (2009) Expression and activity of sulfate transporters and APS reductase in curly kale in response to sulfate deprivation and re-supply. J Plant Physiol 166:168–179

Koralewska A, Stuiver CEE, Posthumus FS, Shahbaz M, Buchner P, Hawkesford MJ, De Kok LJ (2010) Whole plant regulation of sulfate uptake and distribution in *Brassica* species. In: Sequi P, Ferri D, Rea E, Montemurro AV, Fornado F (eds) More sustainability in agriculture: new fertilizers and fertilizer management, 18th international symposium of CIEC, Fertilitas Agrorum, Rome pp 274–280

Shahbaz M, Tseng MH, Stuiver CEE, Koralewska A, Posthumus FS, Venema JH, Parmar S, Schat H, Hawkesford MJ, De Kok LJ (2010a) Copper exposure interferes with the regulation of the uptake, distribution and metabolism of sulfate in Chinese cabbage. J Plant Physiol, Rome 167:438–446

Shahbaz M, Tseng MH, Stuiver CEE, Posthumus FS, Parmar S, Koralewska A, Hawkesford MJ, De Kok LJ (2010b) Impact of copper exposure on physiological functioning of Chinese cabbage (*Brassica pekinensis*). In: Sequi P, Ferri D, Rea E, Montemurro AV, Fornado F (eds)

More sustainability in agriculture: new fertilizers and fertilizer management, 18th international symposium of CIEC, Fertilitas Agrorum, Rome pp 318–324

Stuiver CEE, Koralewska A, Posthumus FS, De Kok LJ (2009) Impact of sulfur deprivation on root formation, and activity and expression of sulfate transporters in Chinese cabbage. In: Sirko A, De Kok LJ, Haneklaus S, Hawkesford MJ, Rennenberg H, Saito K, Schnug E, Stulen I (eds) Sulfur metabolism in plants: regulatory aspects, significance of sulfur in the food chain, agriculture and the environment. Margraf Publishers, Weikersheim, pp 89–92

Part III
Sulfur Metabolism – Food Security and Environmental Interactions

Sulfur Nutrition and Food Security

Robert M. Norton

Abstract Food security is an issue that will continue to challenge agricultural production and the distribution of produce. While access to food has improved in the past couple of years the challenges of a changing climate, urbanization, and resource security will require attention to improving productivity within the current land resource. Fertilizers enable half the world's food to be produced. Sulfur is important as an essential plant nutrient that assists with both the quantity and the quality of a range of plant products. It is also critical for animal nutrition. In crops, nitrogen and sulfur nutrition are closely tied and changes in N/S ratio can result in changes in grain quality. This balance, along with the role of nitrogen and phosphorus, indicate that sulfur is a key part of a balanced plant nutrition program to meet future food security. Grains represent the major depletion of sulfur in a national sulfur audit, followed by livestock, and a national audit suggests an annual sulfur removal of 0.4 kg S ha^{-1}, similar to an audit from 1995. This is balanced by the input of sulfur containing fertilizer as well as the use of agricultural gypsum and manures and inputs from sulfur in irrigation water and the atmosphere. Much of the sulfur in soils is present in organic matter, which must be mineralized before plants can access it. Sulfate ions remain in soil solution and are readily leached, and with changing farming systems, the number of sulfur deficient soils is increasing. As a result, there has been a steady increase in the demand for sulfur for crop nutrition. There is a global supply of around 50 Mt of sulfur, with nearly all recovered from S-rich oil and gas with a growth of around 5% per annum predicted. 85% of sulfur is used for sulfuric acid production including the manufacture of a range of sulfur containing fertilizers such as ammonium sulfate and single superphosphate, although newer sulfur fortified products are entering

R.M. Norton (✉)
International Plant Nutrition Institute, 54 Florence St, Horsham, VIC 3400, Australia
e-mail: rnorton@ipni.net

L.J. De Kok et al. (eds.), *Sulfur Metabolism in Plants*, Proceedings of the International Plant Sulfur Workshop 1, DOI 10.1007/978-94-007-4450-9_24,

the market. Attention to using the appropriate source of sulfur at the right rate and at the right time and place will be key strategies to help with sustainable food production now and in the future.

Introduction

Food security is defined by the Food and Agriculture Organisation (FAO) (2010) as when all people, at all times, have physical, social and economic access to sufficient, safe and nutritious food. While they conclude from the 2010 report that the number and proportion of hungry people are declining, hunger remains as a challenge for nearly one billion people, particularly as food prices rise. Indeed, the spike in food prices during the 2008 global food crisis demonstrated how sensitive the world is to a sudden decline in food availability (Sheales and Gunning-Trant 2009). Even though the food price index eased after the 2008 peak, from October 2010 it has climbed back up and by February 2011 it was 6% higher than the 2008 peak (FAO 2010). The index has stayed above the previous 2008 peak for most of 2011 (FAO 2011). This is a problem that will not easily go away.

The grain production for the 2011/2012 crop season is forecast for a record 689 Mt up 3.5 Mt on the 2009/2010 previous record. Demand is also strong but supply increases have meant that ending stocks will be around 4% up at the end of the 2011/2012 season. The current carryover is around 29% of consumption, and this has steadily built up from a low of around 22% since 2007/2008 (United States Department of Agriculture 2011). In terms of food security, the situation for all grains (coarse grains and wheat) is somewhat worse that just for wheat – with stocks at 20% of consumption. Pre-2000 stocks were never less than 22% of consumption but since 2000, stocks have fallen to less that 20% in 5 of 10 years.

Looking forward to 2050 where the global population will reach over nine billion, ensuring food security is a significant challenge, especially against the other issues facing agriculture and land use across the globe. By 2050, global food demand is expected to double, and this is set against a background of static land area, climate change, demand of land for nature and the growth in non-food uses for crops (Millennium Project 2008).

The objective of this chapter is to provide a link between food production and management of sulfur as a plant nutrient of emerging importance to agriculture. It considers the link between fertilizers and food security and presents some of the current issues relating to sulfur management in crop and pasture production.

Linking Food Security to Fertilizers

Productivity growth in agriculture is the key strategy to meet food security and the impact of science has been to improve grain yield per unit land area, with only limited recourse to expanding production areas. Cassman (1999) termed this need for high productivity from the same area of land as ecological intensification and this has delivered increasing food supply as well as preserving land for wildlife and non-agricultural pursuits. Fertilizers play a key role in supporting ecologically sustainable production; with around half the current crop yield attributable to the use of commercial fertilizer (Stewart et al. 2005).

Just as the demand for food increases, the demand for nutrients is steadily increasing with an annual growth rate of +2.5% per year (Heffer and Prud'homme 2010). This steady growth follows a price spike in fertilizers in 2008 that saw prices for DAP rise from around US$250 to around US$800 (free on board, Middle East). Prices for other fertilizers followed that trend but all quickly slipped back after that peak. For example the price for sulfur rose to >US$450 t^{-1} at port in the Middle East but finished around US$175 t^{-1}. The price for nutrients such as sulfur is driven by a large number of supply and demand factors, including international trade policies and energy costs and as such it is not possible to predict actual prices into the future.

While prices are uncertain and volatile, what is certain is that demand for nutrients to meet the increasing demand for food will continue because yield is directly related to the amount of nutrients taken up by a crop. Fertilizers combined with good soil management, the correct varieties and crop protection programs melded together in a complete agronomic package are required to achieve high and profitable yields. A good example is the progress made in maize production in Malawi, which has had a range of policy initiatives to promote maize production over the past 45 years. An analysis of maize production over that period concluded that maize production and hence food security is positively linked to increased fertilizer use through fertilizer subsidies (Mkware and Marsh 2011).

On a global scale, food security cannot be achieved without the effective use of fertilizer nutrients in combination with other nutrients sources such as residues and manures on-farm. In the absence of fertilizers, nutrients removed by the produce sold off the farm are effectively lost from the production system. Balanced and complete nutrient management programs will help all farmers meet the demands of society for nutritious and available food into the future and attention to sulfur nutrition is emerging as an important issue to consider.

Sources and Uses of Sulfur

Sulfur is the thirteenth most abundant element in the earth's crust and commercial sources are derived from ores mined from surface or shallow deposits or as by-products from sulfide ore or sulfates (*e.g.* gypsum). About 75% of the sulfur that is traded is derived from "sour gas" or sulfur rich oil and gas wells. This is done using the Shell-Paques process, where di-hydrogen sulfide is reacted with sodium hydroxide to produce elemental sulfur ("brimstone"). Most of this sulfur is then converted to sulfuric acid, which is the world's most used chemical and is used in many manufacturing and chemical industries. It is used in superphosphate manufacture as well as for the production of N, K and sulfur fertilizers. Other used include the production of cosmetics, pigments, water treatment, pharmaceutical and steel pickling (The Sulphur Institute 2011).

Table 1 Global sulfur supply and consumption, 2000–2008 (Heffer and Prud'homme 2010)

Region	S production (Mt)	S consumption (Mt)	Annual growth (%)
West Europe	45.7	36.3	0.9
Central Europe	12.2	7.7	−1.2
E. Europe and C. Asia	78.1	42.9	1.9
North America	169.1	113.0	−2.8
Latin America	19.2	32.4	4.4
Africa	3.6	65.8	1.5
West Asia (M. East)	69.4	20.2	−1.1
South Asia	6.5	25.4	2.4
East Asia	46.0	96.3	9.4
Oceania	0.8	8.3	9.9
Various	–	2.0	−24.6
Total	**450.5**	**450.5**	

While current oil and gas may have only 50 years of reserves left, the United States Geological Survey suggests that there are around five billion t of sulfur available in tar sands, and an additional 600 billion t in coal and shale. Mean sulfur production (2002–2008) was 46.5 Mt (Heffer and Prud'homme 2010) mostly from North America (36%) and Eastern Europe and Central Asia (17%). Demand is in balance with supply and around 24% of the global production is used in North America. Half the global sulfur supply is traded internationally, with Canada as the largest exporter and China, Morocco and the USA are the largest importers (Heffer and Prud'homme 2010). The world demand for sulfur is increasing by 0.7% annually, and growth is strongest in South Asia and Latin America (Table 1) where agricultural production is also expanding. International Fertilizer Industry Association also suggests that growth will increase globally around 62 Mt by 2014/2015, which is an annual rate of 2.6%, which is consistent with the growth in demand for N, P and K over the same period (Heffer and Prud'homme 2010).

Changing Role of Sulfur

Sulfur is an essential plant and animal macronutrient, along with N, P, K, Ca and Mg, and over the past few decades it has attracted increasing interest from producers and agronomists. This is due to several factors including higher crop yields leading to more sulfur off-take, reduced use of sulfur containing pesticides and slower organic matter turnover under conservation tillage practices (Blair 2002). There are two key changes that have had the largest impact and these are the change to high analysis fertilizers globally, and in the northern hemisphere particularly, a reduction in sulfur deposition from the atmosphere.

In 1921–1922 it was reported from the US mid-west that sulfur deposition in rainfall was around 55 kg S ha^{-1} in a rural area and over 150 kg S ha^{-1} in Chicago (Eaton and Eaton 1926). Over the last few decades, industrial societies have reduced atmospheric pollution by replacing high-S coal and oil with low-S natural gas,

Table 2 Approximate percentage nutrient composition of some common fertilizer products

Product	N (%)	P (%)	K (%)	S (%)
Sulfur bentonite	0	0	0	90
Elemental sulfur	0	0	0	>85
Ammonium polysulfide	20	0	0	40–45
Ammonium thiosulfate	12.0	0	0	26
Ammonium sulfate	20.2	0	0	24
Langbenite	0	0	17	21
Potassium sulfate	0	0	41	18
Magnesium sulfate	0	0	0	14
Gypsum	0	0	0	12–18
Superphosphate	0	8.8	0	11
DAP	18.0	20.0	0	1.6
MAP	10.0	21.9	0	1.5
Triple superphosphate	0	20.7	0	1.0

nuclear energy, and a range of renewable energy sources. These changes in energy use have resulted in a major reduction in sulfur deposition from atmospheric sources, often termed "acid rain" because of the unnaturally low pH of the rainwater. Atmospheric SO_2 concentration across much of Europe has declined, for example from around 4 μg S m^{-3} to less than 1 μg S m^{-3} between 1980 and 2000 (Norwegian Institute for Air Research 2003, as reported in McNeill et al. 2005). In turn this has led to a reduction in deposition from, from 1 kg S ha^{-1} $year^{-1}$ in rural areas of northern Norway to more than 20 kg S ha^{-1} $year^{-1}$ in industrial areas, with an average of 2–10 kg S ha^{-1} $year^{-1}$ in many areas (McGrath et al. 2002).

The reduced atmospheric input has been coupled with cropping system changes that have resulted in growing crops with a greater sulfur requirement, such as canola (oilseed rape, *Brassica napus*), lucerne (*Medicago sativa*), pulse crops (*eg* peas, *Psium sativum*) or cruciferous forages (*Brassica* spp.). Other crops such as wheat (*Triticum aestivum*), sugar beet (*Beta vulgaris*) and peanut (*Arachis hypogaea*) are generally considered to have a low sulfur demand (Spencer 1975). For example, a 5 tha^{-1} wheat crop will remove around 6 kg S ha^{-1} while a 3 tha^{-1} canola crop will remove 15 kg S ha^{-1} (Reuter and Robinson 1997). In general, crop yields overall have increased as a result of improved agronomic practice and superior varieties, which in turn increases the overall demand on nutrients including S.

There has also been a reduction in the use of sulfur containing pesticides (Blair 2002) and a general decline is soil organic matter and in consequence organic S. The decline in soil organic matter to a new equilibrium is a consequence of accelerated breakdown of humus mainly due to tillage and the removal of N and other nutrients in harvested products (Connor et al. 2011). Soil organic sulfur fractions are highly correlated with organic carbon and organic nitrogen contents and the ratio of organic forms of C/N/S from 976 soils from across the globe was around 100/8.3/1.4 (Kirkby et al. 2011), which was a similar ratio to an earlier report from Scotland (Williams et al. 1960). So, any reduction in soil organic matter is likely to reduce the overall supply of S.

Single superphosphate (9% P, 11% S, Table 2) was first developed by Lawes and Gilbert in 1843 in the United Kingdom (Rothamsted Research 2006) and the process patented by Lawes involved the acidulation of low solubility P sources with sulfuric acid. Even though P was considered the main nutrient supplied, single superphosphate also supplied plant available sulfur and calcium. The use of superphosphate was one of the great innovations and it was pivotal in the development of modern, productive agricultural systems in places such as Australia (Donald 1965). However, with more intensive and efficient farming systems there has been a trend away from using single superphosphate. This trend has been due to a range of factors including a reduction in the availability of phosphate rock suitable for superphosphate production. Current world phosphate fertilizer production and use is mainly ammoniated phosphates in particular mono-ammonium phosphate (MAP) and di-ammonium phosphate (DAP). In Australia, for example, 68% of phosphorus is applied as DAP or MAP and 27% as single superphosphate (Fertilizer Industry Federation of Australia 2010). Indeed most annual crops are now sown with low sulfur fertilizers such as MAP, DAP or triple superphosphate, mainly to increase the efficiency in transport and handling due to their higher nutrient density. As well, the share of ammonium sulfate (20% N, 24% S) in the total European N market has fallen from 7.2% in 1973 to 3% in 1991 (Ceccotti and Messick 1994) and is now a little less than 4% of total N used globally (Heffer and Prud'homme 2010).

In 1995 and 1996 a nutrient balance by region was constructed across Australia. In most regions, sulfur inputs from fertilizer, soil amendments, water and atmosphere was 10–40 kg S ha^{-1} $year^{-1}$ while exports were generally less than 10 kg S ha^{-1} $year^{-1}$. Sulfur balance varied from being slightly positive (2–5 kg S ha^{-1} $year^{-1}$) to being very highly positive (20–50 kg S ha^{-1} $year^{-1}$), but in southern and central Queensland, there were highly negative balances. These values were recalculated for the period 2005–2009 using figures from the state production figures (Australian Bureau of Agriculture and Resource Economics and Science 2010), it was estimated that 113 kt S $year^{-1}$ was removed in farm produce. The state fertilizer use figures (Fertilizer Industry Federation of Australia 2010) indicate that around 245 kt S $year^{-1}$ is applied, so that Australia is notionally in a positive sulfur balance. Based on the areas fertilized, this is a continental surplus of +2.9 kg ha^{-1} $year^{-1}$. Not included in this balance was added sulfur from mined or by-product gypsum (3.5 Mt mined annually, United States Geological Survey 2011), the atmospheric input (4.5 ± 2.1 kg S ha^{-1} $year^{-1}$, National Heritage Trust 2001) or sulfur input from irrigation. For the latter, the actual sulfur input depends on the position of the supply in the water shed, with areas lower down the water shed having significantly more sulfur input that areas at the top of the water shed.

Despite this apparent surplus, the National Land and Water Audit (National Heritage Trust 2001) reported that the amount of plant available sulfur (as measured by KCl40, Blair et al. 1991) was less than a critical value of 5 mg/kg on 11% of soils across Australia, and was below that level on 25% soils in New South Wales and only on 3% of soils in Victoria. More recent data from the Nutrient Advantage soil testing laboratories are shown in Table 3, with over half the samples submitted showing KCl40 levels below 8 mg/kg. While these values are low, they do not account for subsoil sulfur supply, which can mean that a site – even with a low soil test – will not respond to additional S.

Table 3 2010 soil S (KCl-40) test values, top 10 cm from 1,200 samples collected across Victoria, South Australia and New South Wales

KCl-40 (mg kg^{-1})	Crop (%)	Pasture (%)
<8	52	43
8–12	20	30
>12	28	27

Such regional nutrient balances, coupled to regional soil test values give a good picture of the trends in nutrient use, and can be used to highlight areas where excess nutrient is being added or where removals are in excess of the soils capacity to sustain ongoing production. Despite this importance, there are few such regional or national monitoring programs, other than NuGIS (Nutrient Use Geographic Information Systems) managed by the International Plant Nutrition Institute for the continental United States of America (International Plant Nutrition Institute 2010).

Sulfur as a Plant Nutrient

Like all essential nutrients, sulfur is an important part of a balanced plant nutrition program. Crops grown on sulfur deficient soils suffer reduced yield as well as poor product quality. Sulfur is a present in the essential amino acids cysti(e)ne and methionine which cannot be synthesized by higher animals (Brosnan and Brosnan 2006). Methionine is an initiating amino acid in many eukaryotic proteins, and cyst(e)ine plays a critical role in protein structure through the formation of disulfide bonds. Sulfur rich compounds have also been identified as conferring a range of health benefits including the organic compounds in garlic (allyl cysteine sulfopoxide, Rahman 2003) and onion (capenes and thiosulfates, Dorsch and Wagner 1991), and the glucosinolate compounds in brassicas (Lund 2003). They also make things taste good!

Nitrogen and sulfur metabolism are largely interdependent so that plants maintain a relatively constant ratio of organic N to organic sulfur (Zhao et al. 1999). For wheat, an N:S ratio in grain of around 17 can be used to retrospectively diagnose crop sulfur status (Randall et al. 1981). The relative proportions of nitrogen and sulfur determine protein structure and type, and this is one of the main reasons that sulfur nutrition affects grain and baking quality (Moss et al. 1981; Zhao et al. 1999). A higher sulfur supply increases the glutenin content, which in turn affects dough extensibility and bread making quality (Byers et al. 1987; Weegel et al. 1996).

Plants take up sulfur predominantly as sulfate (SO_4^{2-}) anions from the soil solution (Alves and Lavorenti 2004), although some sulfur can be assimilated as gaseous sulfur dioxide (SO_2) of hydrogen sulfide (H_2S) through leaves although it is only a significant nutrient source in industrial areas with air pollution (Leustek and Saito 1999). Sulfate is actively taken into the roots, and then transported through the

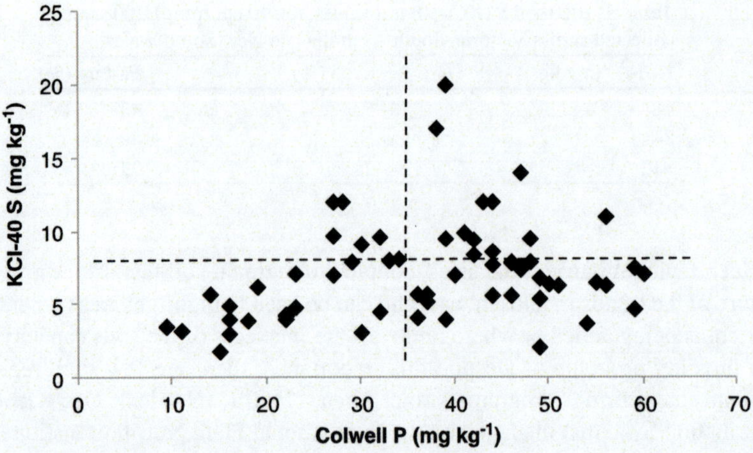

Fig. 1 Soil test values for available sulfur (*KCl-40*) and phosphorus (*Colwell*) in the top 10 cm of soils from the New England Tablelands of New South Wales, Australia. *Horizontal* and *vertical lines* represent the critical soil test values for the KCl-40 and Colwell P tests respectively (Gourley et al. 2007). Data courtesy of Nutrient Advantage Laboratories, 2011

plant mostly unmetabolized and is ultimately either directly incorporated into plant metabolites or reduced and incorporated into cysteine to then go into other metabolic pathways. Under S-limiting conditions, the efficiency of remobilization of vacuole stored sulfate and other pools of reduced sulfur can be low, so that during later growth or grain filling, low tissue sulfur contents may result (Hawkesford 2000). Typically, therefore, the symptoms of sulfur deficiency occur first in the younger tissues and are seen as leaves and veins turning pale green to yellow. These symptoms are similar to those of nitrogen deficiency but symptoms of the latter, because of its higher mobility are more obvious in the older tissues first.

Sulfur in the soil is present in several different pools and the work of Australian researchers A. Ray Till and Graeme Blair has been important in understanding the dynamics and importance of these pools in agricultural production systems. Most of the sulfur is present in organic matter, which has to be mineralized to release sulfate for plant uptake, and the release rate varies with the nature of the particular fraction (Till 2010). Some volcanic soils also contain elemental sulfur (S^0), which must be oxidized by aerobic microbes to sulfate before being available for plant uptake (Till 2010). Also on acid-sulfate soils, sulfides (S^{2-}) can predominate (Rayment and Higginson 1992).

Sulfate forms sparingly soluble gypsum in arid regions but can be leached through soils in wetter environments. In the soil solution, sulfate is highly mobile and only weakly held on colloidal particles. It is easily leached and so can move out of the root zone of crops and pastures, and losses of up to 100 kg S ha^{-1} have been measured at in southern England (Williams 1975). Figure 1 shows the differential retention of sulfur and phosphorus in a set of soil samples taken from pastures in the New England region of Australia that have a history of the application of single superphosphate.

In these samples, 46% of samples were below a critical P value, but 74% showed soil sulfur levels below the critical value. This indicates that much of the sulfur applied has leached and that even if a paddock has a reasonable superphosphate history and a reasonable soil test P level, the sulfur levels can still be low. Sulfate is also taken into microorganisms and plants to become fixed then into soil organic matter.

Because sulfate is an anion, it is not strongly adsorbed in surface soils, and sulfate-S losses in runoff are usually small (White 2006). High sulfate loads in rivers and groundwater can lead to higher sulfur concentrations at the discharge points which can contribute to eutrophication where other nutrients are present. Where iron is poor in these waters and with high organic matter, sulfide can accumulate leading to sulfide toxicity and increased phosphate mobilization in anaerobic sediments, and can also increase the production of sulfides under reducing conditions (Smolders and Roelofs 1993).

Sulfur in Crop Management

Sulfur is only one part of an integrated crop production package that includes other nutrients as well as well adapted varieties and good agronomy, but when combined, can result in significant yield improvements. For example, an integrated agronomy program for maize intensification in Mozambique has demonstrated a fourfold yield increase (Fig. 2, International Fertilizer Development Centre 2009), which in this case included an added sulfur response. The development of fertilizer best

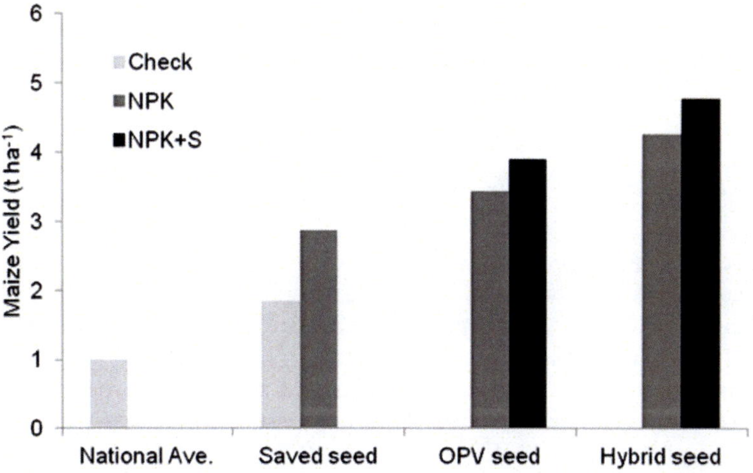

Fig. 2 Agronomic and genetic improvement in maize production in Mozambique. Data are the means from seven locations in 2008–2009. Values are for the national average yield, farmer saved seed, open pollinated varieties (*OPV*) and hybrid varieties. Nutrients applied were 106 kg N ha^{-1}, 32 kg P ha^{-1}, 31 kg K ha^{-1} and 18 kg S ha^{-1} (IFDC 2009)

management practices is a key to achieving those types of results, and this approach looks to apply the right source (or product) at the right rate, right time and right place, a strategy call the "4R's Nutrient Stewardship Framework" (International Fertilizer Industry Association 2009). Under the global 4R's approach, the four "rights" are linked and convey how fertilizer applications can be managed to achieve economic, social and environmental goals. The management of fertilizer sulfur application is no different, and as new products and strategies are developed, they can be considered in the source, rate, time, and place framework (Bruulsema et al. 2009).

Sulfur nutrition is particularly tricky because if the high mobility of the sulfate ion and the number of interactions between sulfur and other nutrients. The co-limitation of nitrogen and sulfur has been mentioned, but sulfur fertilization is also reported to induce molybdenum deficiency due to sulfate-molybdate antagonism during plant uptake (Stout and Meagher 1948; Macleod et al. 1997) due to competition for the same proton coupled symporters (Fitzpatrick et al. 2008; Shinmachi et al. 2010). Sulfur and selenium also are antagonistic for essentially the same reason, such that on soils with high selenium levels, sulfur fertilization reduced the pasture selenium content (Pratley and McFarlane 1974). It is unclear if selenium is an essential element in all plants but selenium is required by animals to prevent muscular dystrophy (Gowariker et al. 2009). Sulfur fertilization is also reported to reduce the uptake of boron, particularly where soil levels of boron are relatively low (Haneklaus et al. 2006). Sulfur fertilization can also stimulate the uptake of some micronutrients such as copper, manganese, nickel and cadmium, probably due to the rhizosphere acidification that can result from the application of elemental sulfur (Haneklaus et al. 2005). McDonald and Moschen (2009) have also indicated that sulfur can enhance the uptake of zinc even under conditions where the latter is adequate for growth.

There is also interest in the interaction of salinity and sulfur uptake, but Grattan and Greive (1999) concluded that the differences in plant responses between chloride and sulfate salinity may depend on the salinity indices used. Many crops are sensitive to high internal chloride levels.

Selecting the Right Rate for Sulfur

The benefits of applying additional sulfur will only accrue when other nutritional and non-nutritional constraints are met, and sulfur becomes the limiting or co-limiting factor. The most common ways to make the assessment of whether sulfur is likely to be limiting is to use a soil test that extracts a fraction of the soil solution and adsorbed sulfur that can then be correlated with plant response to added fertilizer. Various extractants and conditions have been evaluated including monocalcium orthophosphate (MCP), MCP with carbon (CPC) (Peverill and Briner 1974), potassium chloride at two temperatures (KCl at 40°C or at 100°C) or sodium bicarbonate ($NaHCO_3$), potassium phosphate (Watkinson and Kear 1996) as well as assays based on total sulfur content (Rayment and Higginson 1992). Mineralizable organic sulfur can also be assessed using a short-term incubation (Watkinson and

Kear 1996). Under Australian conditions, the KCl-40 test has been the most widely used and is the basis of most soil test recommendations (Blair et al. 1991). The CPC-S is also used diagnostically, although the two sulfur tests are not correlated (Gourley et al. 2007).

There has been some discontent with the KCl-40 test, especially with the need to evaluate potential responses for canola. The current test interpretations are based on a 10 cm sample although it is known that there are large differences in sulfur down a soil profile (Blair et al. 1997). Brennan and Bolland (2006) found that shallow soil tests were overestimating the magnitude of a sulfur deficiency for canola production on the sandy soils in Western Australia because of relatively high levels in the 10–30 cm layers for a significant number of soils. Wong and Wittwer (2009) found that there was significant positive charge and an accumulation of sulfate in Western Australian subsoils. It is therefore recommended to take deeper soil samples to account for stratification of sulfur that leaches down the profile. At present there are no critical values for these deeper tests.

The KCl-40 extraction provides an assessment of ester sulfates (Blair et al. 1991), but it may be that consideration needs to be given to estimating mineralisation of sulfate from organic matter (Watkinson and Kear 1996), similar to the general procedure used to estimate nitrogen demand and supply (Norton 2011). Also, the presence of elemental sulfur may not be adequately considered by the present testing methods. New tests incorporating deeper samples, organic and elemental sulfur pools and better estimates of seasonal demand have been proposed for development in Australia (Grains Research and Development Corporation 2011).

Sulfur uptake and plant sulfur status can be assessed through plant analyses, including the total plant sulfur, sulfate sulfur, total sulfur to sulfate ratio, and the nitrogen to sulfur ratio. Pinkerton et al. (1993) found that for canola, seed yield was well predicted by whole shoot sulfur content, and estimated the critical concentration of the whole shoots to be 0.36– 0.40% S at the buds visible stage. Spencer and Freney (1980) reported the critical sulfur value of whole shoot as 0.30% at mid-tillering, and that these values were likely to be lower in N deficient plants (Reuter and Robinson 1997).

While plant analyses may provide a diagnosis of the nutrient status of a plant at any point in time, the critical values established can change rapidly with crop phenostage, and due consideration should be given to the depth of root penetration when sampled. For sulfate and other mobile nutrients, plants may undergo a temporary shortage of nutrients, but then reach added supply as the roots move deeper into the soil.

Grain analysis is a possible strategy to retrospectively assess the supply of sulfur over the whole growth cycle of the plant, and the ratios of N/S can be a useful indicator of nutrient management strategies employed (Randall et al. 1981). A collection of wheat from two cultivars grown in 70 locations in the National Variety Testing program from south-eastern Australia was assessed for N and sulfur using Inductively Coupled Plasma Optical Emission Spectroscopy and these results showed that – in general – most of the grain was above the critical sulfur content of 0.12% and less than the 15:1 N:S ratio of Randall et al. (1981; Fig. 3). These data

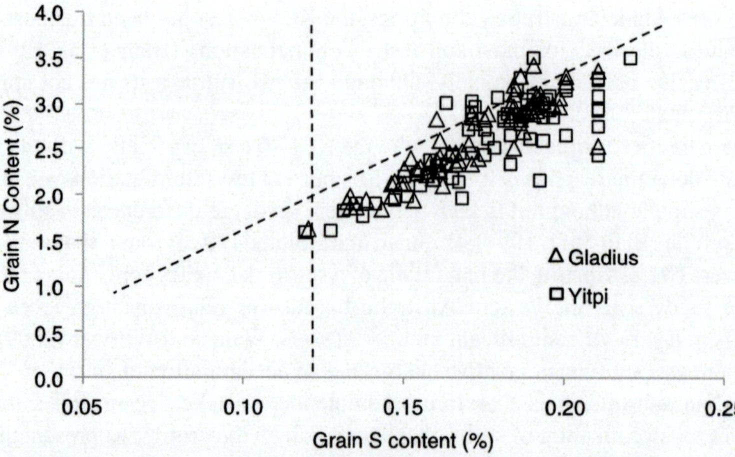

Fig. 3 Grain N and S contents of two wheat cultivars Yitpi and Gladius from 70 sites across south-eastern Australia, graphed along with the critical values for sulfur content (0.12%) and the N/S ratio of 17 proposed by Randall et al. (1981). Data of the author

suggest that normal agricultural practice in the relatively low yielding environments in Australia is generally able to meet the sulfur requirements as well as balancing the sulfur demand with appropriate N management.

Based on experimental work in Australia, Good and Glendinning (1998) suggested that where soil tests were low, 40 kg S ha^{-1} would be adequate for a good canola crop, while with higher soil tests or even with a good history of single superphosphate, 20 kg S ha^{-1} would be recommended.

Selecting the Right Time and Place to Apply Sulfur

Timing and position of nutrients are intimately related. At sowing is really the only time that fertilizer and seed can be placed together, but due consideration should be given to potential phytotoxic effect from the fertilizer, and damage will be worse with high application rates and wide seed rows on poorly buffered and dry soils. Generally, sulfur fertilizers themselves present a moderate damage potential due to the acidifying effect on the soil, and fluid sources such as ammonium thiosulfate or ammonium polysulfide have a high potential to cause crop damage (Holloway et al. 2008) and so are used in fertigation systems rather than as dry materials. Other than care with seed and fertilizer distances, the critical aspect of sulfur placement is to ensure the nutrient is where the roots can access it. As sulfate is very mobile, it will move down the profile and may become depleted in the root zone before the roots proliferate. Controlling release of sulfate will have some benefits in controlling potential losses down the profile, although the release rates need to be in synchrony with plant demand to avoid nutrient stress.

In field grown crops with high sulfur demands such as canola, the symptoms of sulfur deficiency are often not apparent until the period of rapid growth during stem elongation (Pinkerton et al. 1993). Pot grown canola has been shown not to recover from early sulfur stress (Pinkerton and Hocking 1987) although in the field recovery is seen from late applications. Hocking et al. (1996) reported that canola top-dressed with 40 kg S ha^{-1} at rosette stage, flower buds visible or stem elongation produced similar seed yields and oil contents to crops where all the sulfur was applied at sowing. Therefore, timing of sulfur supply is relatively flexible and later applications can result in good yields.

In general, granulated fertilizer such as ammonium sulfate or potassium sulfate can be safely spread over a growing crop. However, application of fluid sources such as ammonium thiosulfate may need to be done with specialized applicator or applied as mixtures with irrigation water through a trickle irrigation system.

Selecting the Right Sulfur Product

Single superphosphate has been the major source of sulfur for many crops and pastures, but its low nutrient density makes if relatively more expensive than higher analysis products due to transport costs. There are a range of sulfur containing evaporites that differ in the proportion and type of associated cations. For example potassium sulfate can be separated from Langbenite (potassium magnesium sulfate), Kieserite (magnesium sulfate) and Sylvenite (sodium and potassium chlorides). These evaporites can be used as mixed fertilizers supplying essential nutrients as well as sulfur. Ammonium sulfate is another soluble sulfur source that is mostly a co-product or by-product from a range of industrial processes. Ammonium sulfate has been a commonly used sulfur source as it supplies both nitrogen and sulfur and has good crop safety properties and when granulated can be spread evenly using mechanical fertilizer spreaders. However, as with all sulfate-sulfur sources, there are likely to be leaching losses in coarse textured soils and it is also a relatively low-analysis product with 24% S (Table 2). Fluid sulfur fertilizers such as the thiosulfates (ammonium, potassium, magnesium or calcium) and ammonium polysulfide have application mainly in fertigation systems or applied directly to soil, although these materials have some functional properties that make them of interest in cropping systems (Holloway et al. 2008). In particular, root zone acidification with these materials could release bound phosphorus and increase supply near the time and place of application. It is also reported that thiosulfates can slow the rate of both urea hydrolysis and nitrification when mixed with urea ammonium nitrate solutions (IPNI 2011), which enhances the efficiency of nitrogen delivery.

Gypsum (calcium sulfate) is often used on sodic soils to supply calcium to reduce the exchangeable sodium percentage and so promote soil structure. Gypsum has a relatively low solubility, so that rates of 500 kg gypsum ha^{-1} (~60 kg S ha^{-1}) will require around 100 mm of rainfall to make the sulfur plant available.

There is also interest in low solubility, pelletized, neutralized low grade sulfuric acid effluent from mineral sands processing (such as from Western Australia). This material has been shown to release sulfur at rates similar to superphosphate and more slowly than coarse gypsum (Summers et al. 2003). This product was evaluated on canola in Western Australia and found to release sulfate at a rate similar to gypsum, but was less prone to leaching (Brennan et al. 2010).

Elemental sulfur the most concentrated form of sulfur, but it is not water soluble and must be oxidized to sulfate before it can be taken up by plant roots. The speed of this microbial process is governed by environmental factors such as soil temperature and moisture, as well as the physical properties of the sulfur (Yang et al. 2010). Oxidation rate is also affected by particle size, and the rate of release can be adjusted by varying the proportions of different sized particles (McCaskill and Blair 1988). Large sulfur particles can take months or even years to oxidize, but very fine particles are difficult to spread and sulfur dust is an explosion hazard during storage and handling.

To enhance the rate of sulfur oxidation, while still maintaining handling and storage properties, a small amount of clay is added to the molten sulfur prior to cooling to form small pellets or pastilles. When wetted in soil, the clay swells and the finely divided sulfur particles oxidize as the pastille disintegrates.

Elemental sulfur has been successfully blended with single superphosphate, triple superphosphate or rock phosphate to provide a slow release sulfate source particularly for pastures on coarse textured soils (Yeates and Clarke 1993). There have also been sulfur coats added to other fertilizers such as urea, but more recent technology has seen very thin layers of elemental sulfur incorporated evenly through the fertilizer granule during manufacturing, so that products like MAP can have sulfate and elemental sulfur incorporated at various ratios for particular situations. The sulfur is oxidized as the granule dissolves, becoming available for plant uptake. A side benefit of this process is that some micronutrients such as iron and zinc become more soluble as the result of the acidifying oxidation process. There are at least two different patented processes that incorporate microfine elemental sulfur and other materials into various fertilizers, which then have designed sulfate release patterns to meet particular crop requirements.

Undoubtedly, new products will come onto the market designed to better match sulfate release to plant demand so reducing losses by leaching as well as reducing losses of sulfate by fixation into organic matter. When this happens, new strategies incorporating the 4R's approach to nutrient stewardship will need to be developed to meet the challenge of high productivity farming systems with minimum environmental impact.

Conclusions

While there have been improvements in food security, continued gains in food production and enhanced food quality are still an imperative of agricultural science. Due to previous inputs from the burning of fossil fuels, sulfur was not often considered as yield limiting nutrient. However as low sulfur fuels develop and farming systems change, it is clear that addressing sulfur nutrition is an important aspect of ensuring food security and this will continue into the future.

There are adequate sulfur resources to meet demand well into the future and there are also good tools to assist with selecting the right rate for applying sulfur, although some adjustments need to be made to account for different soil depths and nutrient mobility. There are a range of sulfur products that are coming onto the market that will enable sulfur to be applied at the right time and in the right place.

Acknowledgements The author would like to acknowledge the soil test data supplied by the Nutrient Advantage Laboratories (Mr. Phil Hoult), and the support and discussions with Mr. Charlie Walker (Incitec Pivot Fertilizers) and Dr. Sam Stacey (ex-Adelaide University). I would also like to acknowledge input through discussion with Prof. Mike McLaughlin (CSIRO/ University of Adelaide) as well as my IPNI colleague Dr. Rob Mikkelsen.

References

Alves ME, Lavorenti A (2004) Sulfate adsorption and its relationships with properties of representative soils of the São Paulo State, Brazil. Geoderma 118:89–99

Australian Bureau of Agricultural and Resource Economics and Sciences (2010) Australian commodity statistics. Australian Bureau of Agricultural and Resource Economics and Sciences, Canberra

Blair GJ (2002) Sulfur fertilisers: a global perspective. Proceedings No.498. International Fertiliser Society, York

Blair GJ, Chinoim N, Lefroy RDB, Anderson GC, Crocker GJ (1991) A soil sulfur test for pastures and crops. Aust J Soil Res 29:619–626

Blair GJ, Anderson G, Crestani M, Lewis D (1997) Soil sulphur status, sulphur fertiliser responses and rainfall sulphur accessions trial sites in the National Reactive Phosphorus Rock Project. Aust J Exp Agric 37:995–1001

Brennan RF, Bell RW, Raphael C, Ealick H (2010) Sources of sulfur for dry matter, seed yield and oil concentration of canola grown in sulfur deficient soils of south-western Australia. J Plant Nutr 33:1180–1194

Brosnan JT, Brosana ME (2006) The sulfur-containing amino acids: an overview. J Nut 136:1636S–1640S

Bruulsema T, Lemunyon J, Herz B (2009) Know your fertilizer rights. Crop Soil 42:13–16

Byers M, Franklin J, Smith SJ (1987) The nitrogen and sulphur nutrition of wheat and its effect on the composition and baking quality of the grain. Aspects Appl Biol, Cereal Qual 15:337–344

Cassman KG (1999) Ecological intensification of cereal production systems: Yield potential, soil quality, and precision agriculture. Proc Natl Acad Sci USA 96:5952–5959

Ceccotti SP, Messick DL (1994) Increasing plant nutrient sulphur use in the European fertilizer industry. Agro-Food-Industry Hi-Tech 5:9–14

Connor DJ, Loomis RS, Cassman KG (2011) Crop ecology: productivity and management in agricultural systems, 2nd edn. Cambridge University Press, Cambridge

Donald CM (1965) The progress of Australian agriculture and the role of pastures in environmental change. Aust J Sci 27:187–198

Dorsch W, Wagner H (1991) New antiasthmatic drugs from traditional medicine? Int Arch Allergy Appl Immunol 94:262–265

Eaton SV, Eaton JH (1926) Sulfur in rainwater. Plant Physiol 1:77–87

Fertilizer Industry Federation of Australia (2010) 2009 fertilizer sales statistics. The Fertilizer No.7, April 2010. FIFA, Canberra

Food and Agriculture Organisation (2010) The state of food insecurity in the world. Food and Agriculture Organization of the United Nations, Rome

Food and Agriculture Organisation (2011) FAO food price index. http://www.fao.org/worldfoodsituation/wfs-home/foodpricesindex/en/. Accessed 22 Dec 2011

Good AJ, Glendinning J (1998) Canola needs sulfur. Better Crop 12:6–8

Gourley CJP, Melland AR, Waller RA, Awty IM, Smith AP, Peverill KI, Hannah MC (2007) Making better fertilizer decisions for grazed pastures in Australia. Victorian Government, Department of Primary Industries, Melbourne, 18 pp

Gowariker V, Krishnamurthy VN, Gowariker S, Dhanorkar M, Paranjape K (2009) The fertilizer encyclopedia. John Wiley & Sons Inc. Hoboken, Jew Jersey

Grains Research and Development Corporation (2011) Investment plan(open and multi-stage tenders), 2012–13. GRDC, Canberra

Grattan SR, Grieve CM (1999) Salinity-mineral nutrient relations in horticultural crops. Sci Hortic 78:127–157

Haneklaus S, Bloem E, Hayfa S, Schnug E (2005) Influence of elemental sulphur and nitrogen fertilisation on the concentration of essential micronutrients and heavy metals in *Eropaeolum majus* L. FAL – Agric Res 286:25–35

Hawkesford M (2000) Plant responses to sulfur deficiency and the genetic manipulation of sulfate transporters to improve S utilization efficiency. J Exp Bot 501:131–138

Heffer P, Prud'homme M (2010) Fertilizer outlook 2010–2014. International Fertilizer Industry Association (IFA), Paris, 11pp

Hocking PJ, Pinkerton A, Good A (1996) Recovery of field grown canola from sulfur deficiency. Aust J Exp Agric 36:79–85

Holloway B, McLaughlin M, McBeath T (2008) Fluid fertilizers: A South Australian manual. Fluid Fertilizer Foundation. http://www.fluidfertilisers.com.au/index.php?page=newsletter. Accessed 22 Dec 2011

International Fertilizer Development Centre (2009) Maize Intensification in Mozambique (MIM) project. Economic analysis 2008–2009. Interim project report prepared for the international plant nutrition, the International Fertilizer Industry Association, the International Potash Institute, and The Sulphur Institute by the International Fertilizer Development Center. October 2009 (unpublished data)

International Fertilizer Industry Association (2009) The global "4R" nutrient stewardship framework. AgCom/09/44, International Fertilizer Industry Association (IFA), 10 pp

International Plant Nutrition Institute (2010) A preliminary nutrient use geographic information system (NuGIS) for the US. IPNI Publication No. 30-3270, Norcross

International Plant Nutrition Institute (2011) Nutrient source specifics, No 8. Thiosulfate. International Plant Nutrition Institute, Norcross

Kirkby CA, Kirkegaard JA, Richardson AE, Wade LJ (2011) C:N:P:S ratios in soil humus and implications for soil organic matter sequestration. In: Proceedings of the Grains Research and Development Corporation Advisor Updates, Adelaide, pp 215–219, 23 & 24 Feb 2011

Leustek T, Saito K (1999) Sulfate transport and assimilation in plants. Plant Physiol 120:637–643

Lund E (2003) Non-nutritive bioactive constituents of plants: dietary sources and health benefits of glucosinolates. Int J Vitam Nutr Res 73:135–143

Macleod JA, Gupta UC, Stanfield B (1997) Molybdenum and sulphur relationships in plants. In: Gupta UC (ed) Molybdenum in agriculture. Cambridge University Press, Cambridge, United Kingdom

McCaskill MR, Blair GJ (1988) Development of a simulation model of sulfur cycling in grazed pastures. Biogeochemisty 5:165–181

McGrath RG, Zhao FJ, Blake-Kalff MMA (2002) History and outlook for sulphur fertilizers in Europe. Proceedings No. 497. International Fertilizer Society, York, UK

McNeill AM, Eriksen J, Bergstrom L, Smith KA, Marstorp H, Kirchmann H, Nilsson I (2005) Nitrogen and sulphur management: challenges for organic sources in temperate agricultural systems. Soil Use Manag 21:82–93

Millennium Project (2008) Global Challenges for Humanity, State of the Future, Millennium Project. http://www.millennium-project.org/millennium/challeng.html. Accessed 22 Dec 2011
Mkware B, Marsh D (2011) Effects of maize fertilizer subsidies on food security in Malawi. Department of Economics working paper in Economics 14/11, Waikaito University, New Zealand, 18 pp
Moss HJ, Wrigley CW, MacRitchie F, Randall PJ (1981) Sulfur and nitrogen fertilizer effects on wheat. II. Influence on grain quality. Aust J Agric Res 32:213–226
National Heritage Trust (2001) Nutrient balance in regional farming systems and soil nutrient status. National Land and Water Resources Audit Project, Version 1.1, May 2001. National Heritage Trust, Canberra, 89 pp
Norton RM (2011) 4R approach to nutrient best management practice, Nitrogen application in wheat in south-eastern Australia. International Plant Nutrition Institute, Norcross, 8 pp
Peverill KI, Briner GP (1974) Correlation for soil tests for sulfur with pasture response in the national soil fertility pasture project in Victoria. In: Australian soil science conference working papers, La Trobe University, Melbourne. Canberra. Soil Science Society of Australia, Canberra, pp 3(a)-1–3(a)-5
Pinkerton A, Hocking PJ (1987) Effects of applying sulfur to sulfur-deficient linseed and rapeseed. In: Proceedings of the 4th Australian agronomy conference. La Trobe University, Melbourne, VC, p 267 (Australian Society of Agronomy, Parkville, VC)
Pinkerton A, Hocking PJ, Good A, Sykes J, Lefroy RBD, Blair GJ (1993) A preliminary assessment of plant analysis for diagnosing S deficiency in canola. In: Proceedings of 9th Australian research assembly on Brassicas, Wagga Wagga, pp 21–28
Pratley JE, McFarlane JD (1974) The effect of sulfate on the selenium content of pasture plants. Aust J Exp Agric Anim Husb 14:533–538
Rahman K (2003) Garlic and aging, new insights into an old remedy. Ageing Res Rev 2:39–56
Randall PJ, Spencer K, Freney JR (1981) Sulfur and nitrogen fertilizer effects on wheat. I. Concentrations of sulfur and nitrogen and nitrogen to sulfur ratio in grain in relation to the yield response. Aust J Agric Res 32:203–212
Rayment GE, Higginson FR (1992) Australian laboratory handbook of soil and water chemical methods. Inkata Press, Melbourne, 314 pp
Research R (2006) Guide to the classical and other long-term experiments, datasets and sample archive. Lawes Agricultural Trust, Harpenden, 52 pp
Reuter DJ, Robinson JB (1997) Plant analysis: an interpretative manual, 2nd edn. CSIRO Publishing, Collingwood
Sheales T, Gunning-Trant C (2009) Global food security and Australia, ABARE report, Canberra, December
Shinmachi F, Buchner P, Stroud JL, Parmar S, Zhao FJ, McGrath SP, Hawkesford MJ (2010) Influence of sulfur deficiency on the expression of specific sulfate transporters and the distribution of sulfur, selenium and molybdenum in wheat. Plant Physiol 153:327–336
Smolders A, Roelofs JGM (1993) Sulphate-mediated iron limitation and eutrophication in aquatic ecosystems. Aquat Bot 46:247–253
Spencer K (1975) Sulfur requirements of plants. In: McLachlan KD (ed) Sulfur in Australasian agriculture. Sydney University Press, Sydney, pp 98–108
Spencer K, Freney JR (1980) Assessing the sulfur status of field grown wheat by plant analysis. Agron J 72:469–472
Stewart WM, Dibb DW, Johnston AE, Smyth TJ (2005). The contribution of commercial fertilizer nutrients to food production. Agron J 97:1–6
Stout PR, Meagher WR (1948) Studies of the molybdenum nutrition of plants with radioactive molybdenum. Science 108:471–473
Summers R, Clarke M, McCafferty P, Pope T (2003) A slowly soluble sulfur fertilizer from a by-product of mineral sands processing. Aust J Soil Res 41:1213–1227
The Sulphur Institute (2011) Sulphur 101, The Sulphur Institute. http://www.sulphurinstitute.org/learnmore/ sulphur101.cfm. Accessed 22 Dec 2011
Till AR (2010) Sulfur and sustainable agriculture. International Fertilizer Industry Association, Paris, France, 71 pp

United States Department of Agriculture (2011) World agricultural supply and food demand estimates (WASDE – 501), Dec 2011 USDA, 40 pp

United States Geological Survey (2011) Gypsum. U.S. Geological Survey, Mineral Commodity Summaries, Jan 2011, p 70

Watkinson JH, Kear MJ (1996) Sulfate and mineralisable organic sulfur in pastoral soils of New Zealand. II. A soil test for mineralisable organic sulfur. Aust J Soil Res 34:405–412

Weegel PL, Hamer RJ, Schofield JD (1996) Functional properties of wheat glutenin. J Cereal Sci 23:1–18

White RE (2006) Principles and practice of soil science: the soil as a natural resource, 4th edn. Blackwell Publishing, Malden, MA, USA

Williams RJB (1975) The chemical composition of water from land drainage at Saxmundham and Woburn (1970–1975). Rothamstead Experimental Station report, Part 2, pp 37–62

Williams CH, Williams EG, Scott NM (1960) Carbon, nitrogen, sulphur and phosphorus in some Scottish soils. J Soil Sci 11:334–346

Wong MT, Wittwer K (2009) Positive charge discovered across Western Australian wheat belt soils challenges key soil and nitrogen management assumptions. Aust J Soil Res 47:127–135

Yang Z-H, Stoven K, Haneklaus S, Singh BR, Schnug E (2010) Elemental sulfur oxidation by *Thiobaccillus* spp and aerobic heterotrophic sulfur-oxidising bacteria. Pedosphere 20:71–79

Yeates JS, Clarke MF (1993) Developing alternatives to phosphate fertilizers of high water solubility. Fert Res 36:141–150

Zhao FJ, Hawkesford MJ, McGrath SP (1999) Sulfur assimilation and effects on yield and quality of wheat. J Cereal Sci 30:1–17

Sulfur Compounds in Multiple Compensation Reactions of Abiotic Stress Responses

Heinz Rennenberg and Cornelia Herschbach

Abstract Plants interact with sulfur in two different ways. As a macronutrient sulfur is needed for growth and development; at the same time, sulfur is an important substrate and reductant during various forms of stresses mediated by the abiotic environment. The use of sulfur compounds as substrate and/or reductant in compensation reactions of abiotic stresses including oxidative stress, heavy metal and xenobiotic exposure is discussed with special emphasis on the S-containing tripeptide glutathione (GSH). The examples shown indicate that individual components of S metabolism are involved in different processes of abiotic stress compensation. In the present review the current knowledge of GSH (i) as reductant in the compensation of oxidative stress, (ii) as reductant as well as a substrate in redox reactions, (iii) its direct and indirect involvement in posttranscriptional modification reactions, and (iv) its constitution as a substrate for chelating heavy metals and for conjugation of xenobiotic is discussed. Competition with plant development and growth is also considered.

Introduction

Plants interact with their abiotic environment in contrasting ways. On the one hand, the abiotic environment provides the resources for, and controls processes of, growth and development of plants. On the other hand, the abiotic environment imposes multiple forms of stress on plants that require compensation mechanisms

---

H. Rennenberg (✉) • C. Herschbach
Institut für Forest Botany and Tree Physiology, Chair of Tree Physiology, Albert-Ludwigs-University Freiburg, Georges-Köhler-Allee 53, 79110 Freiburg, Germany
e-mail: heinz.rennenberg@ctp.uni-freiburg.de

L.J. De Kok et al. (eds.), *Sulfur Metabolism in Plants*, Proceedings of the International Plant Sulfur Workshop 1, DOI 10.1007/978-94-007-4450-9_25,

to achieve proper growth and development. Sulfur interacts with both sides of the abiotic environment. As a constituent of protein and numerous primary and secondary metabolites it is a macro-nutrient essential for plant growth and development (Wirtz and Droux 2005; Kopriva 2006); as substrate and reductant it participates in the compensation of various forms of stresses mediated by the abiotic environment (Noctor 2006; Foyer and Noctor 2009; Buchanan and Balmer 2005; Szalai et al. 2009). Therefore, use of sulfur for stress compensation will withdraw sulfur resources for growth and development and *vice versa* (Cameron and Pakrasi 2010). This requires controlled partitioning of sulfur between these different sinks. Currently, the mechanisms involved in this partitioning and its regulation are not understood.

The use of sulfur compounds as substrates in stress compensation reactions can only be achieved at enhanced substrate production. However, despite enhanced production, intense S-substrate consumption for stress compensation may result in a reduced rather than an enhanced cellular concentration of the respective compound under stress conditions. This has been shown, for example, for heavy metal exposure and GSH (Di Baccio et al. 2005; Ernst et al. 2008). When sulfur compounds are used as co-substrate or reductant in stress compensation reactions, they may rapidly be regenerated, but usually they are not consumed to a considerable extent. Under these conditions its sub-cellular concentration may be enhanced to achieve operation of enzymatic compensation reactions at, or close to saturation. Depending on the particular sub-cellular compartment and the constitutive level in this compartment, this increase may result in significantly enhanced cellular concentrations of the sulfur compounds involved or not. Again, this has been demonstrated for GSH and various forms of abiotic stress (salt stress: Herschbach et al. 2010a; ozone: Haberer et al. 2007; Hofer et al. 2008; heavy metal: Koprivova et al. 2002).

In the present review examples for the use of sulfur compounds in compensation reactions of abiotic stresses as substrates and/or reductant and its significance for sulfur metabolism are discussed with special emphasis on the S-containing tri-peptide glutathione (GSH). The abiotic stresses covered include oxidative stress, a.o. by air pollutants and climate, heavy metal and xenobiotic exposure.

GSH in Redox Reactions for Stress Compensation

At the cellular level abiotic stress often disturbs electron transfer reactions and results in enhanced ROS formation (Gill and Tuteja 2010). Since ROS are involved in the control of a large number of cellular and developmental processes, effective ROS homeostasis under stress conditions is essential for plant growth and development (Noctor 2006). At cellular concentrations observed in plant tissues, ascorbic acid is a superior reductant in chemical ROS detoxification compared to GSH (Polle and Rennenberg 1993). Still threshold amounts of GSH are essentially required for many cellular functions and stress compensation (Noctor 2006). The significance of GSH in redox reactions for stress compensation is emphasized by the observation that accumulation of GSSG as a consequence of stress exposure often results in elevated total glutathione levels as a consequence of maintained GSH levels by *de novo* synthesis. The role of GSH under these conditions seems to be largely dominated by its function as co-substrate in the Foyer-Halliwell-Cycle as a central component of ROS homeostasis (Noctor 2006).

Recently, the involvement of GSH in other enzyme-catalyzed redox reactions has been emphasized. These reactions include glutathionylation and deglutathionylation of proteins (Rouhier et al. 2008) and NO turnover (Arasimowicz and Floryszak-Wieczorek 2007; Wilson et al. 2008; Leitner et al. 2009). Posttranscriptional modification of proteins by glutathionylation and deglutathionylation is likely to be catalyzed by the glutaredoxin (GRX)/ferredoxin thioredoxin reductase (FTR) reaction that uses GSSG (glutathionylation) and GSH (glutathionylation and deglutathionylation) as oxidant and reductant, respectively. For this reaction a monothiol and a dithiol mechanism has been proposed, but the *in vivo* mechanism of this reaction is unknown (Rouhier et al. 2008). Since the total amount of glutathione bound to protein was found to be a few percent of total glutathione only (Rennenberg 1984), glutathionylation and deglutathionylation of proteins cannot be considered as a significant sink of glutathione.

Approximately 150 targets of posttranscriptional modification by glutathionylation have been identified. From these targets it appears that glutathionylation is involved in the regulation of many cellular processes such as glycolysis, signal transduction, protein folding, protein degradation, and intracellular trafficking (Rouhier et al. 2008). In the chloroplast, target proteins of glutathionylation are involved in photosynthesis, ATP metabolism, amino acid metabolism, and oxidative stress reactions and, thus, also appear to be involved abiotic stress response. However, the significance of glutathionylation/deglutathionylation reactions in abiotic stress compensation and responses has been proposed (Rouhier et al. 2008; Szalai et al. 2009) but needs further studies.

Posttranscriptional protein modification is also mediated by NO. This modification is involved in abiotic stress responses, *e.g.* to salinity, temperature, light, and anoxia (Arasimowicz and Floryszak-Wieczorek 2007; Leitner et al. 2009). GSH is involved in NO mediated reaction by the production of *S*-nitrosoglutathione (GSNO) that, besides oxidation of NO to nitrite is considered an important factor in NO turnover and, thus, in NO homeostasis (Wilson et al. 2008). Indirect evidence from systemic responses suggests that GSNO is mobile in the phloem, whereas evidence for xylem transport of GSNO has not been provided (Rustérucci et al. 2007). GSNO also seems to be directly involved in defence gene activation (Leitner et al. 2009). Thus, GSH interactions seem to play a dual role in NO mediated stress responses, indirectly by controlling the NO level by the initiation of NO turnover and directly by GSNO mediated effects on defence gene activity.

Abiotic Stress – Oxidative Stress

Ozone Detoxification

Ozone is a regular constituent of the atmosphere, but its atmospheric gas mixing ratio is increasing due to atmospheric pollution since the beginning of industrialization (Pinto et al. 2010). For plants the primary target of atmospheric ozone is the shoot. Ozone enters the leaf via diffusion through the stomata driven by a concentration gradient between the atmosphere and the substomatal cavity (Laisk et al. 1989) that is maintained by the high reactivity of ozone in the aqueous phase of the apoplastic space and its rapid conversion to reactive oxygen species (ROS) (references in Riikonen et al. 2008). In the apoplastic space the ROS pool is heavily controlled by production and consumption processes (Pignocchi and Foyer 2003), since ROS are involved in a number of developmentally determined processes such as lignin formation by polymerisation of lignin precursor monomers (Almagro et al. 2009). In this control apoplastic ascorbic acid plays an essential role, whereas sulfur compounds are not primarily involved in this detoxification reactions (Pignocchi and Foyer 2003). Dehydroascorbate (DHA) produced in apoplastic ozone consumption reaction cannot be reduced in the apoplastic space, but is exchanged with symplastic ascorbic acid at the plasmembrane by carrier mediated transport (Fig. 1). In the symplastic space of the surrounding cells apoplastic DHA is reduced by the Foyer-Halliwell pathway using glutathione as a reductant. The GSSG produced in this reaction catalyzed by the DHA reductase is regenerated in the cytoplasm by glutathione reductase (GR) at the expense of NAD(P)H oxidation. Thus, GSH is not involved in the primary detoxification reaction of ozone, but acts as co-substrate to regenerate ascorbic acid required for the detoxification of ROS produced from ozone in the apoplastic space. This secondary involvement of GSH in ozone detoxification causes a stimulation of GSH synthesis at the level of expression of genes of S assimilation into glutathione that seems to be induced by elevated jasmonate and can result in enhanced cellular GSH levels in response to elevated atmospheric ozone (Rennenberg et al. 2007; Kopriva and Rennenberg 2004). However, changes in cellular GSH levels strongly depend on the species analysed and the ozone concentration applied (Rennenberg et al. 2007). By contrast, direct oxidation of cysteine residues or methionine requires atmospheric ozone concentrations significantly higher than currently observed in polluted environments and are not even predicted in such environments for the end of the century (Rennenberg et al. 2007).

An early consequence of ozone exposure of leaves is the disturbance of phloem transport (Grantz and Farrar 1999; Topa et al. 2004; Liu et al. 2006). This disturbance may not only result in enhanced accumulation of photosynthate in the leaves and, subsequently, in the inhibition of photosynthetic carbon assimilation, but also in a reduced transport of GSH and ascorbic acid from the leaves to the roots. The GSH and ascorbic acid levels of the roots depend, however, on the delivery of these anti-oxidants by phloem transport (Herschbach et al. 2010b). As a consequence,

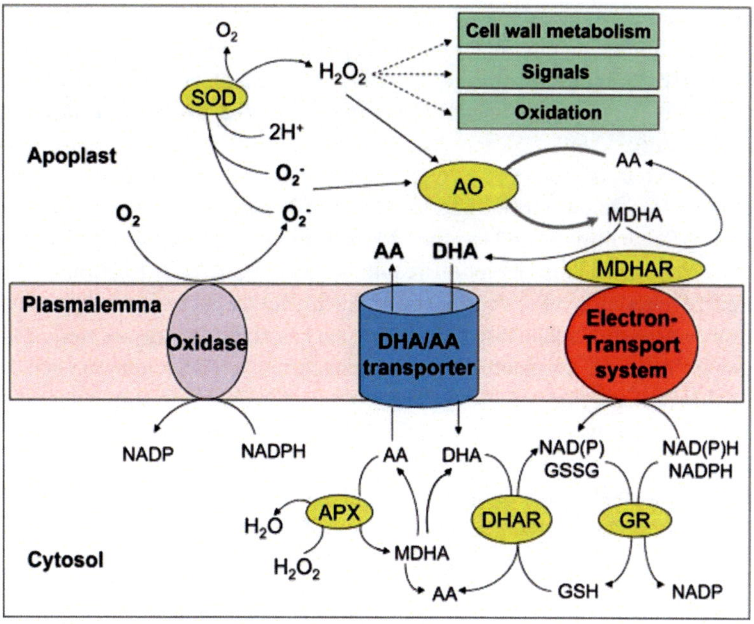

Fig. 1 Control of ROS levels in the apoplastic space. **SOD** superoxide dismutase, **AO** ascorbate oxidase, **MDHAR** monodehydroascorbate reductase, **DHAR** dehydroascorbate reductase, **GR** glutathione reductase, **GSH** reduced glutathione, **GSSG** oxidized glutathione, **AA** ascorbic acid, **DHA** dehydroascorbate, **MDHA** monodehydroascorbate (Redrawn and modified from Pignocchi and Foyer 2003)

the GSH and ascorbic acid level can decline in the roots in response of ozone exposure of the leaves although the roots are not a direct target of ozone exposure (Haberer et al. 2008). The roots are particularly sensitive to a decline of GSH and ascorbic acid because the regular levels are already low compared to the leaves (*e.g.* Herschbach et al. 2010b). Therefore, an ozone mediated decline in anti-oxidants may disturb root developmental processes that depend on ROS control like root hair formation (Sánchez-Fernández et al. 1997; Foreman et al. 2003) and lateral root formation (Potters et al. 2002, 2004; Olmos et al. 2006).

Drought Stress Compensation

Although the roots constitute the obvious primary target of drought stress, changes in soil water contents (SWC) do usually not result is different levels in S compounds such as sulfate and GSH (Seegmüller 1998; Schulte 1998; Herschbach 2003). Different levels of S compounds in roots are also not observed when root tissues of dry and wet years are compared (Haberer et al. 2008; Eimers et al. 2007; Rennenberg et al. 2006). At severe drought, however, sulfur uptake is negatively affected, but this is a general effect on ion uptake rather than a specific effect on S acquisition (Seegmüller 1998; Schulte 1998; Sardans et al. 2008). Apparently, changes in SWC are transferred to reduced root growth rather than reduced nutrient concentration in the roots. Thus, roots are the initial target of drought, but not the initial site for changes in antioxidants.

Drought stress is, however, experienced immediately by the leaves. In response to drought stomata are closed due to ABA synthesised in roots and transported to the leaves (Schachtman and Goodger 2008; Tardieu et al. 2010) and CO_2 inside the leaves is depleted. This depletion results in a reduced CO_2 fixation and, hence, reduced regeneration of electron acceptors in photosynthesis during continuous electron transport. Under these conditions, excess electrons transported in the photosynthetic electron transport chain are transferred to molecular oxygen to yield superoxide radicals that are further converted in the chloroplast by superoxide dismutase to H_2O_2 (Foyer and Noctor 2009; Jaspers and Kangasjarvi 2010). The highly reactive H_2O_2 is detoxified by the reactions of the Foyer-Halliwell pathway in which GSH acts as a co-substrate and reductant. Therefore, it is not surprising that a correlation between anti-oxidant levels, namely of GSH and ascorbic acid, and drought stress susceptibility has frequently been observed in leaves (Schulte 1998; Tausz et al. 2004; Sircelj et al. 2005; Hofer et al. 2008). Apparently, the leaves are the primary target of drought stress to induce changes in anti-oxidants such as GSH and ascorbic acid.

Heavy Metal Chelation

Heavy metal exposure of plants results in up-regulation of almost all steps of assimilatory sulfate reduction and assimilation into cysteine and glutathione (Ernst et al. 2008). Still, not only increased (Koprivova et al. 2002; Jin et al. 2008) but also reduced GSH levels are frequently found in response to heavy metal exposure (Xiang and Oliver 1998; Nocito et al. 2006; Ammar et al. 2008; Anjum et al. 2008; Wojas et al. 2009). This observation is due to the rapid induction of phytochelatin synthase (PCS) by heavy metals (Vatamaniuk et al. 2000; Loscos et al. 2006) and the synthesis of phytochelatins from glutathione by this enzyme. Cd exposure of roots also up-regulates the expression of members of the ABC transporter family (Bovet et al. 2003) that are thought to transport heavy metal phytochelatin complexes into the vacuole (Fig. 2). This view is supported by the observation that over-expression of one of these ABC transporters mediates more effective Cd retention in the roots and enhanced Cd accumulation in leaf vacuoles (Wojas et al. 2009).

Heavy metal phytochelatin complexes are mobile both in the xylem and the phloem. As to be expected from studies on xylem transport of primary metabolites (Heizmann et al. 2001; Herschbach et al. 2011), concentrations of heavy metal phytochelatin complexes in the xylem are low (Mendoza-Cózatl et al. 2008). However, when the *Arabidopsis cad1-3* mutant deficient in PCS (Howden et al. 1995) was complemented by wheat PCS1 expressed under a root specific promotor, heavy

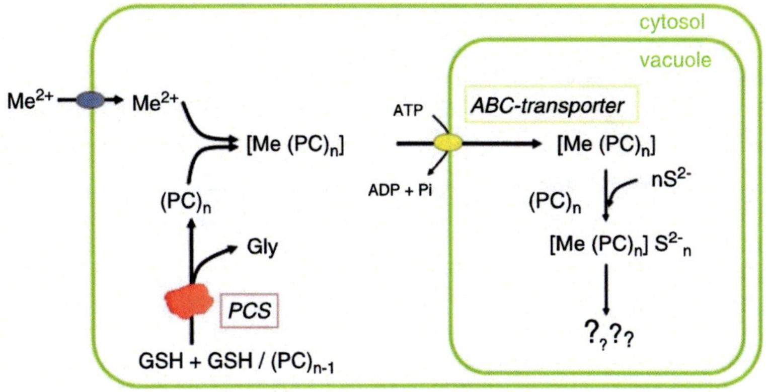

Fig. 2 Heavy metal detoxification and compartmentation by chelation with phytochelatins. **Me** metal ion, **PC** phytochelatin, **PCS** phytochelatin synthase, **S^{2-}** sulfide, **GSH** reduced glutathione

metal sensitivity of the complemented mutant was nullified and in response to heavy metal phytochelatins were not only found in the roots, but also in stems and leaves (Gong et al. 2003). This finding can only be explained by phytochelatin synthesis in the roots of the complemented mutant and subsequent transport of heavy metal phytochelatin complexes in the xylem. Phloem transport of heavy metal phytochelatin complexes has also be concluded from experiments with the *cad1-3* mutant complemented wheat PCS1 in a shoot specific manner (Chen et al. 2006) and, has been verified by the analysis of *Brassica* phloem sap (Mendoza-Cózatl et al. 2008). From these observations it appears that upon heavy metal exposure of the roots the entire plant can be used to store heavy metal phytochelatin complexes in vacuoles. This may be considered a means of dilution to prevent product inhibition in heavy metal chelation by phytochelatins.

Conjugation of Xenobiotics and Oxidants

GSH can be conjugated to a wide range of electrophilic compounds either spontaneously or by enzyme-catalyzed nucleophilic addition mediated by glutathione *S*-transferases (GSTs). GSTs constitute a superfamily of multifunctional enzymes. In plants seven distinct classes of GSTs have been recognized, including the class of dehydroascorbate reductases (DHARs), with four of these classes being specific for plants (Edwards and Dixon 2005). Several GSTs can react with xenobiotics such as herbicides and insecticides (Öztetik 2008; Banerjee and Goswami 2010). Thus, GSTs mediated conjugation is considered a detoxification reaction because toxic xenobiotics and metabolites are converted into non-toxic, hydropholic derivatives that can be degraded in subsequent reactions and can be subjected to internal compartmentation, in particular to sequestration in the vacuole (Edwards et al. 2000). Some other members have an active site chemistry primed for catalyzing reduction

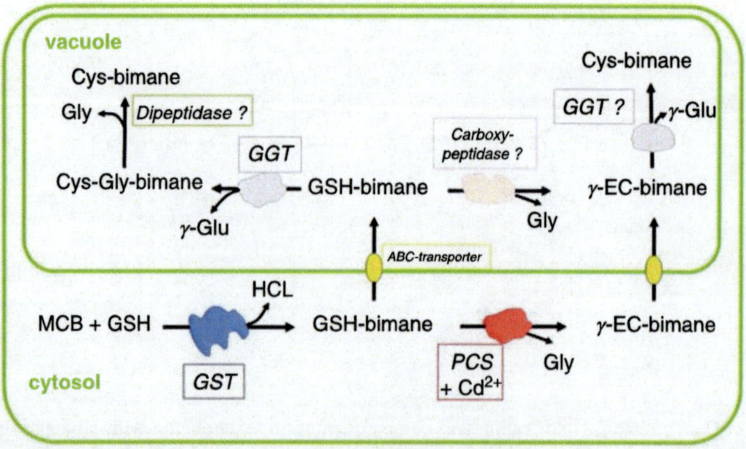

Fig. 3 Conjugation of the xenobiotic monochlorobimane with glutathione and its metabolism. **MCB** monochlorobimane, **GSH** reduced glutathione, **γ-EC** γ-glutamylcysteine, **GST** glutathione *S*-transferase, **PCS** phytochelatin synthase, **GGT** γ-glutamyl transpeptidase. The pathway is deduced from data by Wolf et al. (1996), Steinkamp and Rennenberg (1985), Blum et al. (2007, 2010), Ohkama-Ohtsu et al. (2007), and Grzam et al. (2007)

(these are called GSH peroxidases). Thus, even oxidants originating from the plant´s endogenous metabolism can react with GST's (Edwards and Dixon 2005; Dixon et al. 2010; Lan et al. 2009; Zhao and Zhang 2006). Thus, GSTs play a dual role in plant response to abiotic stress: On the one hand, they can reduce or prevent injury by direct detoxification of man-made compounds taken up from the environment and, on the other hand, they can reduce or prevent oxidative stress indirectly by detoxification of metabolites that are produced inside the plant in response to abiotic stress exposure.

For these detoxification reactions degradation of glutathione conjugates and its sequestration in the vacuole are of particular importance. Figure 3 summarizes current knowledge about these processes using monochlorobimane (MCB) as an example xenobiotic. GST mediated conjugation of MCB in the cytosol will result in the formation of GSH-bimane that will either be transported into the vacuole or, at least in the presence of heavy metals, degraded by PCS to γ-EC-bimane. Such a function of PCS in the degradation of GSH-conjugates has recently been established after blocking its transport into the vacuole, but seems only relevant in the presence of heavy metals (Blum et al. 2007, 2010). The γ-EC-bimane produced might also be sequestered into the vacuole, but this transport seems negligible, because γ-EC-bimane accumulates in the cytosol and is retained therein (Grzam et al. 2007). Inside the vacuole, conversion of GSH-bimane to γ-EC-bimane cannot be achieved by PCS, because PCS is a cytosolic enzyme (Blum et al. 2010), but would requires the action of a carboxy-peptidase that remains to be demonstrated in plant vacuoles. GSH-bimane can be degraded inside the vacuole to Cys-Gly-bimane by vacuolar GGT (Ohkama-Ohtsu et al. 2007; Blum et al. 2010) that may also responsible for the

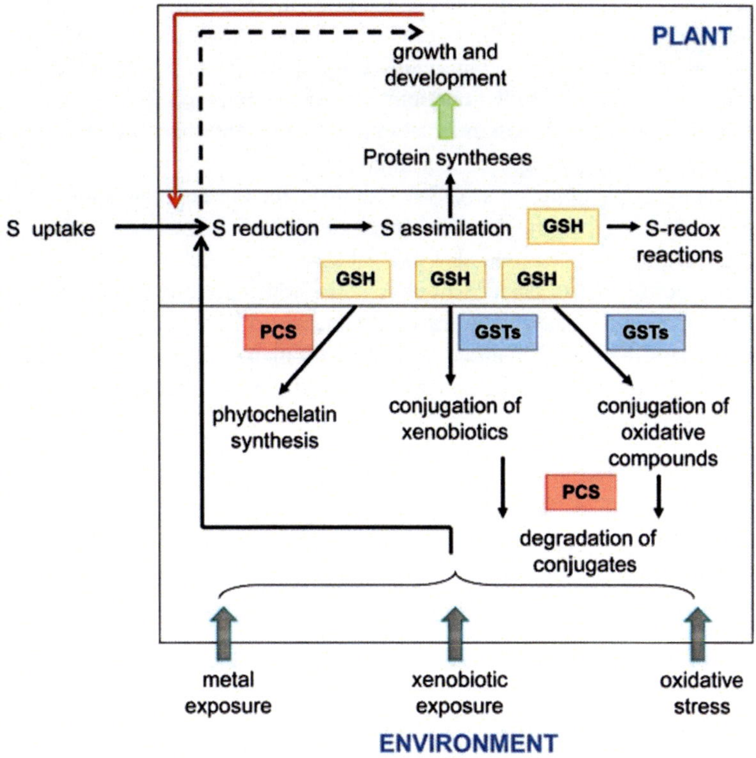

Fig. 4 Interaction of S metabolism with the abiotic environment. **GSH** glutathione, **PCS** phytochelatin synthetase, **GST** glutathione *S*-transferase

conversion of γ-EC-bimane to Cys-bimane. Whether Cys-Gly-bimane is further degraded to Cys-bimane by a vacuolar dipeptidase remains to be elucidated. Thus, the presence of both γ-EC- and Cys-Gly-conjugates, unique in plants (Rennenberg and Lamoureux 1990), can be explained by reactions catalyzed by PCS in the cytosol and by GGT in the vacuole.

Conclusions

These examples show that individual components of S metabolism are involved in different processes of abiotic stress compensation (Fig. 4). GSH acts as co-substrate in the compensation of oxidative stress, participates as co-substrate and substrate in redox reactions, is directly and indirectly involved as substrate in posttranscriptional modification reactions, and constitutes the substrate for chelation of heavy metals and conjugation of xenobiotic. PCS does not only catalyze the formation of PCs responsible for heavy metal detoxification by chelation, but also participates in the degradation of GSH-conjugates. The family of GSH *S*-transferase enzymes does not only catalyze the detoxification of xenobiotics, but is also responsible for the detoxification of oxidative compounds originating from secondary metabolism. This clearly indicates multiple functions of distinct compensation and detoxification reactions in plants adaptation to the abiotic environment. Since plants in their environment are usually not exposed to a single, but to multiple abiotic stress factors, the use of individual components of S-metabolism for the compensation of different forms of abiotic stress requires complex co-ordination and regulation. As plant growth, development, reproduction as well as the plant fitness compete with stress related responses these aspects should also be considered. It will be an important challenge of future research to unravel and characterize this network and to use this knowledge for the selection and development of plants that can better cope with its abiotic environment that confers increasing constrains on plant growth and development due to human activities.

References

Almagro L, Ros LVG, Belchi-Navarro S, Bru R, Ros Barcelo A, Pedren MA (2009) Class III peroxidises in plant defence reactions. J Exp Bot 60:377–390

Ammar WB, Mediouni C, Tray B, Ghorbel MH, Jemal F (2008) Glutathione and phytochelatin contents in tomato plants exposed to cadmium. Biol Plant 52:314–320

Anjum NA, Umar S, Ahmad A, Iqbal M, Khan NA (2008) Sulfur protects mustard (*Brassica campestris* L.) from cadmium toxicity by improving leaf ascorbate and glutathione. Plant Growth Regul 54:271–279

Arasimowicz M, Floryszak-Wieczorek J (2007) Nitric oxide as a bioactive signaling molecule in plant stress responses. Plant Sci 172:876–887

Banerjee S, Goswami R (2010) GST profile expression study in some selected plants: in silico approach. Mol Cell Biochem 336:109–126

Blum R, Beck A, Korte A, Stengel A, Letzel T, Lendzian K, Grill E (2007) Function of phytochelain synthase in catabolism of glutathione-conjugates. Plant J 49:740–749

Blum R, Meyer KC, Wünschmann J, Lendzian KJ, Grill E (2010) Cytosolic action of phytochelatin synthase. Plant Physiol 153:159–169

Bovet L, Eggmann T, Meylan-Bettex M, Polier J, Kammer P, Marin E, Feller U, Martinoia E (2003) Transcript levels of *ATMRPs* after Cadmium treatment: induction of *AtMRP3*. Plant Cell Environ 26:371–381

Buchanan BB, Balmer Y (2005) Redox regulation: a broadening horizon. Annu Rev Plant Biol 56:187–220

Cameron JC, Pakrasi HB (2010) Essential role of glutathione in acclimation to environmental and redox perturbations in the cyanobacterium *Synechocystis* sp. PCC 6803. Plant Physiol 154: 1672–1685

Chen A, Komives EA, Schroeder JI (2006) An improved grafting technique for mature *Arabidopsis* plants demonstrates long-distance shoot-to-root transport of phytochelatins in *Arabidopsis*. Plant Physiol 141:108–120

Di Baccio D, Kopriva S, Sebastiani L, Rennenberg H (2005) Does glutathione metabolism have a role in the defence of poplar against zinc excess? New Phytol 167:73–80

Dixon DP, Skipsey M, Edwards R (2010) Roles for glutathione transferases in plant secondary metabolism. Phytochemistry 71:338–350

Edwards R, Dixon PD (2005) Plant glutathione transferases. Method Enzymol 401:169–186
Edwards R, Dixon DP, Walbot V (2000) Plant glutathione S-transferases: enzymes with multiple functions in sickness and in health. Trends Plant Sci 5:193–198
Eimers MC, Watmough SA, Buttle JM, Dillon PJ (2007) Drought-induced sulphate release from a wetland in south-central Ontario. Environ Monit Assess 127:399–407
Ernst WHO, Krauss G-J, Verkleij JAC, Wesenberg D (2008) Interaction of heavy metals with the sulfur metabolism in angiosperms from an ecological point of view. Plant Cell Environ 31:123–143
Foreman J, Demidchik V, Bothwell JHF, Mylona P, Miedema H, Torres MA, Linstead P, Costa S, Brownlee C, Jones JDG, Davies JM, Dolan L (2003) Reactive oxygen species produced by NADPH oxidase regulate plant cell growth. Nature 422:442–446
Foyer CH, Noctor G (2009) Redox regulation in photosynthetic organisms: signaling, acclimation, and practical implications. Antioxid Redox Signal 11:861–905
Gill SS, Tuteja N (2010) Reactive oxygen species and antioxidant machinery in abiotic stress tolerance in crope plants. Plant Physiol Biochem 48:909–930
Gong J-M, Lee DA, Schroeder JI (2003) Long-distance root-to-shoot transport of phytochelatins and cadmium in *Arabidopsis*. Proc Natl Acad Sci USA 100:10118–10123
Grantz DA, Farrar JF (1999) Acute exposure to ozone inhibits rapid carbon translocation from source leaves of Pima cotton. J Exp Bot 50:1253–1262
Grzam A, Martin MN, Hell R, Meyer AJ (2007) γ-Glutamyl transpeptidase GGT4 initiates vacuolar degradation of glutathione *S*-conjugates in *Arabidopsis*. FEBS Lett 581:3131–3138
Haberer K, Herbinger K, Alexou M, Tausz M, Rennenberg H (2007) Antioxidative defence of old growth beech (*Fagus sylvatica*) under double ambient O-3 concentrations in a free-air exposure system. Plant Biol 9:215–226
Haberer K, Herbinger K, Alexou M, Rennenberg H, Tausz M (2008) Effects of drought and canopy ozone exposure on antioxidants in fine roots of mature European beech (*Fagus sylvatica*). Tree Physiol 28:713–719
Heizmann U, Kreuzwieser J, Schnitzler J-P, Brüggemann N, Rennenberg H (2001) Assimilate transport in the xylem sap of young oak (*Quercus robur*) trees. Plant Biol 3:132–138
Herschbach C (2003) Whole Plant regulation of sulfur nutrition of deciduous trees – influences of the environment. Plant Biol 5:233–244
Herschbach C, Teuber M, Eiblmeier M, Ehlting B, Ache P, Polle A, Schnitzler J-P, Rennenberg H (2010a) Changes in sulphur metabolism of grey poplar (*Populus* x *canescens*) leaves during salt stress: a metabolic link to photorespiration. Tree Physiol 30:1161–1173
Herschbach C, Scheerer U, Rennenberg R (2010b) Redox states of glutathione and ascorbate in root tipps of poplar (*Populus tremula x P. alba*) depend on phloem transport from the shoot to the roots. J Exp Bot 61:1065–1074
Herschbach C, Gessler A, Rennenberg H (2011) Long-distance transport and plant internal cycling of N- and S-compounds. Prog Bot 73:161–188
Hofer N, Alexou M, Heerdt C, Low M, Werner H, Matyssek R, Rennenberg H, Haberer K (2008) Seasonal differences and within-canopy variations of antioxidants in mature spruce (*Picea abies*) trees under elevated ozone in a free-air exposure system. Environ Pollut 154:241–253
Howden R, Goldsbrough PB, Andersen CR, Cobbett CS (1995) Cadmium-sensitive, cad1 mutants of *Arabidopsis thaliana* are phytochelatin deficient. Plant Physiol 107:1059–1066
Jaspers P, Kangasjarvi J (2010) Reactive oxygen species in abiotic stress signaling. Physiol Plant 138:405–413
Jin X, Yang X, Mahmood Q, Islam E, Liu D, Li H (2008) Response of antioxidant enzymes, ascorbate and glutathione metabolism towards cadmium in hyperaccumulator and nonhyperaccumulator ecotypes of *Sedium alfredii* H. Environ Toxicol 23:517–529
Kopriva S (2006) Regulation of sulfate assimilation in *Arabidopsis* and beyond. Ann Bot 97:479–495
Kopriva S, Rennenberg H (2004) Control of sulfate assimilation and glutathione synthesis: interaction with N and C metabolism. J Exp Bot 55:1831–1842

Koprivova A, Kopriva S, Jager D, Will B, Jouanin L, Rennenberg H (2002) Evaluation of transgenic poplars over-expressing enzymes of glutathione synthesis for phytoremediation of cadmium. Plant Biol 4:664–670

Laisk A, Kull O, Moldau H (1989) Ozone concentration in leaf intercellular air spaces is close to zero. Plant Physiol 90:1163–1167

Lan T, Yang Z-L, Yang X, Liu Y-J, Wang X-R, Zeng Q-Y (2009) Extensive functional diversification of the *Populus* glutathione S-transferase supergene family. Plant Cell 21:3749–3766

Leitner M, Vandelle E, Gaupels F, Bellin D, Delledonne M (2009) No signals in the haze. Nitric oxide signalling in plant defence. Curr Opin Plant Biol 12:451–458

Liu X, Rennenberg H, Matyssek R (2006) Effects of elevated pO_3 on carbon cycle between above and belowground organs of trees. J Environ Sci (China) 18:932–936

Loscos J, Naya L, Ramos J, Clemente MR, Matamoros MA, Becana A (2006) A reassessment of substrate specificity and activation of phytochelatin synthases from model plants by physiologically relevant metals. Plant Physiol 140:1213–1221

Mendoza-Cózatl DG, Butko E, Springer F, Torpey JW, Komives AE, Kehr J, Schroeder JI (2008) Identification of high levels of phytochelatines, glutathione and cadmium in the phloem sap of *Brassica napus*. A role for-thiol-peptides in the long-distance transport of cadmium and the effect of cadmium on iron translocation. Plant J 54:249–259

Nocito FF, Lancilli C, Crema B, Fourcroy P, Davidian J-C, Sacchi GA (2006) Heavy metal stress and sulfate uptake in maize roots. Plant Physiol 141:1138–1148

Noctor G (2006) Metabolic signalling in defence and stress: the central roles of soluble redox couples. Plant Cell Environ 29:409–425

Ohkama-Ohtsu N, Zhao P, Xiang C, Oliver DJ (2007) Glutathione conjugates in the vacuole are degraded by γ-glutamyl transpeptidase GGT3 in *Arabidopsis*. Plant J 49:878–888

Olmos E, Kiddle G, Pellny TK, Kumar S, Foyer CH (2006) Modulation of plants morphology, root architecture, and cell structure by low vitamin C in *Arabidopsis thaliana*. J Exp Bot 57:1645–1655

Öztetik E (2008) A tale of plant glutathione *S*-transferases: since 1970. Bot Rev 74:419–437

Pignocchi C, Foyer CH (2003) Apoplastic ascorbate metabolism and its role in the regulation of cell signalling. Curr Opin Plant Biol 6:379–389

Pinto DM, Blande JD, Souza SR, Nerg A-M, Holopainen JK (2010) Plant volatile organic compounds (VOCs) in ozone (O_3) polluted atmospheres: the ecological effects. J Chem Ecol 36:22–34

Polle H, Rennenberg H (1993) Significance of antioxidants in plant adaptation to environmental stress. In: Fowden L, Mansfield T, Stoddard F (eds) Plant adaptation to environmental stress. Chapman & Hall, London, pp 263–273

Potters G, De Gara L, Asard H, Horemans N (2002) Ascorbate and glutathione: guardians of the cell cycle, partners in crime? Plant Physiol Biochem 40:537–548

Potters G, Horemans N, Bellone S, Caubergs RJ, Trost P, Guisez Y, Asard H (2004) Dehydroascorbate influences the plant cell cycle through a glutathione-independent reduction mechanism. Plant Physiol 134:1479–1487

Rennenberg H (1984) The fate of excess sulfur in higher plants. Annu Rev Plant Physiol 35:121–153

Rennenberg H, Lamoureux G (1990) Physiological processes that modulate the concentration of glutathione in plant cells. In: Rennenberg H, Brunold Ch, De Kok LJ, Stulen I (eds) Sulfur nutrition and assimilation in higher plants; fundamental, environmental, and agricultural aspects. SPB Acadamic Publishing, The Hague, pp 53–65

Rennenberg H, Loreto F, Polle A, Brilli F, Fares S, Beniwal RS, Gessler A (2006) Physiological responses of forest trees to heat and drought. Plant Biol 8:556–571

Rennenberg H, Herschbach C, Haberer K, Kopriva S (2007) Sulfur metabolism in plants: are trees different? Plant Biol 9:620–637

Riikonen J, Kets K, Darbah J, Oksanen E, Sober A, Vapaavuori E, Kubiske ME, Nelson N, Karnosky DF (2008) Carbon gain and bud physiology in *Populus tremuloides* and *Betula papyrifera* grown under long-term exposure to elevated concentrations of CO_2 and O_3. Tree Physiol 28:243–254

Rouhier N, Lemaire SD, Jacquot J-P (2008) The role of glutathione in photosynthetic organisms: emerging functions for glutaredoxins and glutathionylation. Annu Rev Plant Biol 59:143–166

Rustérucci C, Espunya MC, Díaz M, Chabannes M, Martínez MC (2007) *S*-Nitrosoglutathione reductase affords protection against pathogens in *Arabidopsis*, both locally and systemically. Plant Physiol 143:1282–1292

Sánchez-Fernández R, Fricker M, Corben LB, White NS, Sheard N, Leaver CJ, Van Montagu M, Inzé D, May MJ (1997) Cell proliferation and hair tip growth in the *Arabidopsis* root are under mechanistically different forms of redox control. Proc Natl Acad Sci USA 94:2745–2750

Sardans J, Peñuelas J, Ogaya R (2008) Drought's impact on Ca, Fe, Mg, Mo and S concentration and accumulation patterns in the plants and soil of a Mediterranean evergreen *Quercus ilex* forest. Biogeochemistry 87:49–69

Schachtman DP, Goodger JQD (2008) Chemical root to shoot signaling under drought. Trends Plant Sci 13:281–287

Schulte M (1998) Der Einfluss von erhöhtem atmospherischem CO_2 auf den Kohlenstoff-, Stickstoff- und Schwefelhaushalt von Eichen. Ph.D thesis Albert-Ludwigs-University, Freiburg

Seegmüller S (1998) Der Einfluss von Mykorrizen, atmospärischem Kohlendioxid und Wassermangel auf das Wachstum und die Schwefel- und Stickstoffernährung der Stieleiche (*Quercus robur* L.). Ph.D thesis Albert-Ludwigs-University, Freiburg

Sircelj H, Tausz M, Grill D, Batic F (2005) Biochemical responses in leaves of two apple tree cultivars subjected to progressing drought. J Plant Physiol 162:1308–1318

Steinkamp R, Rennenberg H (1985) Degradation of glutathione in plant cells: evidence against the participation of a γ-glutamyltranspeptidase. Z Naturforsch 40c:29–33

Szalai G, Kellös T, Galiba G, Kocsy G (2009) Glutathione as an antioxidant and regulatory molecule in plants under abiotic stress conditions. J Plant Growth Regul 28:66–80

Tardieu F, Parent B, Simonneau T (2010) Control of leaf growth by abscisic acid: hydraulic or non-hydraulic processes. Plant Cell Environ 3:636–647

Tausz M, Sircelj H, Grill D (2004) The glutathione system as a stress marker in plant ecophysiology: is a stress-response concept valid? J Exp Bot 55:1955–1962

Topa MA, McDermitt DJ, Yun SC (2004) Do elevated ozone and variable light alter carbon transport to roots in sugar maple? New Phytol 162:173–186

Vatamaniuk OK, Mari S, Lu Y-P, Rea PA (2000) Mechanism of heavy metal ion activation of phytochelatin (PC) synthase. J Biol Chem 275:31451–31459

Wilson ID, Neill SJ, Hancock JT (2008) Nitric oxide synthesis and signaling in plants. Plant Cell Environ 31:622–631

Wirtz M, Droux M (2005) Synthesis of the sulfur amino acids: cysteine and methionine. Photosynth Res 86:345–362

Wojas S, Hennig J, Plaza S, Geisler M, Siemianowski O, Sklodowska A, Ruszczynska A, Bulska E, Antosiewicz DM (2009) Ectopic expression of *Arabidopsis* ABC transporter MRP7 modifies cadmium root-to-shoot transport and accumulation. Environ Pollut 157:2781–2789

Wolf AE, Dietz K-J, Schröder P (1996) Degradation of glutathione *S*-conjugates by a carboxypeptidase in the plant vacuole. FEBS Lett 384:31–34

Xiang CB, Oliver DJ (1998) Glutathione metabolic genes coordinately respond to heavy metals and jasmonic acid in *Arabidopsis*. Plant Cell 10:1539–1550

Zhao FY, Zhang H (2006) Salt and paraquat stress tolerance results from co-expression of the *Suaeda salsa* glutathione *S*-transferase and catalase in transgenic rice. Plant Cell Tissue Organ Cult 86:349–358

Cysteine Synthesis in the Chloroplast Is Not Required for Resistance of *Arabidopsis thaliana* to H_2S Fumigation

Hannah Birke, Markus Wirtz, Luit J. De Kok, and Rüdiger Hell

Abstract Reduced sulfur is necessary for synthesis of various essential compounds in the plant cell including cysteine and glutathione. For this reason sulfide is continuously produced in plastids as intermediate of the assimilatory sulfate reduction pathway. Varying demand for sulfide during development and in response to external stress challenges a plant to rapidly re-organize the entire sulfur metabolism, since high endogenous sulfide levels are potentially toxic for the cellular metabolism. Upon prolonged exposure of *Arabidopsis* to 1 μl l^{-1} H_2S, the endogenous sulfide level in the shoot was kept low, whereas there was a mass increase in thiol content. The cysteine and glutathione content increased approximately 30- and 3.4-fold, respectively. Notably, the dramatic changes on thiol levels were not accompanied by changes in extractable activities of the cysteine synthesizing enzymes, serine acetyltransferase and *O*-acetylserine(thiol)lyase (OAS-TL). It was evident that, despite the production of sulfide in plastids, its consumption by OAS-TL in the same compartment was not crucial for adaptation of *Arabidopsis* to high H_2S levels in the environment.

Sulfur plays an essential role in various physiological processes in plants and is a functional component of the amino acids methionine and cysteine in proteins and of glutathione, a key metabolite in redox homeostasis. Notably, in all these metabolites sulfur is bound in its reduced state, but in general sulfur is taken up by plants in its

H. Birke • M. Wirtz • R. Hell (✉)
Department for Plant Molecular Biology, Centre for Organismal Studies (COS), University of Heidelberg, Im Neuenheimer Feld 360, 69120 Heidelberg, Germany
e-mail: ruediger.hell@cos.uni-heidelberg.de

L.J. De Kok
Laboratory of Plant Physiology, University of Groningen, P.O. Box 11103, 9700 CC Groningen, Netherlands
e-mail: l.j.de.kok@rug.nl

L.J. De Kok et al. (eds.), *Sulfur Metabolism in Plants*, Proceedings of the International Plant Sulfur Workshop 1, DOI 10.1007/978-94-007-4450-9_26,

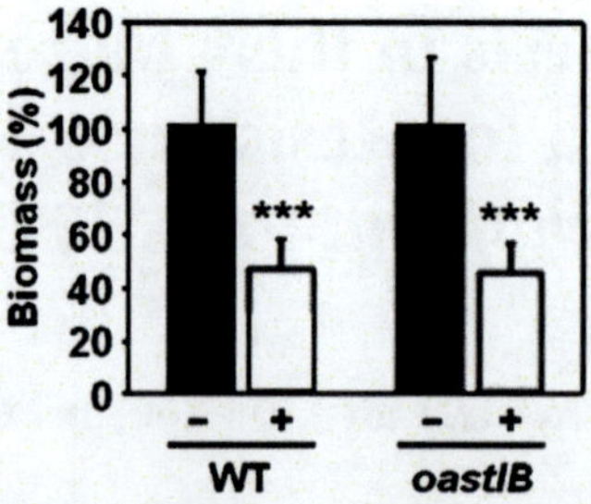

Fig. 1 Shoot biomass production of *Arabidopsis* wild type (*WT*) and *oastlB* mutant plants upon H_2S exposure. Leaves of 5 week old plants (n = 25) exposed to 0 and 1 μl l^{-1} H_2S (–/+) for 2 weeks were harvested and average biomass was determined. Average biomass of plants in the absence of H_2S was set to 100%. *** $p \leq 0.001$, unpaired Student's t-test

oxidized form as sulfate by the root. Sulfate is reduced to sulfide in the plastids in the assimilatory sulfate reduction pathway. Sulfide is subsequently metabolized into cysteine, the major precursor or donor of reduced sulfur for organic sulfur compounds. *O*-acetylserine(thiol)lyase (OAS-TL) uses sulfide to form cysteine by replacing the acetyl group of the intermediate *O*-acetylserine (OAS). OAS is produced by serine acetyltransferase (SAT) that transfers the acetyl moiety from acetyl-CoA to serine. The reduction of sulfate to sulfide is exclusively accomplished in plastids, whereas both SAT and OAS-TL are localized in cytosol, plastids, and mitochondria. SAT and OAS-TL form the cysteine synthase complex, which is suggested to control the cysteine synthesis rate (Wirtz and Hell 2006). However, the three compartments contribute to a surprisingly different extent to total cysteine synthesis of the cell. Whereas OAS is mainly synthesized in the mitochondria, cytosolic OAS-TL activity is primarily responsible for the final step of cysteine synthesis (Haas et al. 2008; Heeg et al. 2008; Watanabe et al. 2008). Although plastidic OAS-TL activity accounts for 45% of total OAS-TL activity, knock out of the respective enzyme (OAS-TL B) has no significant effect on flux of sulfide into cysteine (Heeg et al. 2008). Sulfate reduction in plastids is tightly regulated to balance the demands of mitochondria and cytosol for sulfide and to avoid accumulation of sulfide in plastids to toxic levels. However, so far it is not known whether incorporation of the produced sulfide into OAS in plastids contributes to the regulation of local sulfide levels. Therefore, the function of OAS-TL B in detoxification of excessive cellular sulfide levels was investigated.

Three week old *Arabidopsis thaliana* plants, ecotype Col-0 (WT) and a T-DNA insertion mutant lacking plastidic OAS-TL (*oastlB*) grown in soil were exposed to either 0 or 1 μl l^{-1} H_2S for 2 weeks in fumigation cabinets (2 per treatment) as described by Koralewska et al. (2008). Shoots of 25 plants per genotype and treatment were harvested and used for analysis of sulfide and sulfur metabolites content and enzyme activities.

Exposure of *Arabidopsis* to 1 μl l^{-1} H_2S for 2 weeks, resulted in an approx. 50% decrease in biomass production of the shoot of both WT and *oastlB* mutants (Fig. 1), whereas the shoot sulfide content increased 2.5-fold (Fig. 2). These sulfide contents

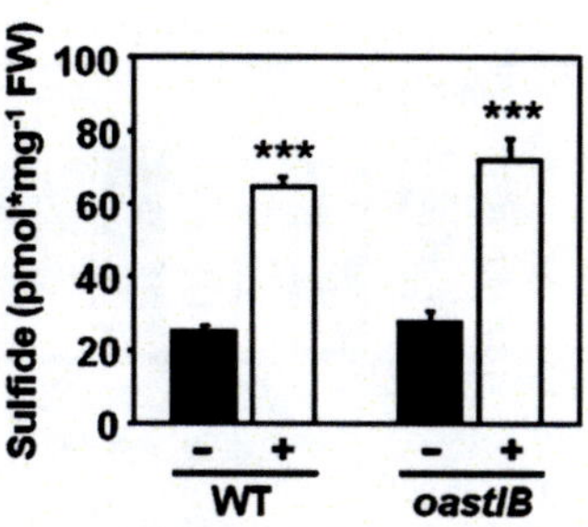

Fig. 2 Endogenous sulfide levels in shoots of H_2S exposed *Arabidopsis* wild type (*WT*) and *oastlB* mutant plants. Sulfide concentration was determined by HPLC according to Völkel and Grieshaber 1994 in 5-week old plants grown in the absence and presence of 1 µl l^{-1} H_2S (–/+) for 2 weeks. Shoots of 25 plants of each genotype and treatment were pooled and three aliquots were used for analysis. *FW* fresh weight; *** $p \leq 0.001$, unpaired Student's t-test

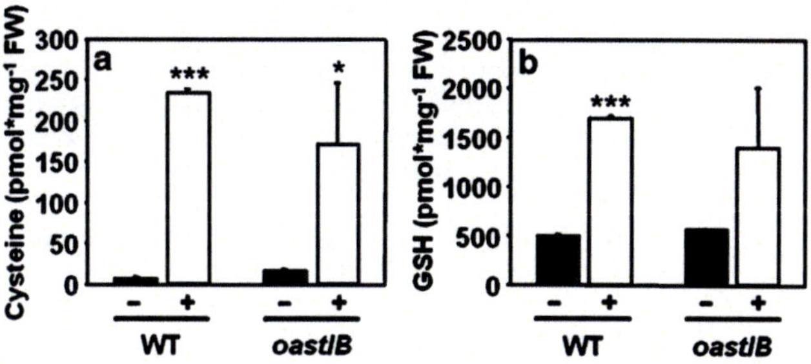

Fig. 3 Thiol levels in shoots of H_2S exposed *Arabidopsis* wild type (*WT*) and *oastlB* mutant plants. Cysteine (**a**) and glutathione (GSH, **b**) were determined by HPLC according to Heeg et al. (2008) in plants exposed to H_2S as described in Fig. 2. *FW* fresh weight; * $p \leq 0.05$, *** $p \leq 0.001$, unpaired Student's t-test

were a factor 100 higher than predicted on basis of the Henry's gas law (De Kok et al. 2007). H_2S exposure resulted in a mass increase in cysteine and glutathione content of the shoots (Fig. 3). Cysteine levels were increased more than 30-fold in H_2S-exposed WT plants (Fig. 3a), whereas that of glutathione increased 3.4-fold (Fig. 3b). H_2S exposure of *oastlB* plants also resulted in similar enhanced cysteine and GSH content, revealing no differences between the mutant and WT plants (Fig. 3). The enhanced cysteine content in the shoot upon H_2S exposure was not accompanied by a substantial change in the activity or protein levels of SAT and OAS-TL (Fig. 4). Total SAT activity was only slightly increased in H_2S-exposed WT and *oastlB* plants compared to the control plants (32% and 24%, respectively, Fig. 4a). However, immunological detection of the most abundant SAT isoform SAT3 and all OAS-TL isoforms did not reveal detectable alterations in protein amount (Fig. 4b). Nevertheless, the complete loss of OAS-TL B in *oastlB* plants was confirmed (Fig. 4b). In contrast to SAT enzymatic activity, OAS-TL enzymatic activity was not altered upon H_2S exposure in WT and *oastlB* plants.

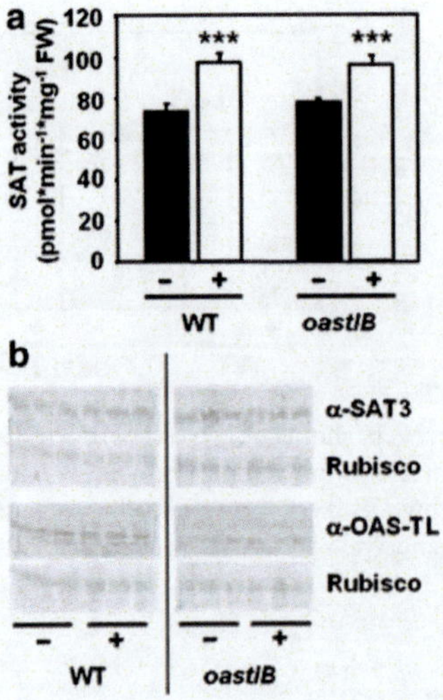

Fig. 4 Abundance and activity of SAT and OAS-TL in H_2S exposed wild type (*WT*) and *oastlB* mutant plants. (**a**) SAT activity was determined in soluble protein extracts according to Heeg et al. 2008 from plants exposed to H_2S as described in Fig. 2. (**b**) Amounts of SAT3 and OAS-TL proteins were determined by immunological detection using the respective primary antibodies according to Heeg et al. (2008) (OAS-TL A: lower signal, OAS-TL B: middle signal, OAS-TL C: upper signal). Rubisco was stained using Ponceau S to serve as a loading control. *FW* fresh weight; *** $p \leq 0.001$, unpaired Student's t-test

H_2S exposure had a substantial negative effect on biomass production of both *Arabidopsis* WT and *oastlB* plants. The comparable impact of H_2S exposure on growth of both the WT and *oastlB* mutant also demonstrated that the absence of OAS-TL B in the plastids did not add to the potential toxicity of sulfide for *Arabidopsis*. Moreover, H_2S exposure yielded in strongly enhanced cysteine and glutathione levels in the shoot, whereas the enzymatic activities of SAT and OAS-TL were not changed substantially. A similar response was also observed in poplar plants showing an increased cysteine content without changes in protein amounts of SAT and OAS-TL upon H_2S exposure for 4 days (Nakamura et al. 2009). It cannot be entirely excluded that moderate H_2S exposure for a shorter period as described here, would reveal differences between the genotypes. A high sulfide concentration, as used here, has the potential to override endogenous differences due to compartment specific perception and detoxification. However, this is unlikely since total sulfide levels did increase only about 2.5-fold in WT and *oastlB* mutants. Therefore, it is suggested, that the metabolic impact of an elevated H_2S level in the environment was not dependent on OAS-TL B activity in plastids but seemed to rely on responses in several subcellular compartments.

Acknowledgments H.B. is affiliated with The Hartmut Hoffmann-Berling International Graduate School of Molecular and Cellular Biology, University of Heidelberg, Germany, and the graduate program ENDS; the authors gratefully acknowledge support by DFG grant He 1848/14-1 and the Schmeil Stiftung Heidelberg.

References

De Kok LJ, Durenkamp M, Yang L, Stulen I (2007) Atmospheric sulfur. In: Hawkesford MJ, De Kok LJ (eds) Sulfur in plants – an ecological perspective. Springer, Dordrecht, The Netherlands pp 91–106

Haas FH, Heeg C, Queiroz R, Bauer A, Wirtz MA, Hell R (2008) Mitochondrial serine acetyltransferase functions as a pacemaker of cysteine synthesis in plant cells. Plant Physiol 148: 1055–1067

Heeg C, Kruse C, Jost R, Gutensohn M, Ruppert T, Wirtz M, Hell R (2008) Analysis of the *Arabidopsis O*-acetylserine(thiol)lyase gene family demonstrates compartment-specific differences in the regulation of cysteine synthesis. Plant Cell 20:168–185

Koralewska A, Stuiver CEE, Posthumus FS, Kopriva S, Hawkesford MJ, De Kok LJ (2008) Regulation of sulfate uptake, expression of the sulfate transporters Sultr1;1 and Sultr1;2, and APS reductase in Chinese cabbage (*Brassica pekinensis*) as affected by atmospheric H_2S nutrition and sulfate deprivation. Funct Plant Biol 35:318–327

Nakamura M, Kasugai I, Abe M, Youssefian S (2009) Increased thiol biosynthesis of transgenic poplar expressing a wheat *O*-acetylserine(thiol) lyase enhances resistance to hydrogen sulfide and sulfur dioxide toxicity. Plant Cell Rep 28:313–323

Völkel S, Grieshaber MK (1994) Oxygen dependent sulfide detoxification in the lugworm *Arenicola marina*. Mar Biol 118:137–147

Watanabe M, Mochida K, Kato T, Tabata S, Yoshimoto N, Noji M, Saito K (2008) Comparative genomics and reverse genetics analysis reveal indispensable functions of the serine acetyltransferase gene family in *Arabidopsis*. Plant Cell 20:2484–2496

Wirtz M, Hell R (2006) Functional analysis of the cysteine synthase protein complex from plants: structural, biochemical and regulatory properties. J Plant Physiol 163:273–286

Is the Eliciting Effect of Stress Metabolites on Pathogenesis in Winter Wheat Linked to the Sulfur Supply?

Elke Bloem, Silvia Haneklaus, and Ewald Schnug

Abstract The effect of a graded sulfur (S) supply to the soil combined with foliar applications of three different defense metabolites (S^0, salicylic acid and cysteine) on the infection rate of winter wheat with powdery mildew (*Blumeria graminis*) was tested in a pot experiment. Spray applications with all three compounds reduced the infection rate in comparison to the control. The strongest effect was obtained expectedly after S^0 application due to its well-established fungicidal effect in this host-pathogen relationship. S^0 reduced the infection rate by 47% and 67% at maximum when 50 and 100 mg S^0 pot^{-1} were applied, respectively 8 days after application in comparison to the control. In case of salicylic acid (SA) and cysteine the lower dose of 50 mg pot^{-1} resulted in a stronger reduction of the infection rate than the higher one when the S supply of the crop was low (5 mg S pot^{-1}). Generally, the infection proceeded slower when plants were fully supplied with S though the initial infection rate was higher in these treatments. In case of spray applications with SA and cysteine the combination effect together with soil-applied sulfate never boosted the effect of a single treatment. Only with S^0 the strongest reduction in fungal infection was observed when a high soil S supply was combined with a high foliar S^0 application. Cysteine and SA trigger pathogen-related response in plants but defense mechanisms were not further enhanced by soil-applied S fertilization in combination with SA and cysteine spray applications, but obviously down-regulated. This suggests feedback regulation between S uptake, S metabolism, infection rate and SA and cysteine concentrations in the plant.

E. Bloem (✉) • S. Haneklaus • E. Schnug
Institute for Crop and Soil Science, Julius Kühn-Institute (JKI), Federal Research Centre for Cultivated Plants, Bundesallee 50, 38116 Braunschweig, Germany
e-mail: elke.bloem@jki.bund.de; ewald.schnug@jki.bund.de

L.J. De Kok et al. (eds.), *Sulfur Metabolism in Plants*, Proceedings of the International Plant Sulfur Workshop 1, DOI 10.1007/978-94-007-4450-9_27,

Field investigations revealed the positive effect of S fertilization on the resistance of crop plants against fungal pathogens (Haneklaus et al. 2007, 2009). The basic mechanisms of the so-called sulfur-induced resistance (SIR) are meanwhile understood and different S metabolites were shown to be involved in resistance mechanisms of crops against pathogens. The significance of these compounds for SIR has already been comprehensively discussed before (Bloem et al. 2005, 2007; Haneklaus et al. 2007). Yet, under field conditions it is not possible to trigger SIR consistently by S fertilizer practices. In this study a new experimental setup was chosen to get deeper insight into the mechanisms involved in S-mediated plant-pathogen response. Winter wheat was supplied with graded S rates and inoculated with powdery mildew (*Blumeria graminis*). Then plants were sprayed with three different stress-related compounds: Salicylic acid was used as a defense elicitor, cysteine as one major stress metabolite involved in SIR and S^0 which has a proven fungicidal effect in the tested host-pathogen relationship and which is discussed as a defense metabolite, too. The compounds were sprayed after symptoms of infection became visible in order to test the individual capacity of these compounds to influence the further course of pathogenesis.

Cysteine is the precursor of all relevant S-containing metabolites involved in SIR. In addition, cysteine can be directly involved in pathogen resistance (Vidhyasekaran 1988). In numerous field experiments it was shown that both, S fertilization and infection by fungal pathogens increased the cysteine content in the leaf tissue significantly whereby this effect was pronounced strongest and significantly in combination (Haneklaus et al. 2009). Besides its nutritional effect, the pesticidal efficacy of S^0 against pests and diseases is well established (Forsyth 1802; Hoy 1987). S^0 has been shown to be most efficient against powdery mildew infections (Hoy 1987; Reuveni 2001). The mode of action of this effective fungicide is still not known, but has been attributed to the lipophilic character of S^0, which may enter into the cell wall of the fungi where it disturbs redox reactions (Börner 1997).

The application of salicylic acid (SA) was tested as SA mediates the plant response to pathogen infection by initiating and maintaining the hypersensitive response and systemic acquired resistance (Malinowski et al. 2007). Synthesis of SA requires Coenzyme A (CoASH) in the β-oxidation pathway (Ryals et al. 1996), which displays a possible link to S metabolism (Haneklaus et al. 2007). During infection the concentration of SA may increase close to infection sites (Malinowski et al. 2007) and it was shown that treatment with exogenous SA induced genes of pathogenesis-related proteins (PR) and resistance indicating to the function as natural transduction signal (Malamy et al. 1990). In this context it is an interesting fact that genes of the primary and secondary S metabolism were up-regulated after pathogen attack (Kruse et al. 2007). In this study the question was addressed if an exogenous spraying of cysteine, S^0 or SA mediates the plant response to pathogenesis and whether such effect is related to the S supply of the crop.

A bi-factorial greenhouse experiment was conducted with three S levels (5, 25 and 50 mg S per pot as K_2SO_4) and seven different treatments of spray applications with stress-related compounds (H_2O as control, 50 and 100 mg pot^{-1} SA, cysteine and S^0). The chemicals were dissolved in 5 mL distilled water containing 0.1% Tween

Table 1 Infection rate (%) of winter wheat flag leaves with powdery mildew (*Blumeria graminis*) in relation to the S supply of flag leaves and older leaves at flowering before foliar application of stress-related metabolites

		Infection rate with powdery mildew[a] (% ± SD)	S content of flag leaves (mg g^{-1})	S content of older leaves (mg g^{-1})
Control	(5 mg S pot^{-1})	23.4 ± 6.8	3.2	2.7
S-1	(25 mg S pot^{-1})	28.9 ± 8.8	4.6	4.2
S-2	(50 mg S pot^{-1})	32.3 ± 9.1	4.8	3.9
$LSD_{5\%}$		*3.9*	*0.9*	*0.8*

[a]Infection rates before spray application are mean values from 28 pots

as detergent and sprayed on the infected plants. Four days after spray application, when first changes caused by the spray application became visible for the first time, the visual scoring of the infection rate was performed and further on after 6, 8 and 11 days. Eight plants of winter wheat (variety *Ritmo,* which has been rated as being medium resistant against powdery mildew) were grown in *Mitscherlich* pots, filled with 8 kg of pure sand. S was fertilized before planting and potassium was balanced with KCl. Plants were inoculated with powdery mildew (*Blumeria graminis f. sp. tritici*) during their main growth at BBCH 39–41 (Stauß et al. 1994). The infection was carried out by dispersing mycelium of severely infected young wheat plants in the headspace of the experimental plants. Three weeks after inoculation at BBCH 59 all plants showed an infection rate of on average 28% with a high variability in relation to the S supply (Table 1). Scoring of the infection rate was performed on the flag leaves before and regularly after foliar application of S compounds. The percentage of leaf area affected by the fungi was visually estimated according to Moll et al. (1996). Each treatment was conducted in fourfold repetition. Eleven days after spray application the plants were harvested and analyzed for total S by employing *Vario Max CNS* equipment (Elementar Analysensysteme GmbH, Hanau, Germany). Two-way ANOVA was used to analyze the results and means were compared by the *Tukey* test at 5% probability level.

A high variability of the infection rate with powdery mildew was observed before stress-related metabolites were foliarly applied (Table 1). Graded S supply to the soil resulted in a significant increase in the S content of young flag leaves as well as older leaves (Table 1). Surprisingly, the infection rate was significantly higher than in control pots when 25 and 50 mg S pot^{-1} were applied to the soil. Reasons for this effect remain at the moment speculative.

In Table 2 the infection rates 4, 6, 8 and 11 days after spraying were compared to the infection before foliar dressings were applied. Spraying of S^0 at the higher dose of 100 mg S per pot resulted in the highest reduction of the infection rate with powdery mildew. Already 4 days after S^0 application the infected leaf area was reduced by 16%, while the infection of control plants did not change during the same time but advanced later on. Over time S^0 dressings at the lower dose retarded mycelial growth compared to the control but yielded only a reduction 4 days after S^0 application. The higher S^0 rate had the strongest effect with a significant decline of

Table 2 Development of the infection rates (%) of winter wheat flag leaves with powdery mildew (*Blumeria graminis*) 4, 6, 8, and 11 days after spray applications with S^0, cysteine, and salicylic acid compared to the respective infection rates prior to foliar treatments

	Change of infection rate (%) after spray application and soil-applied S			
Days after application of metabolites	4	6	8	11
Control	+0.0	+13.8	+35.0	+20.0
50 mg S^0 (S^0-50)	−3.8	**0.0**	**+13.8**	**+6.3**
100 mg S^0 (S^0-100)	**−16.3**	**−6.3**	**−2.5**	**−10.0**
$LSD_{5\%}$	*9.0*	*13.5*	*10.6*	*11.5*
50 mg cysteine (cys-50)	−1.3	**−3.8**	**−8.8**	**−6.25**
100 mg cysteine (cys-100)	−3.8	**−1.3**	**+7.5**	**+2.5**
$LSD_{5\%}$	*8.2*	*10.1*	*8.3*	*11.9*
50 mg SA (SA-50)	−5.0	**+1.3**	**+5.0**	**+1.3**
100 mg SA (SA-100)	+2.5	+6.3	**+11.3**	+7.5
$LSD_{5\%}$	*8.0*	*10.8*	*11.5*	*15.9*
25 mg S pot^{-1} (S-25)	**+13.8**	+13.8	+26.3	+7.5
50 mg S pot^{-1} (S-50)	−5.0	**−5.0**	**−6.3**	**−6.3**
$LSD_{5\%}$	*8.7*	*12.9*	*15.0*	*13.0*

Bold numbers indicate statistically significant differences to control plants; data for spray applications with S^0, cysteine and SA include plants with a soil S supply of 5 mg pot^{-1}

mycelial growth 4 days after spraying, too. Hereafter the reduction of the infection rate kept at a high level (Table 2).

With a view to cysteine spray applications significantly reduced the infection rate at the lower dose at all dates but this effect was far less pronounced when compared to S^0. In comparison, the higher rate of cysteine reduced the infection rate after 4 and 6 days somewhat, while it proceeded thereafter though keeping visibly below the infection rate of the control plants (Table 2).

Application of SA reduced infection rate only at the lower dose and only 4 days after application. Later on fungal growth was retarded and with the lower dose of 50 mg pot^{-1} always significantly slower than in the control plants. Soil-applied S fertilization had a significant effect on the infection rate, too. The higher dose of 50 mg S pot^{-1} significantly and consistently decreased the infection rate. In contrast, the infection rate proceeded at a high level but tended to decrease after 11 days when only 25 mg S pot^{-1} were applied. As the S status of the crop plant was similar in both treatments (Table 1) it may be assumed that an S supply which exceeds the physiological requirement of the plant is beneficial to strengthen the natural resistance against fungal pathogens. Such effect has been proposed previously by Haneklaus et al. (2007, 2009).

For an easier comparison of results data are presented as relative changes after foliar dressings whereby the infection rate before application was set to 100% (Fig. 1). Values of higher than 100% indicate an increase while lower values indicate a reduction of the infection intensity with powdery mildew. Generally, the infection proceeded slower when plants were fully supplied with S though the

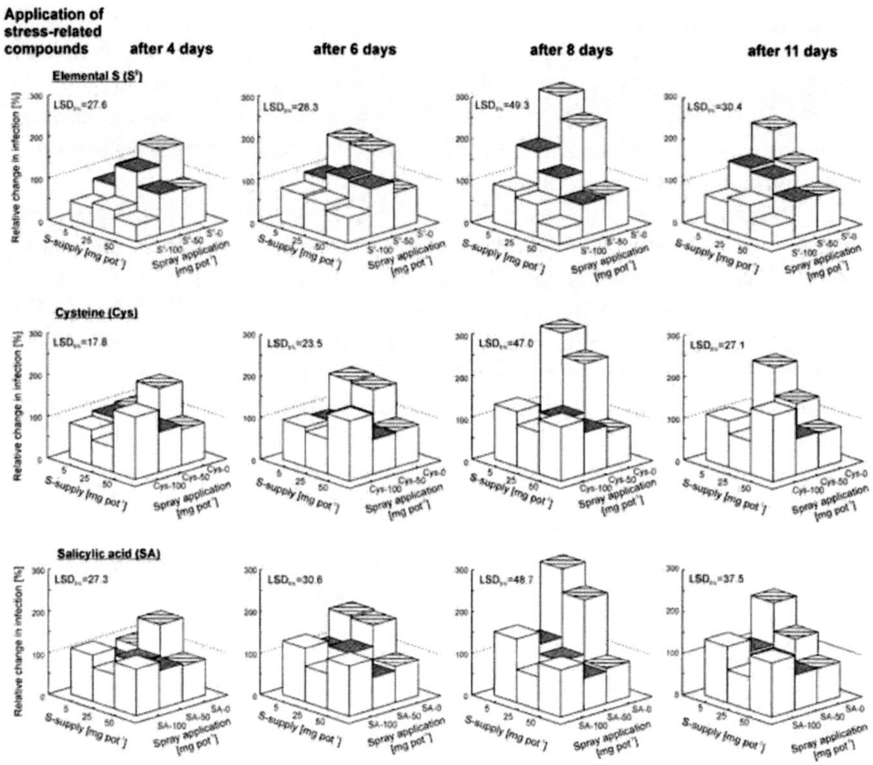

Fig. 1 Relative changes in the infection rate (%) of winter wheat flag leaves with powdery mildew (*Blumeria graminis*) *before* (100%) and *after* application of stress-related metabolites in relation to the external S supply

initial infection rate was higher in these treatments (Fig. 1). The infection rate was influenced strongest 8 days after spraying in relation to foliar dressing and S nutritional status. Spray application of all three stress-related compounds caused a reduction of the fungal infection rate with powdery mildew. Application of S^0 was most efficient with a significant reduction of 47% and 67% at rates of 50 and 100 mg pot^{-1}, respectively at an S supply of 25 mg pot^{-1} (Fig. 1). The application of the higher S dose to the soil (50 mg S pot^{-1}) yielded a reduction of the infection rate that was comparable with that of 50 mg cysteine and 100 mg S^0. Spraying of S^0 reduced the infection additionally by 6% and 56%, at an S^0 application of 50 and 100 mg pot^{-1}, respectively.

The pure SA effect equaled a significant reduction of the infection rate of 47% at the higher SA dose 8 days after spraying (Fig. 1). Soil-applied S fertilization alone caused an even greater reduction of 71% at the same time. Interactions between soil-applied S and foliar dressings proved to be significant. The combined effect of soil-applied sulfate (S-50) and SA spray application (SA-100) yielded a reduction of 60% (Fig. 1) compared to the control. Interactions between both factors resulted in a lower reduction of fungal infection than the pure fertilizer S effect. So, after 6 days the combination effect of SA and soil-applied S yielded only a reduction by 30% in contrast to the single SA (39%) and S treatment (50%) to the soil.

For foliar cysteine applications similar results were determined: 8 days after cysteine was sprayed it caused a reduction of the infection rate by 54%. In comparison, soil-applied S reduced the infection rate by 71%. The combination of S-50 and Cys-100 caused a reduction of still 55%. Six days after spraying this value decreased to only 16% while at the same time soil-borne S reduced the infection intensity by 50% and foliar cysteine by 45%. Just in case of foliar spraying of S^0 the strongest reduction of the infection rate was obtained at the highest S supply of the soil. Eight days after spraying, S^0 reduced the infection rate by 68%, soil-applied S by 71% and the combined application of S^0/S even by 87%.

While S^0 supposedly has a direct toxic effect on the leaf surface (Haneklaus et al. 2009), cysteine and SA are known to trigger defense mechanisms after uptake by the plant. The fact that foliar application of both compounds caused a reduction of the infection rate with powdery mildew while the combination effect of soil-applied sulfate with SA or cysteine dressings never boosted their individual effect supports the assumption that feedback regulation between sulfate uptake/sulfur status and cysteine and SA, and regulation of defense mechanisms exists. A promising compound that is actively involved in induced resistance and that was shown to regulate sulfate uptake is glutathione. However, further studies are required to verify further regulatory functions.

References

Bloem E, Haneklaus S, Schnug E (2005) Significance of sulfur compounds in the protection of plants against pests and diseases. J Plant Nutr 28:763–784

Bloem E, Haneklaus S, Salac I, Wickenhäuser P, Schnug E (2007) Facts and fiction about sulfur metabolism in relation to plant-pathogen interactions. Plant Biol 9:596–607

Börner H (1997) Pflanzenkrankheiten und Pflanzenschutz, 7th edn. Eugen Ulmer, Stuttgart

Forsyth W (1802) A treatise on the culture and management of fruit trees. Nichols and Son, London

Haneklaus S, Bloem E, Schnug E (2007) Sulfur and plant disease. In: Datnoff L, Elmer W, Huber D (eds) Mineral nutrition and plant diseases. APS Press, Saint Paul, pp 101–118

Haneklaus S, Bloem E, Schnug E (2009) Plant disease control by nutrient management: sulphur. In: Walters D (ed) Disease control in crops – biological and environmentally friendly approaches. Wiley/Blackwell, Oxford, pp 221–236

Hoy MA (1987) Sulfur as a control agent for pest mites in agriculture. In: Proceedings of the international symposium elemental sulphur in agriculture, vol 1. Nice, France, 25–27 March 1987, pp 51–61

Kruse C, Jost R, Lipschis M, Kopp B, Hartmann M, Hell R (2007) Sulfur-enhanced defence: effects of sulfur metabolism, nitrogen supply, and pathogen lifestyle. Plant Biol 9:608–619

Malamy J, Carr JP, Klessig DF, Raskin I (1990) Salicylic acid: a likely endogenous signal in the resistance response of tobacco to viral infection. Science 250:1002–1004

Malinowski J, Krzymowska M, Godon K, Hennig J, Podstolski A (2007) A new catalytic activity from tobacco converting 2-coumaric acid to salicylic aldehyde. Physiol Plant 129:461–471

Moll E, Walther U, Flath K, Prochnow J, Sachs E (1996) Methodische Anleitung zur Bewertung der partiellen Resistenz von Sorten bzw. Linien unter Berücksichtigung epidemiologischer Aspekte. Ber Biol Bundesanst Land- Forstwirtsch 12, 7–20

Reuveni M (2001) Activity of trifloxystrobin against powdery and downey mildew diseases of grapevines. Can J Plant Path 23:52–59

Ryals JA, Neuenschwander UH, Willits MG, Molina A, Steiner H-Y, Hunt MD (1996) Systemic acquired resistance. Plant Cell 8:1809–1819

Stauß R, Bleiholder H, van der Boom T, Buhr L, Hack H, Hess M, Klose R, Meier U, Weber E (1994) Einheitliche Codierung der phänologischen Entwicklungsstadien mono- und dikotyler Pflanzen. Ciba-Geigy AG, Basel

Vidhyasekaran P (1988) Physiology of disease resistance in plants, vol II. CRC Press, Inc, Boca Raton, USA

Elevated Atmospheric CO_2 Affects Grain Sulfur Concentration and Grain Nitrogen/Sulfur Ratio of Wheat (*Triticum aestivum* L.)

Nimesha Fernando, Joe Panozzo, Michael Tausz, Robert M. Norton, Glenn Fitzgerald, and Saman Seneweera

Abstract Wheat (*Triticum aestivum* L. cv. Yitpi) was grown under field conditions in the Australian Grains Free-Air Carbon Dioxide Enrichment (AGFACE) facility during the 2008 and 2009 growing seasons. Current atmospheric (384 μmol mol^{-1}) and elevated CO_2 concentration (550 μmol mol^{-1}) were combined with two different times of sowing (TOS). The "normal sowing" date (TOS_1) contrasted with "late sowing" (TOS_2) to provide different growing conditions including higher temperatures during grain filling. Nitrogen (N) and sulfur (S) concentrations and N/S ratios were analyzed in mature grains. Elevated CO_2 concentration significantly reduced grain N by 12.3% and 13.1%, and S concentration by 4.8% and 9.9% in the 2008 and 2009 growing seasons, respectively. The largest reduction in grain N and S concentrations at elevated CO_2 relative to ambient CO_2 concentration was observed

N. Fernando • S. Seneweera (✉)
Department of Agriculture and Food Systems, Melbourne School of Land and Environment, The University of Melbourne, Natimuk Road, Private Box 260 Horsham, VIC 3401, Australia
e-mail: samans@unimelb.edu.au

J. Panozzo • G. Fitzgerald
Department of Primary Industries, Natimuk Road, Private Box 260, Horsham, VIC 3401, Australia

M. Tausz
Department of Forest and Ecosystem Science, Melbourne School of Land and Environment, The University of Melbourne, Water Street, Creswick, VIC 3363, Australia

R.M. Norton
Department of Agriculture and Food Systems, Melbourne School of Land and Environment, The University of Melbourne, Natimuk Road, Private Box 260 Horsham, VIC 3401, Australia

International Plant Nutrition Institute, 54 Florence St, Horsham, VIC 3400, Australia

L.J. De Kok et al. (eds.), *Sulfur Metabolism in Plants*, Proceedings of the International Plant Sulfur Workshop 1, DOI 10.1007/978-94-007-4450-9_28,

at TOS_2. Regardless of CO_2 concentration, TOS_2 increased grain N concentration by 42.7% and 16.5%, and S concentration by 26.7% and 18.8% in 2008 and 2009, respectively. As a result, N/S ratio was reduced by 7.6% under elevated CO_2 concentration in 2008 and the trend was similar in 2009. Overall, our results suggest that both elevated CO_2 concentration and TOS are likely to modify grain S and N concentrations and ratios, which play an important role in determining the nutritive value and baking quality of wheat grain.

The increase in atmospheric concentration of carbon dioxide ([CO_2]), a major greenhouse gas, is one of the most relevant contributing factors to global climate change. Atmospheric [CO_2] is expected to increase from currently 384 μmol mol^{-1} to 550 μmol mol^{-1} by the middle of the twenty-first century (Carter et al. 2007). It is predicted that global temperatures will concomitantly increase by an average of 1.5–4.5°C by 2100 and there may be more frequent occurrences of extreme climatic events such as heat waves and/or droughts (Carter et al. 2007). On the other hand, elevated [CO_2] can have a positive effect on plant growth, yield formation and development, because C_3 photosynthesis is limited by current atmospheric [CO_2] (Seneweera et al. 1996). In addition, elevated [CO_2] is likely to alter plant carbon (C) and nitrogen (N) metabolism from the cellular to the whole plant level leading to changes in chemical composition in vegetative plant parts (Loladze 2002; Seneweera et al. 2005) and in grains (Erbs et al. 2010; Högy et al. 2009). Moreover, it has been reported that high [CO_2] tends to reduce grain sulfur (S) concentration in wheat (Erbs et al. 2010). Although wheat requires a relatively low amount of S, sulfur can exert a large influence on yield and grain quality (Zhao et al. 1999). Sulfur is contained in essential amino acids such as cysteine and methionine, hence functionally important in many proteins. Wheat proteins, especially metabolic and several types of gluten proteins are rich in S containing amino acids (Shewry and Halford 2002). Thus, sulfur deficiency can lead to reduced synthesis of gluten and metabolic proteins. Strong correlations between S and rheological properties of dough have been well documented (Vensel et al. 2005). It has been well established that elevated [CO_2] results in a reduction of N and protein concentration in both vegetative parts and grains (Taub et al. 2008). However, the mechanism/s by which [CO_2] enrichment reduces grain protein content is not well understood. There is a strong inter-dependence of N and S metabolism in crop plants (Stewart and Porter 1969), but there is little information available on how elevated [CO_2] and variable growth conditions interact on grain N and S concentrations and its ratios. This paper addresses this open question in a Free Air CO_2 Enrichment (FACE) facility.

Wheat (*Triticum aestivum* L. cv. Yitpi, a hard wheat, bread-making cultivar) was grown under field conditions at the Australian Grains FACE (AGFACE) facility, at a satellite site in Walpeup, Victoria, Australia (35°12′S, 142°00′E) during the 2008 and 2009 growing seasons. Ambient [CO_2] (384 μmol mol^{-1}) and elevated [CO_2] (550 μmol mol^{-1}) were combined with two different times of sowing (TOS): A "normal sowing" date (TOS_1) contrasted with "late sowing" (TOS_2) to investigate this interaction with [CO_2]. Growth conditions of TOS_1 and TOS_2 in both growing

Table 1 Growing conditions during the experiment

	Total water supplied (R+I) mm	Days T 30–35°C	Days T >35°C	TMAX during grain filling (°C)	TMAX during growing season (°C)	Grain filling days
2008-TOS1	37.4	4	3	25.5	19.0	43
2008-TOS2	37.4	4	4	26.6	20.2	37
2009-TOS1	97.8	2	3	26.2	20.3	67
2009-TOS2	0	3	12	35.5	22.8	19

TMAX mean daily maximum temperature, *R*, rainfall, *I* irrigation, *T* temperature, *TOS* time of sowing

seasons were as described in Table 1. The experimental design was a completely randomized block design with 16 rings of 4 m diameter each. Each treatment combination was replicated four times. CO_2 enrichment started at plant emergence and continued until grain maturity stage. Grains were harvested at physiological maturity. Harvested ears were dried at room temperature and grains were separated and aspirated (Vaccum separator, Kimseed, Australia). Total grain N concentration was measured by near infrared reflectance spectroscopy (NIR, Foss, Sweden) based on 11% moisture content. Followed by an acid digestion, grain S concentration was analyzed using inductively-coupled plasma atomic emission spectrometry (Applied Research Laboratories, 3580B, Switzerland). Mineral concentration (g kg^{-1}) was expressed on a grain dry weight basis. N/S ratio was calculated on mass basis.

Grain N concentration was 26.8 gkg^{-1} (2008) and 27.2 gkg^{-1} (2009) under ambient [CO_2] and 23.5 gkg^{-1} (2008) and 23.7 gkg^{-1} (2009) under elevated [CO_2], showing 12.3% and 13.1% reduction at elevated [CO_2] in both years. A greater reduction in grain N concentration at elevated [CO_2] relative to ambient [CO_2] was observed at TOS_2. Regardless of [CO_2] treatment, grain N concentration was higher at TOS_2 in both years (29.6 gkg^{-1}, 27.4 gkg^{-1}) than at TOS_1 (20.8 gkg^{-1}, 23.5 gkg^{-1}), which means an increase at TOS_2 relative to TOS1 by 42.7% and 16.5% in 2008 and 2009, respectively (Table 2).

Grain S concentration significantly declined by 9.9% at elevated [CO_2] in 2009. In contrast, smaller reduction of 4.8% was observed in 2008. A greater reduction in grain S concentration at elevated [CO_2] was observed at TOS_2 in both years (Table 2). However, grain S concentration increased at TOS_2 relative to TOS_1 by 26.7% and 18.8% in 2008 and 2009 respectively. The N/S ratio was significantly reduced under elevated [CO_2], by 7.6% in 2008 and by 3.7% in 2009. Reduction of N/S ratio at elevated [CO_2] was greater in TOS_1 than at TOS_2 in both years. In contrast, increase in N:S ratio was observed at TOS_2 compared to TOS_1, by 26.7% and 18.8% in 2008 and 2009, respectively (Table 2). Increases in grain nutrient concentrations at TOS_2 versus TOS_1 are probably related to lower gain yield (data not shown) under TOS_2 conditions.

Approximately 80% of N and S are invested in protein in many plant species. The maintenance of N and S ratio at an optimum level is central to protein synthesis

Table 2 Grain N and S concentrations and N/S ratio of *Triticum aestivum* L. cv. Yitpi grown under two CO_2 concentrations (384 and 550 μmol mol^{-1}) and two times of sowing (TOS_1 and TOS_2) during 2008 and 2009 growing seasons

Main effects						
2009	aCO_2	eCO_2	% Change	TOS_1	TOS_2	% Change
Grain N (g kg^{-1})	27.2	23.7	−13.1	23.5	27.4	16.5
Grain S (g kg^{-1})	1.83	1.65	−9.9	1.6	1.9	18.8
N/S ratio	14.9	14.4	−3.7	14.8	14.5	−1.7
2008						
Grain N (g kg^{-1})	26.8	23.5	−12.3	20.8	29.6	42.7
Grain S (g kg^{-1})	1.75	1.66	−4.8	1.5	1.9	26.7
N/S ratio	15.2	14.1	−7.6	13.8	15.5	12.5
Interactive effects						
2009	TOS_1 aCO_2	TOS_1 eCO_2	% Change	TOS_2 aCO_2	TOS_2 eCO_2	% Change
Grain N (g kg^{-1})	24.8	22.2	−10.6	29.7	25.2	−15.2
Grain S (g kg^{-1})	1.7	1.5	−7.1	2.0	1.8	−12.1
N/S ratio	15.1	14.5	−3.8	14.8	14.3	−3.5
2008						
Grain N (g kg^{-1})	22.0	19.5	−11.3	31.7	27.5	−13.1
Grain S (g kg^{-1})	1.52	1.49	−2.3	2.0	1.8	−6.7
N/S ratio	14.4	13.2	−8.8	16.0	15.0	−6.6
ANOVA						
2009		CO_2		TOS		$CO_2 \times$ TOS
Grain N (g kg^{-1})		*		*		ns
Grain S (g kg^{-1})		**		***		ns
N/S ratio		ns		*		ns
2010						
Grain N (g kg^{-1})		*		***		ns
Grain S (g kg^{-1})		ns		***		ns
N/S ratio		**		***		ns

Summary of two-way ANOVA results are shown
ns not significant ($p>0.05$)
$*p<0.05$; $**p<0.01$; $***p<0.001$

(Zhao et al. 1999). If the N/S ratio is not maintained at an optimum of around 15/1 (w/w), then protein synthesis will be suppressed (Zhao et al. 1999). Our results demonstrate that N/S ratio was reduced by 7.6% and 3.7% at elevated [CO_2] compared to the optimum value (15/1) in both the 2008 and 2009 growing season. Reduction in N/S ratio at elevated [CO_2] was due to greater reduction of N concentration relative to the reduction in S concentration under elevated [CO_2]. Reduction in grain N concentration was almost similar in both growing seasons, but the reduction in grain S concentration was greater in 2009 than in 2008. This reduction in grain S concentration in the 2009 growing season could be due to lower S uptake possibly linked to accelerated plant development, due to both higher growth temperature and drought stress that plants experienced during grain filling in 2009 (Table 1).

Experimental evidence suggests that grain N concentration is generally decreased at elevated [CO_2] (Taub et al. 2008). In addition, a strong inter-dependence of N and S metabolism has been reported (Stewart and Porter 1969). Although N and S metabolism are closely related, insufficient S supply can lead to transient and steady state nitrate accumulation and/or perturbations in specific amino acid pools, because S deficiency may have a greater effect on N assimilation than on N uptake (Zhao et al. 1999). It is likely that S deficiency could further suppress N assimilation at elevated [CO_2].

Wheat grains consist of 8–20% protein, which is classified into three main groups – structural, metabolic and storage proteins (Delmolino et al. 1988). Metabolic proteins are rich in S-containing amino acids cysteine (Cys) and methionine (Met) (Woodman and Engledow 1924). Storage proteins are considered gluten proteins, which form viscoelastic networks during dough mixing and correlate with rheological properties (Vensel et al. 2005). Gluten proteins differ greatly in the content of cysteine residues and therefore are classified as S-poor (ω5-, ω1,2-gliadins), S-rich (α- and γ-gliadins and low molecular weight glutenin subunits) and intermediate S containing gluten protein fractions (high molecular weight glutenin subunits; Shewry and Halford 2002). It has been suggested that elevated CO_2 concentration reduces ω5-, ω1,2-gliadins and HMW glutenin subunits which contain relatively low S content, more than other types of proteins (Wieser et al. 2008). Comparison of the total N and S ratio to the ratio of protein types (S-poor, S-rich) would provide more information on how N and S metabolism is modified under elevated [CO_2].

This study concludes that elevated [CO_2] decreased the N/S ratio in wheat grain which is probably linked to whole plant N and S transport and/or assimilation. The overall reduction of N/S ratio was greater at the earlier sowing dates, suggesting that elevated [CO_2] mediated reduction in N and S concentration is dependent on growing conditions, among them temperatures during grain filling is critical in determining the S:N ratio.

Acknowledgements Research at the Australian Grains Free Air CO_2 Enrichment (AGFACE) facility is jointly run by the Victorian State Department of Primary Industries and the University of Melbourne. The authors gratefully acknowledge financial support by the Australian Commonwealth Department of Agriculture, Fisheries and Forestry (DAFF), The Grains Research and Development Corporation (GRDC), and the International Plant Nutrition Institute. We thank Mahabubur Mollah, research engineer, for operating the AGFACE facilities, and Ivan Mock for site management in Walpeup. Peter Howie, Cassandra Walker, Nathan Neumann, and Russel Argall and the AGFACE team are gratefully acknowledged for their technical support.

References

Carter TR, Jones RN, Lu X (2007) New assessment methods and the characterisation of future conditions. In: Parry ML, Canziani OF, Palutikof JP, van der Linden PJ, Hanson CE (eds) Climate change (2007) Impacts, adaptation and vulnerability. IPCC, Cambridge University Press, Cambridge, pp 133–171 (Contribution of working group II to the fourth assessment report of the intergovernmental panel on climate change)

Delmolino IMM, Rojo B, Martinezcarrasco R, Perez P (1988) Amino-acid composition of wheat-grain. 1. Changes during development. J Sci Food Agric 42:29–37

Erbs M, Manderscheid R, Jansen G, Seddig S, Pacholski A, Weigel HJ (2010) Effects of free-air CO_2 enrichment and nitrogen supply on grain quality parameters and elemental composition of wheat and barley grown in a crop rotation. Agric Ecosyst Environ 136:59–68

Högy PH, Wieser P, Kohler K, Schwadorf J, Breuer J, Erbs M, Weber S, Fangmeier A (2009) Does elevated atmospheric CO_2 allow for sufficient wheat grain quality in the future? J Appl Bot Food Qual 82:114–121

Loladze I (2002) Rising atmospheric CO_2 and human nutrition: toward globally imbalanced plant stoichiometry? Trends Ecol Evolut 17:457–461

Seneweera S, Blakeney A, Milham P, Basra AS, Barlow EWR, Conroy J (1996) Influence of rising atmospheric CO_2 and phosphorus nutrition on the grain yield and quality of rice (*Oryza sativa* cv. Jarrah). Cereal Chem 73:239–243

Seneweera S, Makino A, Mae A, Basra AS (2005) Response of rice to p(CO_2) enrichment: the relationship between photosynthesis and nitrogen metabolism. J Crop Improv 13:31–53

Shewry PR, Halford NG (2002) Cereal seed storage proteins: structures, properties and role in grain utilization. J Exp Bot 53:947–958

Stewart BA, Porter LK (1969) Nitrogen-sulfur relationships in wheat (*Triticum aestivum* L), corn (*Zea mays* L), and beans (*Phaseolus vulgaris* L). Agron J 61:267–271

Taub DR, Miller B, Allen H (2008) Effects of elevated CO_2 on the protein concentration of food crops: a meta-analysis. Global Change Biol 14:565–575

Vensel WH, Tanaka CK, Cai N, Wong JH, Buchanan BB, Hurkman WJ (2005) Developmental changes in the metabolic protein profiles of wheat endosperm. Proteomics 5:1594–1611

Wieser H, Manderscheid R, Erbs M, Weigel HJ (2008) Effects of elevated atmospheric CO_2 concentrations on the quantitative protein composition of wheat grain. J Agr Food Chem 56:6531–6535

Woodman HE, Engledow KL (1924) A chemical study of the development of the wheat grain. J Agr Sci 14:563–586

Zhao FJ, Hawkesford MJ, McGrath SP (1999) Sulphur assimilation and effects on yield and quality of wheat. J Cereal Sci 30:1–17

Sulfate and Nitrate Assimilation in Leaves of *Quercus ilex* and *Quercus pubescens* Grown Near Natural CO_2 Springs in Central Italy

Cornelia Herschbach, Monika Schulte, Peter von Ballmoos, Christian Brunold, and Heinz Rennenberg

Abstract The effect of long-term exposure to elevated pCO_2 concentrations on sulfate and nitrate assimilation was studied under field conditions using leaves from *Quercus ilex* and *Quercus pubescens* trees growing with ambient or elevated CO_2 concentrations in the vicinity of three natural CO_2 springs, Bossoleto, Laiatico and Sulfatara, in Tuscany, Italy. The activity of the key enzymes of sulfate assimilation, adenosine 5′-phosphosulfate reductase (APR) and nitrate assimilation, nitrate reductase (NR), were measured together with the levels of acid soluble thiols, and soluble non-proteinogenic nitrogen compounds. Whereas NR activity remained unaffected in *Q. ilex* or increased *Q. pubescence*, APR activity decreased in the area of CO_2 springs. The latter changes were often accompanied by increased GSH concentrations, apparently synthesized from H_2S and SO_2 present in the gas mixture emitted from the CO_2 springs. Thus, the diminished APR activity in leaves of *Q. ilex* and *Q. pubescence* from spring areas can best be explained by the exposure to gaseous sulfur compounds. Although the concentrations of H_2S and SO_2 in the gas mixture emitted from the vents at the CO_2 springs were low at the Bossoleto and Laiatico spring, these sulfur gases pose physiological effects, which may override consequences of elevated pCO_2.

C. Herschbach (✉) • M. Schulte • H. Rennenberg
Institute of Forest Botany and Tree Physiology, Chair of Tree Physiology, Albert-Ludwigs-University Freiburg, Georges-Köhler-Allee 053/054, 79085 Freiburg im Breisgau, Germany
e-mail: cornelia.herschbach@ctp.uni-freiburg.de

P. von Ballmoos • C. Brunold
Institut für Plant Sciences, University Bern, Altenbergrain 21, 3013 Bern, Switzerland

L.J. De Kok et al. (eds.), *Sulfur Metabolism in Plants*, Proceedings of the International Plant Sulfur Workshop 1, DOI 10.1007/978-94-007-4450-9_29,

The current increase in atmospheric pCO_2 may affect the physiology, the development and the growth of plants. Wealth of information concerning these effects has been published from experiments, in which plants were exposed to elevated pCO_2 under controlled conditions in different experimental set ups starting from controlled growth chambers, open top chambers and Free-Air CO_2 Enrichment (FACE) experiments. Within these experiments the pCO_2 concentration was enhanced from ambient to elevated pCO_2 levels within one step. This approach does not allow long-term acclimation to rising pCO_2 concentrations over several generations, especially relevant for plant communities and trees. Vegetation, that grows for generations in a naturally pCO_2-enriched atmosphere around pCO_2 springs, provide a unique opportunity to address questions of long-term adaptation. At these springs, pCO_2 of geological origin is continuously released to the atmosphere through natural vents, thus producing local increases in the atmospheric pCO_2 concentrations as expected in a future pCO_2-enriched world (Miglietta and Raschi 1993; Schulte et al. 1999). These conditions are ideal for assessing physiological responses (Körner and Miglietta 1994) and micro-evolutionary adaptations (Schulte et al. 2002) to long-term elevated pCO_2 that cannot be simulated experimentally.

At three selected spring sites, *i.e.* near Laiatico, Bossoleto and Solfatara, the pCO_2 enrichment was comparable to the pCO_2 concentrations expected at the end of this century (Miglietta et al. 1993; Körner and Miglietta 1994). However, the gas mixture emitted from these vents contains SO_2 and H_2S in addition to CO_2 (Table 1, Schulte et al. 1999), pollutants which have been shown to affect the physiology of plants (De Kok et al. 1998) and may even be phytotoxic (De Kok 1990). The average H_2S and SO_2 concentrations did not reach levels thought to mediate visible symptoms of injury (De Kok 1990; Schulte et al. 1999) and, indeed, symptoms of H_2S and SO_2 damage were not observed at the field sites. Nevertheless, it can be assumed that atmospheric sulfur is taken up by the leaves and is used as an additional sulfur source (Brunold and Erismann 1975; De Kok 1990). Exposure of the shoot to atmospheric H_2S or SO_2 normally enhances the GSH content of leaves and roots (De Kok 1990), but high levels of cysteine were also detected (Buwalda et al. 1988; De Kok et al. 1988). This increase in thiols in the presence of SO_2 and H_2S may be explained by a direct incorporation of the sulfur into cysteine and the usage of cysteine for the formation of GSH which have been shown to decrease the activity of ATP sulfurylase (Lappartient and Touraine 1996; Lappartient et al. 1999) and APS reductase (APR, Westerman et al. 2001; Vauclare et al. 2002).

In *A. thaliana* shoots, not only the content of water-soluble non protein-SH compounds, but also the organic N in the shoot increased significantly by SO_2 exposure (Van der Kooij et al. 1997), indicating an interaction between sulfur and nitrogen metabolism. Such an interaction has been established for a long time (reviewed in Brunold 1993; Brunold et al. 2003; Kopriva and Rennenberg 2004) and has also been described at the whole plant level (Kruse et al. 2007). An interaction between sulfur and carbon metabolism is evident from experiments with *Arabidopsis* and *Lemna*. Here APR activity depends on light, is increased by sucrose and glucose treatment and is strongly reduced when CO_2 is omitted (Kopriva et al. 1999, 2002; Hesse et al. 2003). Also nitrate reductase (NR), the key enzyme of nitrate assimilation,

Table 1 Summary of the mean CO_2, SO_2 and H_2S concentration measured at the three natural CO_2-springs (Schulte et al. 1999)

	Geographical coordinates	CO_2 ($\mu l\ l^{-1}$)	H_2S ($nl\ l^{-1}$)	SO_2 ($nl\ l^{-1}$)	Time of emission
Bossoleto (Rapolano)	43°17′N; 11°35′E	1,074	22	12	Night
Laiatico	43°24′N; 10°50′E	795	22	4	Day
Solfatara	42°30′N; 12°08′E	797	245	18	Day and Night

interacts with carbon metabolism. Elevated pCO_2 reduced the decline of NR activity during the second part of the photoperiod and partially reversed the dark inactivation in herbaceous plants (Scheible et al. 1997; Geiger et al. 1998). The increased stability of NR at elevated CO_2 might be a result of the higher level of sugars at elevated pCO_2 (Kaiser and Huber 1994). In tobacco elevated pCO_2 increased the turn-over of amino N, most remarkable Glu and Ala, in mature leaves indicating enhanced nitrate reduction (Kruse et al. 2003).

Effects of elevated pCO_2 in oak depend on the origin of *Q. ilex* acorns (Schulte et al. 2002). Elevated pCO_2 decreased APR activity and thiol levels in *Q. ilex* leaves when acorns were collected from control area. However, although thiols were reduced at elevated pCO_2, when acorns were collected from the Laiatico spring area, APR activity was not (Schulte et al. 2002). NR increased at elevated pCO_2 when *Q. ilex* trees were grown from acorns originating from the spring area. As these results indicate long-term acclimation to elevated pCO_2, micro evolutionary adaptation of sulfur and nitrogen nutrition as a consequence to elevated pCO_2 in *Q. ilex* and *Q. pubescens* are expected. To test this assumption, three natural springs with elevated pCO_2 and corresponding control areas with ambient atmospheric pCO_2 were selected in Tuscany, Italy. The key enzymes of both pathways, *i.e.* APR and NR, were investigated in combination with analyses of reduced S and N compounds.

Mature leaves were randomly harvested between 10 am and 14 pm in June 1995 and 1996 from 2 years old twigs of six *Q. ilex* and *Q. pubescens* trees grown at three different natural CO_2-springs or at the corresponding control areas each. The Bossoleto (Rapolano) spring is situated at about 40 km SE of Siena at 250 m (Körner and Miglietta 1994) and released pCO_2 mainly in the night (Table 1, Schulte et al. 1999). Thereby the atmospheric pCO_2 ranged from 325 $\mu l\ l^{-1}$ up to peak values of 8,097 $\mu l\ l^{-1}$ in the night. The emitted gas mixture also contained trace amounts of H_2S (2–83 $nl\ l^{-1}$ H_2S) and SO_2 (5–23 $nl\ l^{-1}$ SO_2, Schulte et al. 1999). The spring is surrounded by a typical Mediterranean forest composed mainly of *Q. ilex* and *Q. pubescens*. The control site was at a distance of 5 km from the source at Poggia San Cecilia, where the vegetation, the climate and the soil were comparable to the Bossoleto spring (A. Raschi, personal communication). The Laiatico spring is situated on the slope of a hill near the village of Laiatico, which is at 10 km W from the city of Volterra. CO_2 is emitted from one major and a number of smaller vents at the foot of the slope mainly during the day. The atmospheric pCO_2 ranged from 209 up to 2,815 $\mu l\ l^{-1}$ (Schulte et al. 1999). The gas mixture at the Laiatico spring was comparable with that observed at the Bossoleto spring (Table 1). During the day the atmospheric H_2S concentration ranged from 2 to 65 nl l^{-1}, the atmospheric SO_2 concentration from 0.25 to 4 nl l^{-1}. The slope is covered with a dense, 20 year old forest consisting mainly of *Q. ilex* and *Q. pubescens* and a few *Fraxinus excelsior* trees. Leaves from trees at distances up to 200 m from the main hole of the vent were harvested for analyses. The Solfatara spring is situated close to the village Grotte San Stefano, which is located near the city of Viterbo. At the Solfatara site there are a large number of holes along a brook emitting gas during the day and the night with peak values of atmospheric pCO_2 of 2,044 μl l^{-1} (Miglietta et al. 1993; Schulte et al. 1999). The gas mixture emitted at the Solfatara site contained sulfur gases in much higher concentrations compared to the two other sites (Table 1). pH_2S showed peak values up to 850 nl l^{-1} (Schulte et al. 1999), the atmospheric pSO_2 concentration increased up to 140 nl l^{-1}. The control site was at 400 m E from these holes. The vegetation, the climate and the soil were similar in the whole transect.

Thiols were extracted from approx. 50 mg powdered leaf material. Extraction, reduction and derivatization were performed as described (Schulte et al. 2002). Thiol derivatives were separated and quantified after HPLC analysis by fluorescence detection (Schupp and Rennenberg 1988). Powdered leaf samples were taken for the extraction of soluble amino compounds based on Winter et al. (1992) and were separated by ion exchange chromatography as previously described (Schneider et al. 1996). After post-column derivatization with ninhydrin absorption of the amino-ninhydrin derivatives was measured at 440 and 570 nm. For measurement of nitrate reductase and APR activity, leaves stored in liquid nitrogen were homogenised in the frozen state using a micro-dismembrator (B. Braun, Melsungen AG, Melsungen, Germany). The resulting frozen powder was transferred into phosphate buffer (0.1 M, pH 7.7) containing 1% (v/v) Tween 80, 4% (w/v) PVPP K30, 5 mM EDTA, 10 mM DTE, 10 mM L-cysteine and 20 μM FAD, using 10 parts (w/v) of buffer per one part of frozen leaf material. The resulting suspension was homogenized using a Polytron (Kinematica, Littau, Switzerland). The homogenates were made cell-free by passing through two layers of viscose fleece (Millette, Migros, Switzerland). An aliquot of 190 μl of this crude extract was added to 810 μl of the assay mixture for NR composed of 25 mM phosphate buffer, pH 7.5, 3.5 mM KNO_3 and 0.15 mM of each NADPH and NADP (Neyra and Hagemann 1975). These reductants were omitted from the blanks. The NR activity was linear at 30°C for at least 20 min of incubation. After 20 min the reaction was stopped by adding 400 μl of 125 mM zinc acetate. The nitrite produced in an aliquot was detected spectrophotometrically at 540 nm after deazotization in a 1:1 mixture of 1% (w/v) sulfanylamide in 1.5 M HCl and 0.02% (w/v) N-(1-naphtyl) ethylene diamine dihydrochloride. APR activity was determined according to Brunold and Suter (1990) by measuring the acid volatile radioactivity produced from $AP^{35}S$ using DTE as the reductant (Suter et al. 2000). Radioactivity was determined by liquid scintillation spectrometry using a Betamatic instrument (Kontron Instruments, Zürich, Switzerland). Proteins in the crude extracts were determined according to Bradford (1976) after precipitation with 10% trichloric acid and subsequent solubilization in 0.1 M KOH using BSA as a standard.

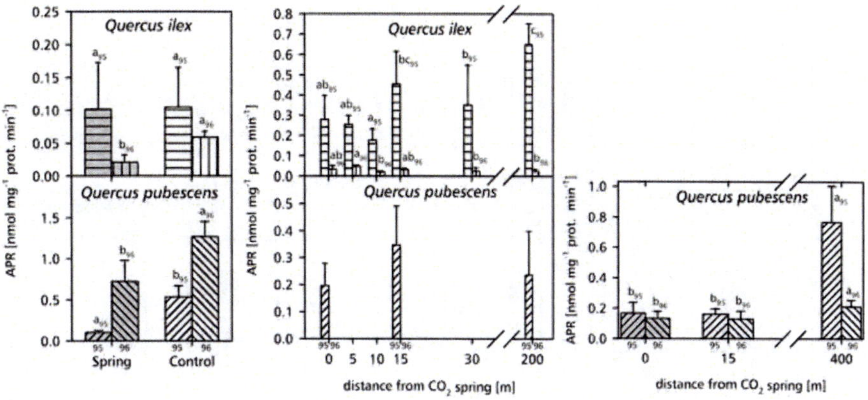

Fig. 1 Activity of APR in leaves of *Q. ilex* and *Q. pubescens* trees growing at the CO_2 springs at Bossoleto (left), Laiatico (middle), and Sulfatara (right) and at corresponding control sites. Mean values from six leaves collected from six different trees (±SD) are presented for 1995 (*striped diagonal up* for *Q. pubescence*; *striped horizontal* for *Q. ilex*) and 1996 (*striped diagonal down* for *Q. pubescence*; *striped vertical* for *Q. ilex*). Values carrying different letters are different at $P \leq 0.05$ within one sampling date. APR was not measured in *Q. pubescens* from the Laiatico spring in 1996

Data shown represent means (±SD) of measurements from individual mature leaves from different trees (n=6) of comparable size. Statistical analysis of the results was done using the Student's t-test after testing for normality and equal variance and the multifactorial Duncan test (SPSS for Windows, 7.0). If normality was not present, a Mann–Whitney rank sum test was used.

Comparison of the activity of APR from leaves of oak trees growing near the springs with those from various control sites results in two patterns: Either the levels were the same (*Q. ilex,* Bossoleto 1995 and Laiatico 1996; *Q. pubescens,* Laiatico 1995) or APR levels at the spring area were significantly lower than at the control areas (*Q. ilex,* Bossoleto 1996, Laiatico 1995; *Q. pubescens,* Bossoleto 1995 and 1996, Solfatara 1995 and 1996) (Fig. 1). In the leaves of *Q. pubescens* from the Bossoleto and the Solfatara site APR activities increased with increasing distance from the spring area in parallel with decreasing NR activities (Fig. 2). At the Laiatico site, NR activity of *Q. ilex* leaves was not significantly different between the spring area, the control area at 200 m distance, and areas of intermediate distances from the spring (Fig. 2). In *Q. pubescens* leaves a significantly lower NR activity was measured at the spring than at 15 m distance and at the control area in 1995 (Fig. 2). NR activities of *Q. pubescens* leaves were much lower in 1995 than in 1996. The same pattern was found in leaves of *Q. pubescens* from the Bossoleto site, indicating that an additional ecological factor was regulating NR activity in 1995. At the Bossoleto site, the level of NR activity was typically higher at the spring than at the control area. However, significant differences were only found in *Q. pubescens* leaves in 1996 (Fig. 2). At Solfatara site, NR activity in leaves of *Q. pubescens* was

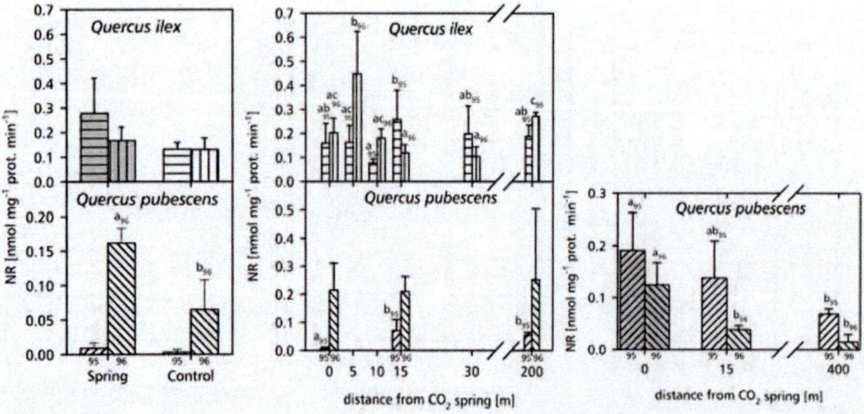

Fig. 2 Activity of NR in leaves of *Q. ilex* and *Q. pubescens* trees growing at the CO_2 springs near Bossoleto (left), Laiatico (middle) and Sulfatara (right) and at corresponding control sites. Mean values from six leaves collected from six different trees (±SD) are presented for 1995 (*striped diagonal up* for *Q. pubescence*; *striped horizontal* for *Q. ilex*) and 1996 (*striped diagonal down* for *Q. pubescence*; *striped vertical* for *Q. ilex*). Values carrying different letters are different at $P \leq 0.05$ within one sampling date. At the Solfatara spring only *Q. pubescens* was present

always higher at the spring than at the control area at 400 m distance from the spring in both years, 1995 and 1996 (Fig. 2). Even at intermediate distance from the spring, NR activity was lower than at the spring area.

At the Bossoleto and the Solfatara site, the level of cysteine and GSH measured in the leaves of both species was typically higher at the spring than at the control area (Fig. 3). Only in 1995 *Q. ilex* leaves contained comparable levels of cysteine and GSH at the Bossoleto spring and the control area (Fig. 3). A higher level of γ-EC was detected at the Solfatara spring in *Q. pubescens* leaves harvested in 1995 (Fig. 3). At the Laiatico site, significant differences in cysteine, γ-EC or GSH contents between leaves from the control and the spring area for both oak species were not observed (Fig. 3). The lower levels of cysteine and GSH in the leaves of *Q. ilex* in 1996 (Fig. 3) corresponded to lower levels of APR activity in that year (Fig. 1). In the leaves of *Q. pubescens*, TSNN, main amino acids and ammonium contents were typically not significantly different in leaves from the three springs and their corresponding control areas (Fig. 4). Only at the Bossoleto site the ammonium content was significantly lower at the spring than at the control area (Fig. 4). *Q. ilex* leaves from the Laiatico spring area contained more ammonium than at the control area (Fig. 4). At the Bossoleto site, *Q. ilex* leaves contained significantly more alanine and arginine at the control than at the spring area (Fig. 4) despite similar NR activities (Fig. 2). All other main amino acids and TSNN were comparable at the Laiatico and Bossoleto springs and their corresponding control area (Fig. 4).

When the results of the present field study were compared with a cross exchange experiment under controlled conditions with acorns of *Q. ilex* collected from the Laiatico spring (Schulte et al. 2002), some important differences could be identified.

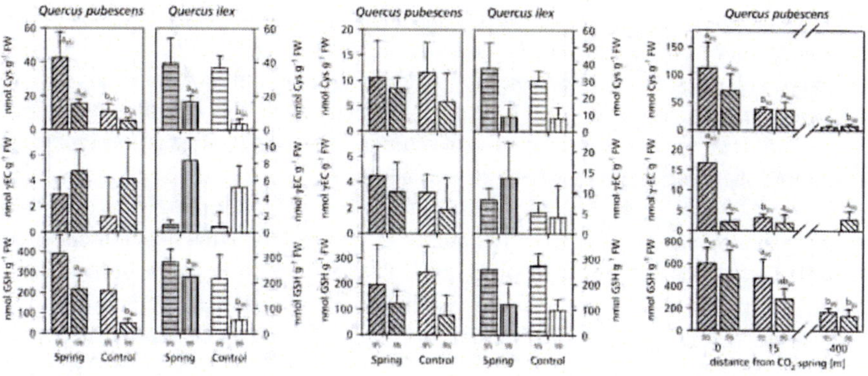

Fig. 3 Cysteine (Cys), γ-EC and glutathione (GSH) in leaves of *Q. ilex* and *Q. pubescens* trees growing at the CO_2 springs at Bossoleto (left two columns), Laiatico (middle two columns), and Sulfatara (right column) and at corresponding control sites. Mean values from six leaves collected from six different trees (±SD) are presented for 1995 (*striped diagonal up* for *Q. pubescence*; *striped horizontal* for *Q. ilex*) and 1996 (*striped diagonal down* for *Q. pubescence*; *striped vertical* for *Q. ilex*). Values carrying different letters are different at $P \leq 0.05$ within one sampling date

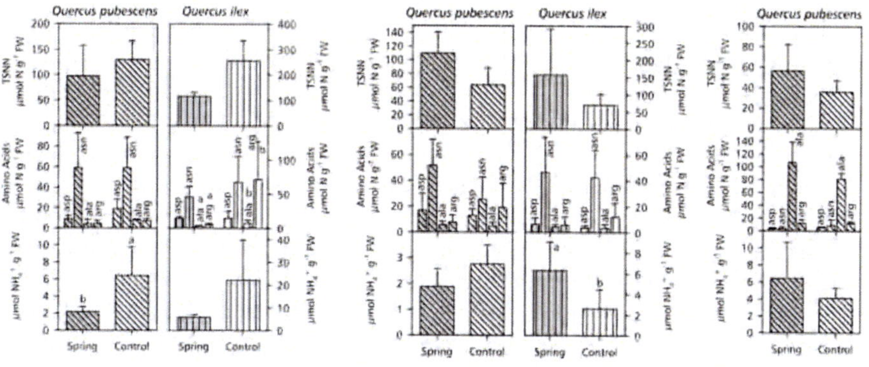

Fig. 4 Total soluble non-proteinogenic nitrogen compounds (TSNN), main amino acids and ammonium (NH_4^+) content in leaves of *Q. ilex* and *Q. pubescens* trees growing at the CO_2 springs at Bossoleto (left two columns), Laiatico (middle two columns), and Sulfatara (right column) and at corresponding control sites. Mean values from six leaves collected from six different trees in 1996±S.D. (*striped diagonal down* for *Q. pubescence*; *striped vertical* for *Q. ilex*) are presented. Values carrying different letters are different at $P \leq 0.05$

When the acorns collected from the Laiatico spring area were cultivated at elevated pCO_2 under controlled conditions, APR activity was comparable to that determined in *Q. ilex* leaves at ambient pCO_2 (Schulte et al. 2002). In contrast, APR activity was down regulated in *Q. ilex* leaves of the spring area in the field in 1995, whereas thiol levels were not affected. Under controlled conditions Cys and GSH contents were reduced in leaves of *Q. ilex* plants cultivated at elevated pCO_2 (Schulte et al. 2002). This clearly indicates that the sulfur metabolism reacts differently under controlled and field condition, which was not observed for nitrogen metabolism. TSNN in

Q. ilex leaves from the Laiatico spring was slightly enhanced at the spring area in the present field study and also under control conditions at elevated pCO_2 (Schulte et al. 2002). This goes parallel with unchanged levels of major amino compounds in *Q. ilex* leaves under field (present study) and controlled conditions (Schulte et al. 2002). Also total leaf nitrogen concentration of *Q. ilex* and *Q. pubescens* was not significantly different between plants growing at elevated pCO_2 close to the spring or at atmospheric pCO_2 (Körner and Miglietta 1994). Thus, effects from elevated pCO_2 on nitrogen metabolism were not observed for *Q. ilex* neither in the field nor under controlled conditions. In contrast, sulfur assimilation from atmospheric sulfur gases in the field seems to counteract the effects observed under controlled conditions.

From several studies with herbaceous (for review see De Kok 1990, further literature: Herschbach et al. 1995a, b; Durenkamp and De Kok 2004; Westerman et al. 2000, 2001; Yang et al. 2006; Koralewska et al. 2007) and deciduous (Herschbach et al. 2000) plants it is well known that sulfur gases effects thiol contents and, as a consequence, APR activity (Westerman et al. 2001; Lang et al. 2007; Durenkamp et al. 2007; Koralewska et al. 2008). APR activity tends to decrease in leaves collected at the spring area of the Bossoleto and Sulfatara site, although not at any time and in any oak species, and also in *Q ilex* leaves from the Laiatico spring in 1996. Cysteine and GSH levels were higher in leaves from trees at the spring area of the Bossoleto and Sulfatara site, but unaffected at the Laiatico site. These results indicate a strong correlation between thiol contents and APR activity also under field conditions and are consistent with current knowledge of feedback inhibition of APR (for review see Kopriva and Rennenberg 2004; Kopriva 2006; Davidian and Kopriva 2010).

When the observed results from the three locations are related to the sulfur gas concentrations of the emitted gas mixture and to the diurnal course of emission controversial results obtained. The Sulfatara site exhibits a higher atmospheric pH_2S concentration compared to the Bossoleto and Laiatico spring that correlates with the highest Cys, γ-EC and GSH content in *Q. pubescence* leaves. Hence at the Sulfatara site a clear correlation between gas emission, thiol increment and APR activity was observed that agrees with the common knowledge (Kopriva and Rennenberg 2004; Kopriva 2006).

The mean value of the sulfur gas concentration of the Bossoleto and Laiatico spring was comparable. Thus, differences in sulfur metabolism observed between these sites could not be explained by differences in the gas composition, but probably by the time of sulfur gas emission. At the Bossoleto site the gas is mainly emitted during the early morning with peak values up to 8,097 $\mu l\ l^{-1}$ pCO_2 (Schulte et al. 1999). During daytime, *i.e.* between 9 am and 6 pm, mean pCO_2 was $368 \pm 54\ \mu l\ l^{-1}$. Net photosynthesis of *Q. ilex* and *Q pubescence* leaves at this site was higher at the spring area compared to the control area (Tognetti et al. 1998; Blaschke et al. 2001) and leaf conductance was reduced at the spring site resulting in higher water use efficiency (Tognetti et al. 1998, 1999). At the Laiatico spring, where the gas is emitted during the day (Schulte et al. 1999), *Q. pubescence* leaves showed higher net photosynthesis and slightly reduced leaf conductance (Stylinski et al. 2000). As neither H_2S nor SO_2 seem to penetrate the cuticula in appreciable amounts (Taylor and Tingey 1983; De Kok et al. 1991) and stomatal conductance is generally around zero during the night, uptake of H_2S and SO_2 should be restricted to the daytime when concentrations of both sulfur gases were low at the Bossoleto site. However, at the Bossoleto spring cysteine and GSH contents were enhanced indicating uptake and usage of the emitted sulfur gases. From these considerations the observed results seem contradictory. However, also other environmental factors such as drought can influence GSH contents in leaves. At drought stress Marabottini et al. (2001) observed higher GSH contents in *Q. pubescence* leaves collected from the Bossoleto spring area than from the control area and unchanged GSH contents in *Q. ilex*. Although meteorological data are not available, it can be assumed from observations at the Bossoleto site that a drought period prior the harvest in 1996 had influenced the GSH content in *Q. pubescence* leaves.

On the other hand, elevated sulfur availability by sulfur gases emitted from the vents at the spring areas itself could function as an environmental factor influencing nitrogen assimilation. As nitrate reductase activity increased, decreased or remained constant at elevated pCO_2 (reviewed in Stitt and Krapp 1999, see also Natali et al. 2009) the findings of increased NR activity in *Q. pubescence* leaves from the Bossoleto and Sulfatara site as well as its decrease at the Laiatico spring in 1995 are not surprising. However, nitrogen assimilation and, thus, NR activity also depend on the sulfur supply (Brunold 1993). Sulfur deficiency diminishes NR activity (Brunold 1993), but information of a surplus of sulfur on NR activity is scarce. In *Q. pubescence* leaves from the Bossoleto and the Sulfatara spring area NR was higher compared to the control areas, whereas APR activity was lower. As the lower APR activity correlates with enhanced reduced sulfur levels, elevated NR activity may be an indication of an enhanced demand for reduced nitrogen for protein synthesis. This view is supported by the decreased leaf NH_4^+ content of *Q. pubescence* at the Bossoleto spring. In conclusion, the presented results support the diversity of C, N and S interactions and, additionally, the complex environmental influences on these interactions.

Acknowledgements This study was financially supported by the European Commission (MAPLE – Project EV5VCT940432), and the Swiss National Science Foundation. The authors thank M. Stalder for excellent technical assistance.

References

Blaschke L, Schulte M, Raschi A, Slee N, Rennenberg H, Polle A (2001) Photosynthesis, soluble and structural carbon compounds in two Mediterranean oak species (*Quercus pubescens* and *Q. ilex*) after lifetime growth at natural elevated CO_2 concentrations. Plant Biol 3:288–298

Bradford MM (1976) A rapid and sensitive method for the quantification of microgram quantities of protein utilizing the principle of protein-dye binding. Anal Biochem 72:248–254

Brunold C (1993) Regulatory interactions between sulfate and nitrate assimilation. In: De Kok LJ, Stulen I, Rennenberg H, Brunold C, Rauser WH (eds) Sulfur nutrition and sulfur assimilation in higher plants: regulatory agricultural and environmental aspects. SPB Academic Publishing, The Hague, pp 61–75

Brunold C, Erismann KH (1975) H_2S as sulfur source in *Lemna minor* L.: direct incorporation into cysteine and inhibition of sulfate assimilation. Experientia 31:508–509

Brunold C, Suter M (1990) Adenosine 5′-phosphosulfate sulfotransferase. In: Lea P (ed) Methods in plant biochemistry. Academic, London, pp 339–343

Brunold C, von Ballmoos P, Hesse H, Fell D, Kopriva S (2003) Interactions between sulfur, nitrogen and carbon metabolism. In: Davidian J-C, Grill D, De Kok LJ, Stulen I, Hawkesford MJ, Schnug E, Rennenberg H (eds) Sulphur transport and assimilation in plants. Backhuys Publishers, Leiden, pp 45–56

Buwalda F, De Kok LJ, Stulen I, Kuiper PJC (1988) Cysteine, γ-glutamylcysteine and glutathione contents of spinach leaves as affected by darkness and application of excess sulfur. Physiol Plant 74:663–668

Davidian JC, Kopriva S (2010) Regulation of sulfate uptake and assimilation – the same or not the same? Mol Plant 3:314–325

De Kok LJ (1990) Sulfur metabolism in plants exposed to atmospheric sulfur. In: Rennenberg H, Brunold C, De Kok LJ, Stulen I (eds) Sulphur nutrition and sulphur assimilation in higher plants. SPB Academic Publishing, The Hague, pp 111–130

De Kok LJ, Buwalda F, Bosma W (1988) Determination of cysteine and its accumulation in spinach leaf tissue upon exposure to excess sulfur. J Plant Physiol 331:502–505

De Kok LJ, Rennenberg H, Kuiper PJC (1991) The internal resistance in spinach leaves to atmospheric H_2S deposition is determined by metabolic processes. Plant Physiol Biochem 29:463–470

De Kok LJ, Stuiver CEE, Stulen I (1998) Impact of atmospheric H_2S on plants. In: De Kok LJ, Stulen I (eds) Responses of plant metabolism to air pollution and global change. Backhuys Publishers, Leiden, pp 51–63

Durenkamp M, De Kok LJ (2004) Impact of pedospheric and atmospheric sulphur nutrition on sulphur metabolism of *Allium cepa* L., a species with a potential sink capacity for secondary sulphur compounds. J Exp Bot 55:1821–1830

Durenkamp M, De Kok LJ, Kopriva S (2007) Adenosine 5′-phosphosulphyte reductase is regulated differently in *Allium cepa* L. and *Brassica oleracea* L. upon exposure to H_2S. J Exp Bot 58:1571–1579

Geiger M, Walch-Piu L, Hernecker J, Schulze E-D, Ludwig F, Sonnewald U, Scheible W-R, Stitt M (1998) Enhanced carbon dioxide leads to a modified diurnal rhythm of nitrate reductase activity in older plants, and a large stimulation of nitrate reductase activity and higher levels of amino acids in higher plants. Plant Cell Environ 21:253–268

Herschbach C, De Kok LJ, Rennenberg H (1995a) Net uptake of sulfate and its transport to the shoot in spinach plants fumigated with H_2S or SO_2: does atmospheric sulfur affect the 'inter-organ' regulation of sulfur nutrition. Bot Acta 108:41–46

Herschbach C, De Kok LJ, Rennenberg H (1995b) Net uptake of sulfate and its transport to the shoot in tobacco plants fumigated with H_2S or SO_2. Plant Soil 175:75–84

Herschbach C, van der Zalm E, Schneider A, Jouanin L, De Kok LJ, Rennenberg H (2000) Regulation of sulfur nutrition in wild-type and transgenic poplar over-expressing γ-glutamylcysteine synthetase in the cytosol as affected by atmospheric H_2S. Plant Physiol 124:461–473

Hesse H, Trachsel N, Suter M, Kopriva S, von Ballmoos P, Rennenberg H, Brunold C (2003) Effect of glucose on assimilatory sulfate reduction in *Arabidopsis* roots. J Exp Bot 54:1701–1709

Kaiser WM, Huber SC (1994) Correlation between apparent activation state of nitrate reductase in higher plants. Plant Physiol 106:817–821

Kopriva S (2006) Regulation of sulfate assimilation in *Arabidopsis* and beyond. Ann Bot 97:479–495

Kopriva S, Rennenberg H (2004) Control of sulphate assimilation and glutathione synthesis: interaction with N and C metabolism. J Exp Bot 55:1831–1842

Kopriva S, Muheim R, Koprivova A, Trachsel N, Catalano C, Suter M, Brunold C (1999) Light regulation of assimilatory sulfate reduction in *Arabidopsis thaliana*. Plant J 20:37–44

Kopriva S, Suter M, von Ballmoos P, Hesse H, Krähenbühl U, Rennenberg H, Brunold C (2002) Interaction of sulfate assimilation with carbon and nitrogen metabolism in *Lemna minor*. Plant Physiol 130:1406–1413

Koralewska A, Posthumus FS, Stuiver CEE, Buchner P, Hawkesford MJ, De Kok LJ (2007) The characteristic high sulfate content in *Brassica oleracea* is controlled by the expression and activity of sulfate transport. Plant Biol 9:654–661

Koralewska A, Stuiver CEE, Posthumus FS, Kopriva S, Hawkesford MJ, De Kok LJ (2008) Regulation of sulfate uptake, expression of the sulfate transporters Sultr1;1 and Sultr1;2 and APS reductase in Chinese cabbage (*Brassica pekinensis*) as affected by atmospheric H_2S nutrition and sulfate deprivation. Funct Plant Biol 35:318–327

Körner C, Miglietta F (1994) Long term effects of naturally elevated CO_2 on Mediterranean grassland and forest trees. Oecologia 99:343–351

Kruse J, Hetzger I, Haensch R, Mendel R-R, Rennenberg H (2003) Elevated pCO_2 affects C and N metabolism in wild type and transgenic tobacco exhibiting altered C/N balance in metabolite analysis. Plant Biol 5:540–549

Kruse J, Kopriva S, Haensch R, Krauss G-J, Mendel R-R, Rennenberg H (2007) Interaction of sulfur and nitrogen nutrition in tobacco (*Nicotiana tabacum*) plants: significance of nitrogen source and root nitrate reductase. Plant Biol 9:638–646

Lang C, Popko J, Wirtz M, Hell R, Herschbach C, Kreuzwieser J, Rennenberg H, Mendel RR, Hänsch R (2007) Sulphite oxidase as key enzyme for protecting plants against sulphur dioxide. Plant Cell Environ 30:447–455

Lappartient AG, Touraine B (1996) Demand-driven control of root ATP sulfurylase activity and SO_4^{2-}uptake in intact canola. The role of phloem-translocated glutathione. Plant Physiol 111:147–157

Lappartient AG, Vidmar JJ, Leustek T, Glass ADM, Touraine B (1999) Inter-organ signalling in plants: regulation of ATP sulfurylase and sulfate transporter genes expression in roots mediated by phloem-translocated compound. Plant J 18:89–95

Marabottini R, Schraml C, Paolacci AR, Sorgona A, Raschi A, Rennenberg H, Badiani M (2001) Foliar antioxidant status of adult Mediterranean oak species (*Quercus ilex* L. and *Q. pubescens* Willd.) exposed to permanent CO_2-enrichment and to seasonal water stress. Environ Pollut 115:413–423

Miglietta F, Raschi A (1993) Studying the effect of elevated CO_2 in the open in a naturally enriched environment in Central Italy. Vegetatio 105:391–400

Miglietta F, Raschi A, Bettarini I, Resti R, Selvi F (1993) Natural CO_2 springs in Italy: a resource for examining long-term response of vegetation to rising atmospheric CO_2 concentrations. Plant Cell Environ 16:873–878

Natali SM, Sañudo-Wilhemy SAS, Lerdau MT (2009) Effects of elevated carbon dioxide and nitrogen fertilization on nitrate reductase activity in sweetgum and loblolly pine trees in two temperate forests. Plant Soil 314:197–210

Neyra CA, Hagemann RH (1975) Nitrate uptake and induction of nitrate reductase in excised corn roots. Plant Physiol 56:692–695

Scheible W-R, Gonzales-Fontes A, Morcuende R, Lauerer M, Geiger M, Glaab J, Gojon R, Schulze E-D, Stitt M (1997) Tobacco mutants with a decreased number of functional nia genes compensate by modifying the diurnal regulation of transcription, post-translational modification and turnover of nitrate reductase. Planta 203:304–319

Schneider S, Gessler A, Weber P, von Sengbusch D, Hanemann U, Rennenberg H (1996) Soluble N compounds in trees exposed to high loads of N: a comparison of spruce (*Picea abies*) and beech (*Fagus sylvatica*) grown under field conditions. New Phytol 134:103–114

Schulte M, Raiesi FG, Papke H, Butterbach-Bahl K, van Breemen N, Rennenberg H (1999) CO_2 concentration and atmospheric trace gas mixing ratio around natural CO_2 vents in different Mediterranean forests in central Italy. In: Raschi A, Vaccari FP, Miglietta F (eds) Ecosystem response to CO_2: the maple project results. Office for Official Publications of the European Communities, Luxembourg, pp 168–188

Schulte M, von Ballmoos P, Rennenberg H, Herschbach C (2002) Life-long growth of *Quercus ilex* L. at natural CO_2 springs acclimates sulphur, nitrogen and carbohydrate metabolism of the progeny to elevated pCO_2. Plant Cell Environ 25:1715–1727

Schupp R, Rennenberg H (1988) Diurnal changes in the glutathione concentration of spruce needles (*Picea abies* L.). Plant Sci 57:113–117

Stitt M, Krapp A (1999) The interaction between elevated carbon dioxide and nitrogen nutrition: the physiological and molecular background. Plant Cell Environ 22:583–621

Stylinski CD, Oechel WC, Gamon JA, Tissue DT, Miglietta F, Raschi A (2000) Effects of lifelong [CO_2] enrichment on carboxylation and light utilization of *Quercus pubescens* Willd. Examined with gas exchange, biochemistry and optical techniques. Plant Cell Environ 23:1353–1362

Suter M, von Ballmoos P, Kopriva S, Op den Camp R, Schaller J, Kuhlemeier C, Brunold C (2000) Adenosine 5′-phosphosulfate sulfotransferase and adenosine 5′-phosphosulfate reductase are identical enzymes. J Biol Chem 275:930–936

Taylor GE, Tingey DT (1983) Sulphur dioxide flux into leaves of *Geranium carolinianum* L. Plant Physiol 72:237–244

Tognetti R, Johnson JD, Michelozzi M, Raschi A (1998) Response of foliar metabolism in mature trees of *Quercus pubescens* and *Quercus ilex* to long-term elevated CO_2. Environ Exp Bot 39:233–245

Tognetti R, Longobucco A, Miglietta F, Raschi A (1999) Water relations, stomatal response and transpiration of *Quercus pubescens* trees during summer in a Mediterranean carbon dioxide spring. Tree Physiol 19:261–270

van der Kooij TAW, De Kok LJ, Haneklaus S, Schnug E (1997) Uptake and metabolism of sulphur dioxide by *Arabidopsis thaliana*. New Phytol 135:101–107

Vauclare P, Kopriva S, Fell D, Suter M, Sticher L, von Ballmoos P, Krähenbühl U, Op den Camp R, Brunold C (2002) Flux control of sulfate assimilation in *Arabidopsis thaliana*: Adenosine 5′-phosphosulfate reductase is more susceptible to negative control by thiols than ATP sulfurylase. Plant J 31:729–740

Westerman S, De Kok LJ, Stuiver CEE, Stulen I (2000) Interaction between metabolism of atmospheric H_2S in the shoot and sulfate uptake by the roots of curly kale (*Brassica oleracea*). Physiol Plant 109:443–449

Westerman S, Stulen I, Suter M, Brunold C, De Kok LJ (2001) Atmospheric H_2S as sulfur source for *Brassica oleracea*: consequences for the activity of the enzymes of the assimilatory sulfate reduction pathway. Plant Physiol Biochem 39:425–432

Winter H, Lohaus G, Heldt H-W (1992) Phloem transport of amino acids in relation to their cytosolic levels in barley leaves. Plant Physiol 99:996–1004

Yang L, Stulen I, De Kok LJ (2006) Impact of sulfate nutrition on the utilization of atmospheric SO_2 as sulfur source for Chinese cabbage. J Plant Nutr Soil Sci 169:529–534

Gamma-Glutamyl Cycle in Plants: Possible Implications in Apoplastic Redox Control and Redox Sensing

Antonio Masi

Abstract There is now increasing evidence of a gamma-glutamyl cycle occurring in plants that closely resembles what happens in animals, which consists of glutathione extrusion to the extracellular space, sequential degradation to its constituent amino acids by gamma-glutamyl transferase (GGT) and Cys-Gly dipeptidase (CD) activity, reuptake by amino acid transporters and glutathione reassembly inside the cell. Here we demonstrate that the GGT bound to the cell wall in *Arabidopsis* leaves recovers extracellular glutathione that is probably extruded to counteract conditions of oxidative stress due to UV-B exposure. Experiments on barley roots provide evidence of the existence of two different Cys-Gly dipeptidases: one is associated with the cell wall and has a higher affinity for reduced Cys-Gly; the other is presumably bound to the plasma membrane. The different site of CD isoforms with a different specificity for the reduced and oxidized substrate forms implies that reduced glutathione and Cys-Gly must spread through the apoplastic space before they are cleaved by the corresponding enzyme; in so doing, these thiols scan the extracellular space and eventually react with oxidants or sensitive components bound to the plasma membrane acting as sulfur switches. Taken together, these findings suggest that the gamma-glutamyl cycle is implicated in apoplastic redox control and redox sensing.

It has recently been demonstrated (Ferretti et al. 2009; Ohkama-Ohtsu et al. 2007; Martin et al. 2007) that plants have a gamma-glutamyl cycle functional to the retrieval of extracellular glutathione (GSH), in which glutathione is extruded to the apoplast and sequentially degraded to its constituent amino acids by gamma-glutamyl

A. Masi (✉)
Department of Agricultural Biotechnology, University of Padua,
Viale Università 16, 35020 Legnaro, Padua, Italy
e-mail: antonio.masi@unipd.it

L.J. De Kok et al. (eds.), *Sulfur Metabolism in Plants*, Proceedings of the International Plant Sulfur Workshop 1, DOI 10.1007/978-94-007-4450-9_30,

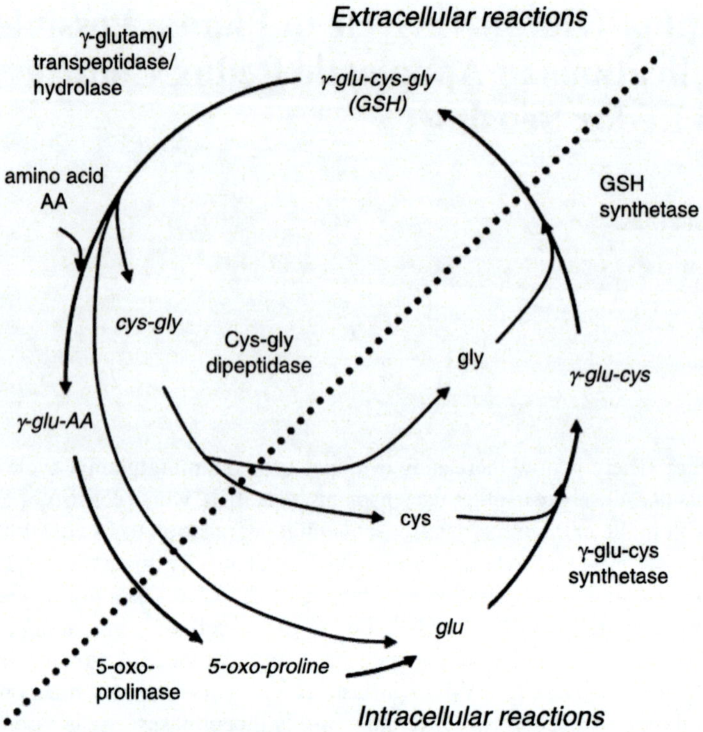

Fig. 1 Gamma-glutamyl cycle in animals, as described by Meister and Tate (1976). This cycle of intracellular glutathione synthesis and its extracellular degradation is now believed to occur in plant cells too, the main difference being that GGT is located on the plasma membrane in animal cells, while it is anchored to the cell wall in plant cells

transferase (GGT) and Cys-Gly dipeptidase activity (Fig. 1). Amino acids are then taken up and glutathione is reassembled inside the cell.

The exact significance of this cycle in plant physiology remains obscure. Several hypotheses may be advanced, relating to inter-organ GSH and cys redistribution, phloem downloading, apoplastic redox control, reuptake of extruded GSH, uptake of GSH from the soil, molecular exchange with mycorrhizae. In *Arabidopsis thaliana*, GGT is represented by a small family of four genes; two different GGT isoforms (GGT1, *At39640*; GGT2, *At39650*) are reportedly located in the apoplastic space and capable of cleaving the gamma-glutamyl bond, thus releasing glu and Cys-Gly. GGT1 is bound to the cell wall and is constitutively expressed in most tissues, whereas the membrane-bound GGT2 isoform is expressed in roots and reproductive tissues (Destro et al. 2011). The vacuolar isoform GGT4 (*At4g29210*) assists in the degradation of GSH conjugates (Grzam et al. 2007). All the enzymes in the gamma-glutamyl cycle have now been identified in plants, except for cys-cly dipeptidase. The significance of the glutathione synthesis and degradation coordinated between

Table 1 Cysteinyl-glycine and glutathione content in apoplastic fluids

Genotype	Treatment	Cys-Gly (nmol ml^{-1} ECWF)	Glutathione (nmol ml^{-1} ECWF)
wt	PAR only	10.4±0.7a	77.6±3.7a
wt	PAR+UV-B	13.0±1.8b	78.7±6.7a
ggt1	PAR only	1.6±0.5c	89.9±11.7b
ggt1	PAR+UV-B	1.6±0.3c	101.1±8.7b

Arabidopsis thaliana plants at the stage of fully-expanded leaves were treated with supplemental UV-B radiation for 1 day. Following extraction with the infiltration/centrifugation technique, thiols were detected fluorimetrically by HPLC. *ECWF* extra-cellular washing fluid from leaves. Data represent the mean of five measurements (±SD). A one-way analysis of variance (ANOVA) was applied to the data; differences between means were evaluated for significance by using Duncan's multiple range test (*DMRT*) ($P<0.05$); means with the same letter are not significantly different. Statistical analyses were performed with SPSS 10.0

the cytoplasm and extracellular space, and the role of this cycle in plant physiology remains largely unknown.

The significance of the gamma-glutamyl cycle in the redox response to oxidative stress conditions was explored in *A. thaliana* wild type and *ggt1* knockout mutants grown in control conditions or in enhanced oxidative conditions generated by supplemental UV-B exposure (8.35 kJ m^{-2} d^{-1} biologically effective irradiance, UVB-BE). Apoplastic glutathione and Cys-Gly concentration were measured in both genotypes and in the presence of UV-B treatment using the infiltration-centrifugation technique (Lohaus et al. 2001); thiols were measured by HPLC following derivatization with the fluorophore SBD-F (Masi et al. 2002).

Cys-Gly was much lower in the *ggt1* genotype, demonstrating the role of apoplastic, cell-wall-bound GGT1 in glutathione degradation. The glutathione concentration was approximately 30% higher in the ggt1 mutant than in the wild type, but only under UV-B exposure (Table 1). These findings point to a role for GGT1 in the recovery of extracellular glutathione, which is extruded more actively to counteract oxidative stress conditions in the extracellular space. Photo-oxidative UV-B conditions are known to coincide with damage to apoplastic components, mainly the plasma membranes, which are a primary target for this harmful radiation (Murphy 1983). On the other hand, the strong similarity between the apoplastic glutathione concentrations in the wild type and *ggt1* mutant under physiological conditions suggests that there are also alternative means for glutathione content buffering in the apoplast, but they are insufficient under oxidative stress.

The existence of a Cys-Gly dipeptidase in the gamma-glutamyl cycle is implicit, but not demonstrated, and indeed very little is known about this enzyme. The site and substrate specificity of Cys-Gly dipeptidase(s) was investigated in barley because previous observations indicated that intact barley roots are able to cleave both reduced and oxidized Cys-Gly supplemented in the growth medium, yielding cyst(e)ine and glycine, which are taken up by amino acid transporters (Ferretti et al. 2009). Given the location of an apoplastic GGT isoform on the cell wall, it was hypothesized that Cys-Gly dipeptidase could also be located in this

Table 2 Degradation of cysteinyl-glycine by dipeptidase activity in cell-wall-enriched fractions

Treatment	Cys-Gly degradation (%)
100 μM reduced Cys-Gly (water)	59.3 ± 3.7
100 μM Cys-Gly (citrate buffer, pH 5)	3.6 ± 0.9
50 μM oxidized Cys-Gly	99.3 ± 1.7

Pellets from centrifuged barley root extracts were incubated with 100 μM reduced Cys-Gly (in water or 50 mM citrate buffer, pH 5, to prevent oxidation) or 50 μM oxidized Cys-Gly for 15 min. Degradation of Cys-Gly is expressed as % of the control and represent the mean of five measurements (±SD)

compartment. Following extraction with 40 mM Tris–HCl pH 8, 1% TritonX-100 from 5-day-old barley roots, cell-wall-enriched fractions were obtained by collecting the pellet resulting from extract centrifugation at 15,000 g for 10 min. After rinsing, this fraction was incubated with a solution containing reduced or oxidized Cys-Gly in water, or reduced Cys-Gly in 50 mM citrate buffer pH 5 to prevent oxidation. As shown in Table 2, the cell-wall-associated dipeptidase was unable to cleave oxidized Cys-Gly, but 60% of the reduced Cys-Gly was degraded after incubation for 15 min. When spontaneous oxidation was prevented by using a citrate buffer, however, over 95% of Cys-Gly was degraded, which goes to show that the thiol group in Cys-Gly is highly susceptible to oxidation.

These findings indicate that a cell-wall-bound Cys-Gly dipeptidase exists and shows specificity to the reduced form of Cys-Gly. It has already been demonstrated (Ferretti et al. 2009) that oxidized Cys-Gly can be cleaved extracellularly in barley roots, giving rise to a parallel release of cystine into the external medium. These findings lead to the conclusion that there are two different apoplastic Cys-Gly dipeptidases in barley roots: one is associated with the cell wall and has a higher affinity for reduced Cys-Gly; the other is probably bound to the plasma membrane. Hence the revised gamma-glutamyl cycle operating in plant tissues as shown in Fig. 2.

The differential site of the GGT and CD isoforms with a different specificity for the reduced and oxidized substrate forms implies that reduced GSH and Cys-Gly spread through the apoplastic space before they are cleaved by the corresponding enzyme; in so doing, in their reactive and reduced form, they can scan the redox state in the extracellular space. Cysteine or cystine will ultimately result from this process, which may be indicative of the extracellular redox state and is possibly a signal for sensitive plasma-membrane bound components, acting as sulfur switches.

While it is generally accepted in the literature that a Cys-Gly dipeptidase activity exists in the apoplast, it has so far been impossible to conclusively assign it to a gene name in *Arabidopsis*; several dipeptidases might be involved, since their specificity has often not been fully characterized, especially when it comes to the substrate's redox state. Future experimentation consequently needs to focus primarily on the biochemical and molecular characterization of Cys-Gly dipeptidase(s), which will involve purifying the enzyme from cell wall fractions, sequencing and gene identification, and defining the kinetic parameters.

Verifying the intriguing hypothesis of a role for the gamma-glutamyl cycle in redox sensing will probably require different strategies. First of all, the sensor needs

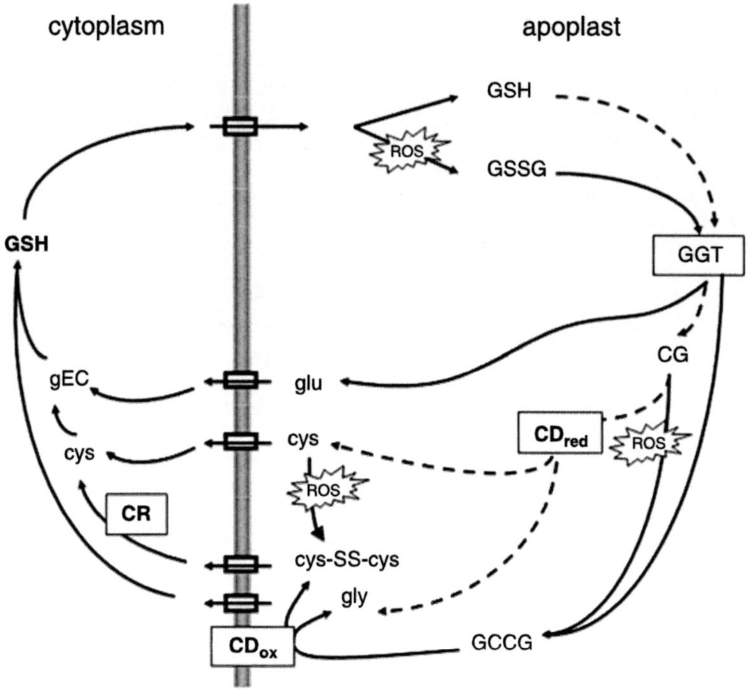

Fig. 2 Schematic representation of the steps in the gamma-glutamyl cycle in plants. Glutathione is synthesized intracellularly, extruded and sequentially hydrolyzed in the apoplast by gamma-glutamyl transferase (*GGT*) and Cys-Gly dipeptidase (*CD*) activity to its constituent amino acids, which are taken up and used for intracellular GSH re-synthesis. There is evidence of the existence of two dipeptidases: one (CD_{red}) is bound to the cell wall and is specific for reduced Cys-Gly, the other (CD_{ox}) is probably bound to the plasma membrane and is capable of cleaving oxidized Cys-Gly. *Dashed line*: degradation pathway for the reduced form of glutathione and cysteinyl-glycine. *CR* cystine reductase, *CG* reduced Cys-Gly, *GCCG* oxidized Cys-Gly, *gEC* gamma-glutamyl-cysteine, *ROS* reactive oxygen species

to be identified. This is most likely located on the plasma membrane, so purified plasma membrane should contain components sensitive to cys or Cys-Gly; this aspect should also be investigated. Although both cys and Cys-Gly are potential candidates as redox-sensing molecules in the gamma-glutamyl cycle, it is worth bearing in mind that, while cys can result from the degradation of different cys-containing peptides, Cys-Gly is the more specific product of GGT degradation activity. Together with the fact that Cys-Gly is highly reactive and liable to oxidation, this makes Cys-Gly a better candidate as a redox-sensing molecule metabolically related to glutathione. Secondly, some physiological conditions, *e.g.* reducing or hypoxia conditions, could be tested to shed light on the role of Cys-Gly dipeptidase in redox sensing. Thirdly, once the Cys-Gly dipeptidase gene in *Arabidopsis* has been identified, mutant lines could be tested to identify possible phenotypes from which gene functions could be inferred.

References

Destro T, Prasad D, Martignago D, Lliso Bernet I, Trentin AR, Renu IK, Ferretti M, Masi A (2011) Compensatory expression and substrate inducibility of γ-glutamyl transferase GGT2 isoform in *Arabidopsis thaliana*. J Exp Bot 62:805–814

Ferretti M, Destro T, Tosatto SCE, La Rocca N, Rascio N, Masi A (2009) Gamma-glutamyl transferase in the cell wall participates in extracellular glutathione salvage from the root apoplast. New Phytol 181:115–126

Grzam A, Martin MN, Hell R, Meyer AJ (2007) γ-Glutamyl transpeptidase GGT4 initiates vacuolar degradation of glutathione *S*-conjugates in *Arabidopsis*. FEBS Lett 581:3131–3138

Lohaus G, Pennewiss K, Sattelmacher B, Hussmann M, Muehling KH (2001) Is the infiltration-centrifugation technique appropriate for the isolation of apoplastic fluid? A critical evaluation with different plant species. Physiol Plant 111:457–465

Martin MN, Saladores PH, Lambert E, Hudson AO, Leustek T (2007) Localization of members of the γ-glutamyl transpeptidase family identifies sites of glutathione and glutathione S-conjugate hydrolysis. Plant Physiol 144:1715–1732

Masi A, Ghisi R, Ferretti M (2002) Measuring low-molecular-weight thiols by detecting the fluorescence of their SBD derivatives: application to studies of diurnal and UV-B induced changes in *Zea mays* L. J Plant Physiol 159:499–507

Meister A, Tate SS (1976) Glutathione and related gamma-glutamyl compounds: biosynthesis and utilization. Annu Rev Biochem 45:559–604

Murphy TM (1983) Membranes as targets of ultraviolet radiation. Physiol Plant 58:381–388

Ohkama-Ohtsu N, Radwan S, Peterson A, Zhao P, Badr AF, Xiang C, Oliver DJ (2007) Characterization of the extracellular γ-glutamyl transpeptidases, GGT1 and GGT2, in *Arabidopsis*. Plant J 49:865–877

Sulfate Transport Processes Under Drought Stress: Interaction with Mycorrhization and Elevated pCO_2 in Young Pedunculate Oak (*Quercus robur* L.) Trees

Stefan Seegmüller and Heinz Rennenberg

Abstract Drought stress mediated changes in sulfate transport processes, *i.e.* sulfate uptake and xylem loading of sulfate in the roots, and its dependency on mycorrhization and pCO_2 were analyzed in young pedunculate oak (*Quercus robur* L.) trees. Withdrawal of water supply caused strong negative values of pre-dawn shoot water potential, indicating severe water stress. Elevated pCO_2, but not mycorrhization transiently improved the water status of the trees. Sulfate uptake was largely independent from pre-dawn shoot water potential irrespective of mycorrhization and pCO_2. In contrast, xylem loading of sulfate decreased with decreasing shoot water potential. Mycorrhization and – to a lesser extent – also elevated pCO_2 improved xylem loading under drought stress. As a consequence, also relative xylem loading, *i.e.* the % amount of the sulfate taken up that was loaded into the xylem, was slightly improved. The positive effect of mycorrhization on xylem loading of sulfate under drought stress is surprising, since plants were inoculated with the ectomycorrhizal fungus *Laccaria laccata* that is not supposed to enter the vascular bundle of the root. Therefore, signaling by the fungus across the bundle sheath has to be assumed.

Sulfur is available to plants mostly in form of sulfate in the soil. Sulfate is acquired from the soil by active, carrier-mediated uptake using low- (LAUS) and high-affinity (HAUS) transport systems (Clarkson et al. 1993). Sulfate taken up by the roots can either be used for sulfur assimilation in the roots, storage in root vacuoles, or can be loaded into the xylem and allocated with the transpiration stream to the leaves, the predominant site of sulfur assimilation (Rennenberg 1984; Brunold 1990).

S. Seegmüller • H. Rennenberg (✉)
Institute of Forest Botany and Tree Physiology, Chair of Tree Physiology,
University of Freiburg, Georges-Köhler-Allee 53, 79110 Freiburg, Germany
e-mail: heinz.rennenberg@ctp.uni-freiburg.de

L.J. De Kok et al. (eds.), *Sulfur Metabolism in Plants*, Proceedings of the International Plant Sulfur Workshop 1, DOI 10.1007/978-94-007-4450-9_31,

All these processes are highly controlled by the sulfate transporter family (Hawkesford and De Kok 2006; Buchner et al. 2004). At the whole plant level this control is thought to be driven by the demand for reduced sulfur (Herschbach et al. 2000; Hartmann et al. 2004; van der Zalm et al. 2005). Sulfate uptake capacity of the roots is largely regulated at the level of transcription. Thus, the expression of sulfate transporter genes in the roots is highly dependent on sulfate availability with high expression at low and low expression at high sulfate concentrations at the site of uptake (Buchner et al. 2004). Whereas the regulatory cross-talk of sulfate uptake and assimilation with carbon and nitrogen metabolism has been studied in detail (Kopriva and Rennenberg 2004), information on the effects of other environmental factors than sulfate availability in the soil and the atmosphere on sulfate transport processes in the roots is scarce. In pedunculate oak (*Quercus robur* L.) neither mycorrhization nor elevated atmospheric pCO_2 improved the rate of sulfate uptake, but total sulfate uptake per plant was enhanced as a consequence of increased root biomass (Seegmüller and Rennenberg 1994; Seegmüller et al. 1996). However, the percentage of the sulfate taken up per root biomass that was loaded into the xylem was enhanced by mycorrhization and elevated atmospheric pCO_2. Thus, these environmental factors increased the sulfate supply of the shoot by enhanced xylem loading as may be required for increased growth of the plants (Seegmüller et al. 1996). Recently, sulfate in the xylem sap received additional attention in the context of drought stress. A number of experiments showed that a decrease in stomatal aperture is among the first responses of plants to soil drying and becomes obvious before any changes in shoot water potential can be measured (Gollan et al. 1986; Goodger et al. 2005). This rapid response has been attributed to chemical or hydraulic signals transmitted from the roots to the leaves (Schachtman and Goodger 2008; Christmann et al. 2007). Analysis of the xylem sap revealed that sulfate was the only compound measured that showed constant enhanced concentrations at early and later stages of soil water stress, already before expression of ABA biosynthetic genes in the roots was enhanced and additional ABA was allocated from the roots to the leaves (Ernst et al. 2010). These observations provide circumstantial evidence that sulfate may be an early root-to-shoot signal of water stress and that this signaling effect is mediated by enhanced xylem loading of sulfate. The present experiments were performed to test whether improved xylem loading of sulfate can also by observed under prolonged drought stress and if the response of xylem loading of sulfate to drought is modulated by mycorrhization and elevated atmospheric pCO_2. For this purpose, young mycorrhizal and non-mycorrhizal pedunculate oak trees were exposed to drought stress at ambient and elevated atmospheric pCO_2 and the uptake and xylem loading of sulfate were determined on excised roots.

Acorns of pedunculate oak (*Quercus robur* L., provenance no. 54 'Manoncourt en Woëvre') were used for the present study. The parent stand grows at 250 m above sea level about 10 km northwest of Nancy in Lorraine, France. The climate is characterized by *c.* 750 mm precipitation, which is evenly distributed over the year and exceeds evapotranspiration by about one-third. The average yearly temperature is 9–10°C, with relatively warm winters (*c.* 5°C in January) and relatively cool summers (*c.* 14°C in July; climate station Nancy, personal communication). Acorns were collected in November, exposed in well-aerated water to 41°C for 3 h to remove surface pathogens, disinfected with Rhodiasan (Rhone-Poulenc, Paris, France) to prevent fungal infections, and stored at 4°C. Immediately before use, acorns were peeled to identify damaged seeds and to allow faster germination, and were soaked in tap water for 2 day at ambient temperature. To enable undisturbed growth of the taproot, acorns were germinated in plastic pipes of 40 cm in length and 8 cm in diameter. The substrate used consisted of autoclaved blond peat (1 part, v/v), a silica sand mixture (19 parts), and slow releasing fertilizer (Nutricote 100, Chisso-Asahi Fertilizer, Tokyo; 4.5 $g\,dm^{-3}$ substrate). Peat was used as an adsorbent of nutrients and to improve the growth of the mycorrhizal fungus. The substrate used provided an adequate supply of plant available water, good aeration of the roots and sufficient nutrients for 1 year's growth (E. Dreyer, INRA Nancy, personal communication). Plants were grown for *c*. 3 months in environmental growth chambers (HPS 1500, Heraeus Vötsch, Hanau) with two replicate chambers for each pCO_2 level (see below) to avoid chamber effects. The photoperiod lasted for 14 h at a temperature of 20°C and 70% relative humidity. Darkness lasted for 10 h at 15°C and 90% relative humidity. Photosynthetically active radiation was maintained at 225–250 $\mu mol\,m^{-2}\,s^{-1}$ at plant level. Light, temperatures and humidity were consistent with the summer climate naturally preferred by pedunculate oak (Dengler and Röhrig 1980; Ellenberg 1986). Half of the plants were inoculated with *Laccaria laccata* (Scop. Ex Fr.) Berk. & Bert. by substituting the autoclaved blond peat with blond peat bearing *L. laccata* (Moser 1958). The success of mycorrhization was determined by visual inspection and exemplary microscopical inspection at harvest. For both treatments (with and without inoculation of *L. laccata*) half of the plants were germinated and cultivated at ambient pCO_2 (359 $\mu l\,l^{-1}$ air), the other half at elevated pCO_2 (1,000 $\mu l\,l^{-1}$). Trees were watered daily with 20 cm^3 de-ionized water per tree. This water supply corresponding to 1,000 mm precipitation $year^{-1}$ guaranteed sufficient water availability. Drought stress was achieved by withdrawing the water supply for 21 or 28 days prior to harvest, whereas controls were watered until harvest as described above. Each water regime was applied to five randomly selected trees in all treatments. At the end of the treatment periods, water conditions were assessed by determination of the pre-dawn shoot water potential with a pressure vessel by the modification of the technique of Scholander et al. (1965) described by Rennenberg et al. (1996). For this purpose plant shoots were dissected from the roots at the shoot bases directly before the start of the photoperiod. The bark was removed from the cut end at a length of 20 mm with a razor blade. The prepared shots were inserted into the pressure vessel with 10 mm of peeled shoot basis extruding. The pressure within the vessel was elevated at 0.1–0.2 MPa min^{-1} until the meniscus of xylem sap appeared at the cut end. This pressure was recorded as pre-dawn shoot water potential.

Uptake and xylem loading of sulfate were studied on excised lateral roots as previously described (Herschbach and Rennenberg 1991; Seegmüller et al. 1996). For this purpose, the root systems of the oak trees were removed from the plastic pipes and carefully washed with M+S nutrient solution (Murashige and Skoog 1962), pH 6.4, containing 0.5 mM sulfate. Lateral roots of 6–10 cm in length were cut with a razor blade in the same nutrient solution. The roots of each tree were used for three independent experiments. For each experiment six lateral roots were placed in an incubation chamber consisting of separate compartments for uptake, buffer and exudation (Herschbach and Rennenberg 1991). The roots were sealed between the compartments of the incubation chamber with Vaseline grease. Root tips in the uptake compartment were incubated at room temperature with M + S nutrient solution, pH 6.4, containing 500 μM sulfate. In order to prevent rhizospheral nutrient depletion, the nutrient solution was stirred by a magnetic stir bar during the entire experiment. The root sections in the buffer and the exudation compartments were exposed to transport medium containing 5 mM bis-tris-propane buffer (Sigma, St. Louis, USA), pH 6.4, and 750 μM calcium chloride. These conditions were applied to achieve optimum sulfate transport rates (Jones and Smith 1981). Each compartment was covered with glass slides. Experiments were started by adding 1,000 kBq carrier-free [^{35}S]-sulfate (NEN-Dupont, Dreieich, Germany) to the uptake compartment. Addition of radioactivity did not markedly increase the sulfate concentration in the uptake compartment. Experiments were terminated after 7 h by removing the nutrient and transport media from each compartment and rinsing the roots three times for 0.5 min with 100 cm^{-3} non-radioactive transport medium. The root tips in the uptake compartment were dissected from the rest of the roots. Root sections were weighed separately and powdered in a mortar under liquid nitrogen. Aliquots of 100 mg of the powder were transferred into 20 cm^3 vials and solubilized in 1 cm^3 Soluene-350 tissue solubilizer (Canberra-Packard, Frankfurt a. M., Germany) at 40°C in a drying oven for 24 h. In order to reduce quenching of radioactivity, samples were bleached with 200 mm^3 30% H_2O_2 at room temperature for 24 h. Radioactivity in the samples was determined by liquid scintillation counting (Wallac Systems 1409, Wallac Oy, Turku, Finland) after addition of 14 cm^3 scintillation fluid (OptiPhase HiSafe II; Wallac Oy, Turku, Finland). Samples were counted at efficiencies of 70–95%, and were corrected for quenching. To determine exudation from the cut ends of the roots, a 500 mm^3 aliquot was taken from the transport medium in the exudation compartment upon termination of the experiment. The aliquot was transferred to a 6 cm^3 vial and mixed with 4 cm^3 scintillation fluid (OptiPhase HiSafe II, Wallac Oy, Turku, Finland). The radioactivity was determined as described above. The rate of sulfate uptake, xylem loading and relative xylem loading (*i.e.* the portion of the sulfate taken up that was loaded into the xylem) were calculated from the radioactivity in the root sections and the radioactivity of the transport medium in the exudation compartment as previously described (Herschbach and Rennenberg 1991). For these calculations dilution of radioactivity by intracellular sulfate pool(s) was neglected. Sulfate uptake and xylem loading rates were correlated with the pre-dawn shoot water potentials of the respective trees. With these data sets regression analyses including non linear equation fitting were carried out using the stasy-500 statistical software (Forstliche Versuchsanstalt, Rheinland-Pfalz, Germany). Regression equations were analysed for significant differences by comparing slopes and y-intersects through t-tests with a probability of error $< 0.05\%$.

When watered daily the pre-dawn shoot water potential of oak trees was similar, irrespective of mycorrhization or pCO_2, and well above −1 MPa, indicating sufficient water supply (Fig. 1). Twenty-one days after the water supply has been withdrawn,

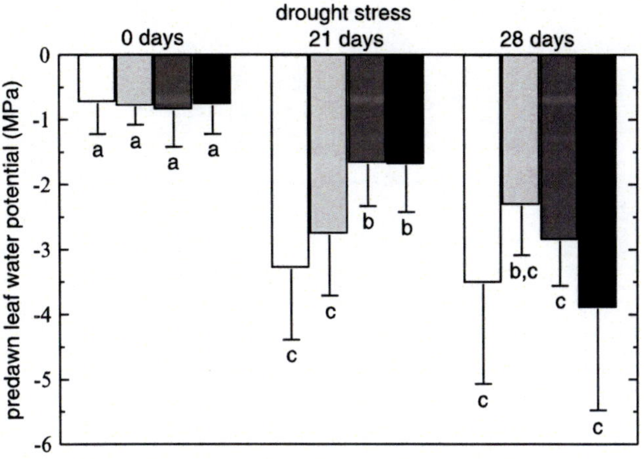

Fig. 1 Predawn leaf water potential of mycorrhizal and non-mycorrhizal pedunculate oak trees (*Quercus robur* L.) at ambient (359 µl l^{-1}) and elevated (1,000 µl l^{-1}) pCO_2 after 0, 21 and 28 days of drought stress. *White bars*: ambient CO_2, non-mycorrhizal; *light grey bars*: ambient CO_2, mycorrizal; *dark grey bars*: non-mycorrhizal, elevated CO_2; *black bars*: mycorrhizal, elevated CO_2

pre-dawn shoot water potential indicated drought stress for all trees analysed, but the extent of drought stress differed between the treatments. Irrespective of mycorrhization pre-dawn shoot water potential showed less negative values for trees grown at elevated than at ambient pCO_2. Mycorrhization did not significantly reduce drought stress. Twenty-eight days after the water supply has been withdrawn, severe drought stress was observed irrespective of the treatment and pre-dawn shoot water potential was not significantly different between the treatments (Fig. 1).

Sulfate uptake of mycorrhizal and non-mycorrhizal roots varied between 25 and 39 nmol sulfate g^{-1} FW h^{-1}, but a dependency on pre-dawn shoot water potential was not found (Fig. 2a). In non-mycorrhizal roots xylem loading of sulfate tended to increase as pre-dawn shoot water potential got less negative than −1 MPa. A similar effect was observed for mycorrhizal roots already when pre-dawn shoot water potential became less negative than −3 MPa (Fig. 2b). As a consequence, xylem loading of sulfate was enhanced by mycorrhization at pre-dawn shoot water potentials less negative than −3 MPa, but similar in mycorrhizal and non-mycorrhizal roots at more negative values. When trees were sufficiently watered, xylem loading of sulfate was about 3-fold higher in mycorrhizal compared to non-mycorrhizal roots. Relative xylem loading, *i.e.* the proportion of the sulfate taken up that was loaded into the xylem, tended to decrease as pre-dawn shoot water potential became more negative and tended to be higher in mycorrhizal than in non-mycorrhizal roots (Fig. 2c), but a clear-cut correlation between relative xylem loading and pre-dawn shoot water potential was not found.

The effect of elevated pCO_2 on sulfate transport processes in mycorrhizal roots is shown in Fig. 3. Sulfate uptake of mycorrhizal roots was similar at ambient and

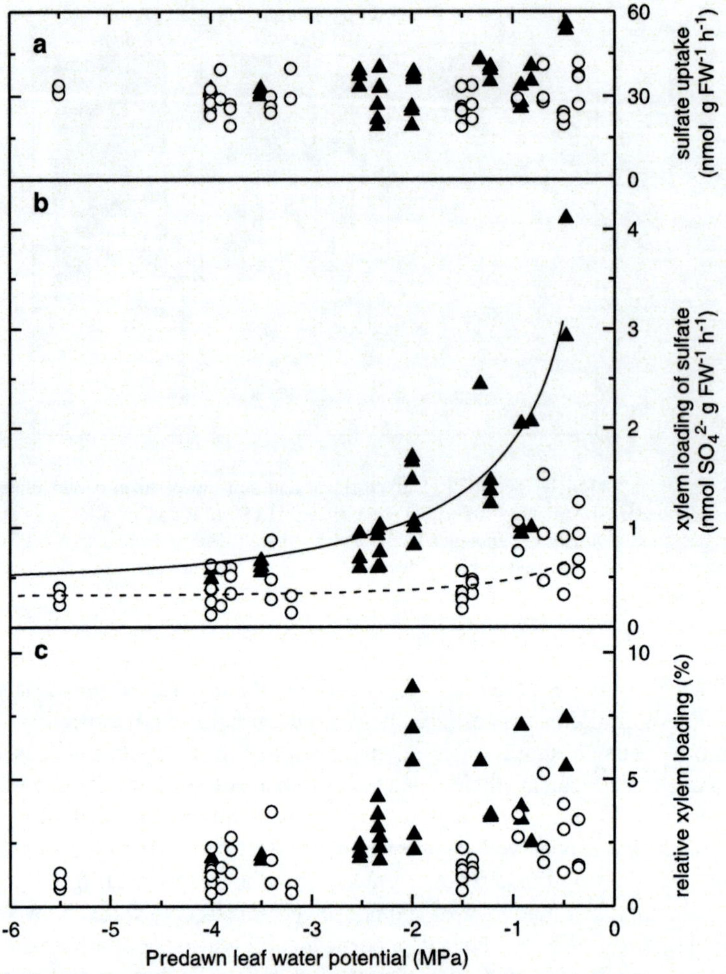

Fig. 2 Sulfate uptake (**a**), xylem loading of sulfate (**b**) and relative xylem loading of sulfate (**c**) of excised lateral roots of mycorrhizal (▲) and non-mycorrhizal (○) pedunculate oak trees (*Quercus robur* L.) under ambient CO_2 at different predawn leaf water potentials. *FW* root fresh weight

elevated pCO_2 and varied between 25 and 65 nmol sulfate g^{-1} FW h^{-1}. A dependency of sulfate uptake on pre-dawn shoot water potential was not observed (Fig. 3a). At both ambient and elevated pCO_2 xylem loading of sulfate increased as pre-dawn shoot water potential became less negative than −3 MPa with slightly higher values at elevated than at ambient pCO_2 (Fig. 3b). Similar results were obtained for relative xylem loading of sulfate, but a clear-cut correlation with pre-dawn shoot water potential was not found because of a high variation of the data (Fig. 3c). In non-mycorrhizal trees neither sulfate uptake (Fig. 4a), nor xylem loading of sulfate (Fig. 4b), or relative xylem loading of sulfate (Fig. 4c) depended on pre-dawn shoot water potential or differed between ambient and elevated pCO_2.

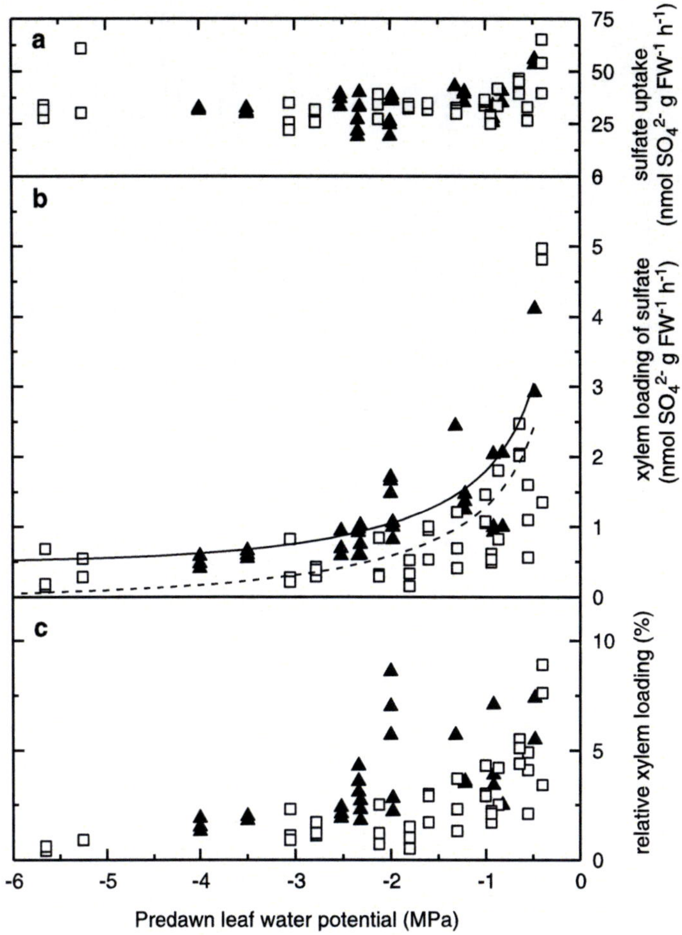

Fig. 3 Sulfate uptake (**a**), xylem loading of sulfate (**b**) and relative xylem loading of sulfate (**c**) of excised lateral roots of mycorrhizal pedunculate oak trees (*Quercus robur* L.) under ambient (359 μl l^{-1}, ▲) and elevated (1,000 μl l^{-1}, □) CO_2 at different pre-dawn leaf water potentials

These results indicate that prolonged drought does not enhance xylem loading of sulfate, which is in contrast to findings based on the sulfate content of the xylem and from the SULTR expression in the elongation zone of maize plants in response to drought (Ernst et al. 2010). Moreover, reduced xylem loading of sulfate and, hence reduced sulfate supply of the shoot were found with oak. Since sulfate uptake was not affected by reduced water supply, it may be assumed that this process is strongly regulated by the growth of the oak trees as also concluded from previous experiments (Seegmüller et al. 1996). In oak, negative drought effects and partial compensation of these effects on sulfate transport processes by mycorrhization and elevated

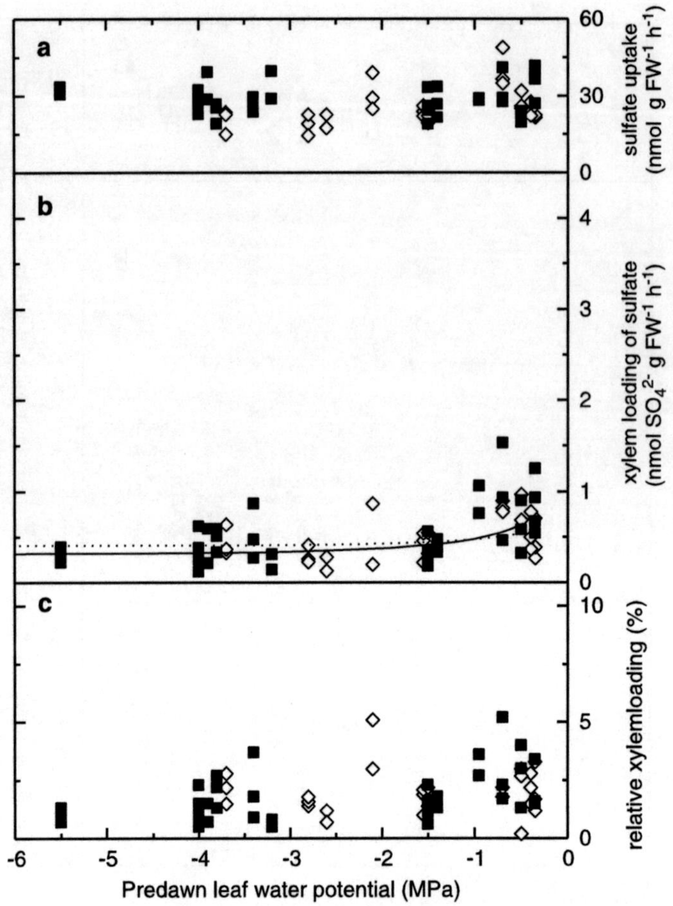

Fig. 4 Sulfate uptake (**a**), xylem loading of sulfate (**b**) and relative xylem loading of sulfate (**c**) of excised lateral roots of non mycorrhizal pedunculate oak trees (*Quercus robur* L.) under ambient (359 µl l⁻¹, ■) and elevated (1,000 µl l⁻¹, ◊) CO_2 at different pre-dawn leaf water potentials

atmospheric pCO_2 are restricted to xylem loading of sulfate. Compensation of the negative effect of drought on xylem loading of sulfate by mycorrhization is surprising, since the oak trees were inoculated with the ectomycorrhizal fungus *Laccaria laccata* that is not supposed to enter the bundle of the root and therefore cannot directly interact with xylem loading. Thus, signaling by the fungus across the bundle sheath has to be assumed.

Acknowledgements Financial support by the commission of the EU under contract no. EC EV5V-CT92-0093 and by the State of Baden-Württemberg is gratefully acknowledged.

References

Brunold C (1990) Reduction of sulfate to sulfide. In: Rennenberg H, Brunold Ch, De Kok LJ, Stulen I (eds) Sulfur nutrition and assimilation in higher plants; fundamental, environmental and agricultural aspects. SPB Academic Publishing, The Hague, pp 13–31

Buchner P, Takahashi H, Hawkesford MJ (2004) Plant sulphate transporter: co-ordination of uptake, intracellular and long-distance transport. J Exp Bot 55:1765–1773

Christmann A, Weiler EW, Steudle E, Grill E (2007) A hydraulic signal in root-to-shoot signalling of water shortage. Plant J 52:167–174

Clarkson DT, Hawkesford MJ, Davidian J-C (1993) Membrane and long-distance transport of sulfate. In: De Kok LJ, Stulen I, Rennenberg H, Brunold C, Rauser WE (eds) Sulfur nutrition and assimilation in higher plants; regulatory, agricultural and environmental aspects. SPB Academic Publishing, The Hague, pp 3–19

Dengler A, Röhrig E (1980) Waldbau auf ökologischer Grundlage. Erster Band: Der Wald als Vegetationstyp und seine Bedeutung für den Menschen. Verlag Paul Parey, Hamburg

Ellenberg H (1986) Vegetation Mitteleuropas mit den Alpen. Verlag Eugen Ulmer, Stuttgart

Ernst L, Goodger JQD, Alvarez S, Marsh EL, Berla B, Lockhart E, Jung J, Li P, Bohnert HJ, Schachtman DP (2010) Sulphate as a xylem-borne chemical signal precedes the expression of ABA biosynthetic genes in maize roots. J Exp Bot 61:3395–3405

Gollan T, Passioura JB, Munns R (1986) Soil water status affects the stomatal conductance of fully turgid wheat and sunflower leaves. Aus J Plant Physiol 13:459–464

Goodger JQD, Sharp RE, Marsh EL, Schachtman DP (2005) Relationships between xylem sap constituents and leaf conductance of well-watered and water-stressed maize across three xylem sap sampling techniques. J Exp Bot 56:2389–2400

Hartmann T, Mult S, Suter M, Rennenberg H, Herschbach C (2004) Regulation of sulphate assimilation by glutathione in poplars (*Populus tremula x P. alba*) of wild type and overexpressing γ-glutamylcysteine synthetase in the cytosol. J Exp Bot 55:837–845

Hawkesford MJ, De Kok LJ (2006) Managing sulfur metabolism in plants. Plant Cell Environ 29:382–395

Herschbach C, Rennenberg H (1991) Influence of glutathione (GSH) on sulphate influx, xylem loading and exudation in excised tobacco roots. J Exp Bot 42:1021–1029

Herschbach C, van der Zalm E, Schneider A, Jouanin L, De Kok LJ, Rennenberg H (2000) Regulation of sulphur nutrition in wild-type and transgenic poplar over-expressing γ-glutamylcysteine synthetase in the cytosol as affected by atmospheric H_2S. Plant Physiol 124:461–474

Jones SL, Smith IK (1981) Sulfate transport in cultured tobacco cells. Effects of calcium and sulfate concentration. Plant Physiol 67:445–448

Kopriva S, Rennenberg H (2004) Control of sulfate assimilation and glutathione synthesis: interaction with N and C metabolism. J Exp Bot 55:1831–1842

Moser M (1958) Die künstliche Mykorrhizaimpfung an Forstpflanzen. II. Die Torfstreukultur von Mykorrhizapilzen. Forstwiss Cbl 77:273–278

Murashige T, Skoog F (1962) A revised medium for rapid growth and bioassays with tobacco tissue cultures. Physiol Plant 15:473–497

Rennenberg H (1984) The fate of excess sulfur in higher plants. Annu Rev Plant Physiol 35:121–153

Rennenberg H, Schneider S, Weber P (1996) Analysis of uptake and allocation of nitrogen and sulphur compounds by trees in the field. J Exp Bot 47:1491–1498

Schachtman DP, Goodger JQD (2008) Chemical root to shoot signaling under drought. Trends Plant Sci 13:281–287

Scholander PF, Hammel HAT, Bradstreet ED, Hemmingsenn EA (1965) Sap pressure in vascular plants. Science 148:339–346

Seegmüller S, Rennenberg H (1994) Interactive effects of mycorrhization and elevated carbon dioxide on growth of young pedunculate oak (*Quercus robur* L.) trees. Plant Soil 167:325–329

Seegmüller S, Schulte M, Herschbach C, Rennenberg H (1996) Interactive effects of mycorrhization and elevated atmospheric CO_2 on sulphur nutrition of young pedunculate oak (*Quercus robur*) trees. Plant Cell Environ 19:418–426

van der Zalm E, Schneider A, Rennenberg H (2005) Regulation of sulphate uptake and xylem loading in poplar roots (*Populus tremula* × *P. alba*). Trees 19:204–212

The Role of Cyclophilin CYP20-3 in Activation of Chloroplast Serine Acetyltransferase Under High Light Stress

Anna Speiser, Agata Kurzyk, Anna Wawrzynska, Markus Wirtz, Agnieszka Sirko, and Rüdiger Hell

Abstract High light induces the production of cysteine and glutathione (GSH) as part of the oxidative stress response triggered by reactive oxygen species. Cysteine is produced in a two-step process, in which serine acetyltransferase (SAT) catalyses the first reaction. In *Arabidopsis thaliana* cyclophilin CYP20-3 was reported to promote plastidic SAT1 activity and thus cysteine and GSH production under high light conditions (Dominguez-Solis et al., Proc Natl Acad Sci USA 105:16386–16391, 2008). Challenging the loss of function mutant *cyp20-3* with high light however did not reveal any hypersensitivity of the mutant questioning the function of CYP20-3 in regulation of cysteine synthesis. Although physical interaction of CYP20-3 and SAT1 could be confirmed by yeast two-hybrid analysis, plastidic cysteine synthesis induced by oxidative stress is probably regulated by mechanisms other than CYP20-3 function.

Biosynthesis of cysteine is the unique entry point of reduced sulfur into plant metabolism. Reduction of sulfate to sulfide exclusively takes place in plastids. The enzyme *O*-acetylserine-(thiol)-lyase (OAS-TL) catalyses cysteine synthesis by fixation of sulfide into the carbohydrate backbone of *O*-acetylserine (OAS). Serine acetyltransferase (SAT) transfers an acetyl moiety from acetyl-CoA to serine and forms OAS. Both enzymes form the hetero-oligomeric cysteine synthase complex (CSC). Dissociation

A. Speiser • M. Wirtz • R. Hell (✉)
Centre for Organismal Studies, University of Heidelberg, Heidelberg, Germany
e-mail: ruediger.hell@cos.uni-heidelberg.de

A. Kurzyk • A. Wawrzynska • A. Sirko
Institute for Biochemistry and Biophysics, Polish Academy of Sciences, Warsaw, Poland

L.J. De Kok et al. (eds.), *Sulfur Metabolism in Plants*, Proceedings of the International Plant Sulfur Workshop 1, DOI 10.1007/978-94-007-4450-9_32,

of the CSC inactivates SAT due to the sensitivity of free SAT to cysteine feedback inhibition. Several lines of evidence demonstrate that SAT activity limits cysteine synthesis. For that reason the dissociation state of the CSC regulates the flux of cysteine synthesis in response to its intermediates OAS and sulfide. In contrast to sulfate reduction cysteine synthesis takes place in the cytosol, the mitochondria and the plastids in *Arabidopsis thaliana*. Compartment specific OAS-TL and SAT proteins are encoded by small gene families of nine and five members, respectively. OAS-TL B and SAT1 represent the plastidic isoforms. Recently AtCYP20-3 was suggested to act as an inducer of plastidic cysteine biosynthesis by promotion of plastidic CSC formation in response to high light stress. The basis for this suggestion was the interaction of CYP20-3 with SAT1 *in vitro* (Dominguez-Solis et al. 2008). Cyclophilins belong to the plant superfamily of immunophilins and assist mainly in protein folding and assembly by peptidyl-prolyl *cis-trans* isomerase (PPI) activity (Romano et al. 2004). They are involved in various cellular processes, *e.g.* development, signal transduction and stress response. Treatment with sodium chloride, rose bengal or high light (300 μmol m^{-2} s^{-1}) causes a retardation of growth of the CYP20-3 knock-out mutant (*cyp20-3*) in comparison to the wildtype (WT). Oxidative stress induced by sodium chloride, rose bengal or high light share the ability to produce reactive oxygen species (ROS), which are scavenged in cytosol, mitochondria, peroxisomes and plastids by the ascorbate-glutathione cycle. An important feature of this cycle is the capacity of glutathione (GSH) to act as a redox buffer. The first step of GSH synthesis is exclusively localized in plastids and is limited by the availability of cysteine. Hence the conclusion was drawn that CYP20-3 assists in folding or assembly of SAT1 protein under oxidative stress and thereby triggers the association of SAT1 and OAS-TL B to form the plastidic CSC in order to produce cysteine (Dominguez-Solis et al. 2008).

The aim of this study was to clarify the role of plastidic cysteine synthesis in *Arabidopsis* upon oxidative stress conditions. To prove the proposed physical interaction of CYP20-3 and SAT1 a yeast two-hybrid assay was applied (Fig. 1). The yeast screening procedure was performed essentially as described in the Yeast MATCHMAKER System 3 manual (Clontech, Palo Alto, CA). Yeast cells carrying the BD-CYP20-3 and AD-SAT1 plasmids with the full-length cDNA of both genes were able to grow on selective media and express the reporter gene. Yeast strains transformed with empty pBD and AD-SAT1 or pAD and BD-CYP20-3 vectors as a negative control failed both selection properties. These results verified the *in vitro* interaction of CYP20-3 and SAT1. *Arabidopsis* WT and *cyp20-3* plants were challenged with high light (300 μmol m^{-2} s^{-1}) to elucidate the *in vivo* role of CYP20-3 regarding cysteine synthesis under ROS inducing conditions. Both WT and *cyp20-3* show no symptoms under control conditions (Fig. 2a, c). Under high light conditions WT plants exhibit a characteristic stress phenotype, which includes stunted leaves, growth retardation and the accumulation of anthocyanins (Fig. 2b). *cyp20-3* plants show the same stress symptoms as WT, revealing no increased sensitivity towards high light (Fig. 2d). CYP20-3 was proposed to promote SAT activity to adjust cysteine production for increasing demand of GSH upon oxidative stress (Dominguez-Solis et al. 2008). To test this hypothesis, thiols were extracted from

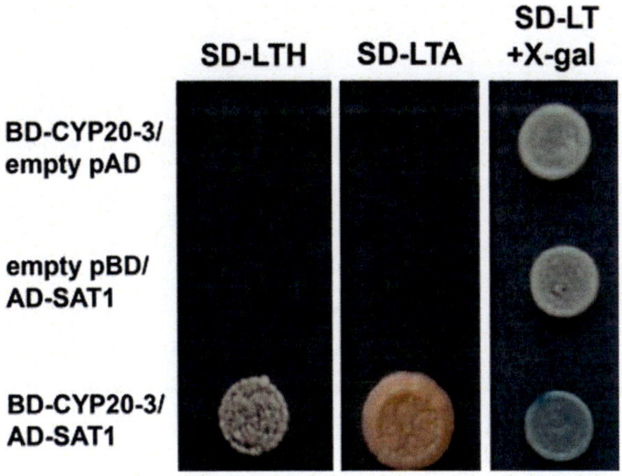

Fig. 1 Interaction of CYP20-3 and SAT1 proved by yeast two-hybrid. Yeast strains co-transformed with indicated BD- and AD-plasmids were grown on selective media without leucine, tryptophan and histidine (*SD-LTH*), leucine, tryptophane and adenine (*SD-LTA*) and additionally screened for α-galactosidase expression (*SD-LT+X-gal*). pBD and pAD empty vectors were used as negative controls

Fig. 2 High light stress experiment using wildtype (WT) and *cyp20-3* plants. WT and *cyp20-3* plants were grown at short day conditions (8 h light, 90 μmol m^{-2} s^{-1}) for 3 weeks and subsequently transferred to long day conditions (14 h light) for another 3 weeks either under control light (90 μmol m^{-2} s^{-1}; **a**, **c**) or high light (300 μmol m^{-2} s^{-1}; **b**, **d**)

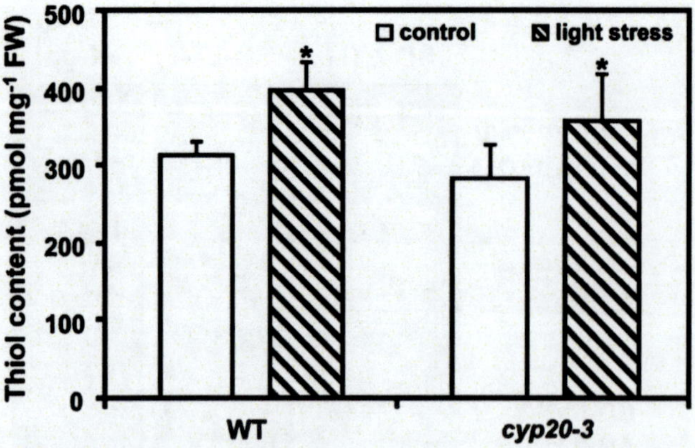

Fig. 3 Determination of thiol levels (total cysteine and glutathione) in wildtype (WT) and *cyp20-3* plants upon high light stress. Thiols were extracted from 6 weeks old plants and detected after derivatization with monobromobimane by HPLC according to Heeg et al. (2008). Plants were grown as described in Fig. 2 (n=3–5). *FW* fresh weight; *$p \leq 0.05$, unpaired Student's t-test

Arabidopsis leaves and detected according to Heeg et al. (2008). The steady state thiol levels in WT and *cyp20-3* increased up to 27% and 26%, respectively, confirming that the applied high light levels caused oxidative stress (Fig. 3). The *cyp20-3* mutant displays indistinguishable levels of GSH in comparison to the wild type, which makes a significant contribution of CYP20-3 for activation of cysteine synthesis in response to high light unlikely.

No hypersensitivity of *cyp20-3* towards oxidative stress, caused by high light, could be observed in the experimental set-up used in this study. The results demonstrate that elevation of thiol steady state levels does not require per se induction of plastidic SAT activity by CYP20-3. Several lines of evidence support this conclusion. A reverse genetics approach demonstrates that SAT1 contributes insignificantly to total SAT activity in leaves. A T-DNA insertion mutant of SAT1 (*serat2;1*) had wild type levels of OAS and cysteine and showed no distinct phenotype (Watanabe et al. 2008). A matter of caution must be added to the last statement, since *serat2;1* has not been tested under high light conditions. In contrast mitochondrial SAT3 represents approximately 80% of total SAT activity in leaves (Watanabe et al. 2008) and consequently triggers the flux of cysteine synthesis under normal growth conditions (Haas et al. 2008). In addition, the plastidic OAS-TL loss-of-function mutant (*oastlB*) neither exhibits any growth phenotype nor shows significant changes in cysteine and GSH levels (Heeg et al. 2008). The wild type like phenotypes of *serat2;1* and *oastlB* proof the existence of a plastid import system for OAS and cysteine. Transport of OAS and cysteine likely contributes to meet the demands of the plastids for efficient GSH synthesis. The high activity of SAT in mitochondria furthermore questions a significant contribution of SAT1 to foliar OAS formation.

The high abundance of CYP20-3 in comparison to SAT1 in the stroma of chloroplast makes an exclusive action of CYP20-3 on SAT1 unlikely, but indicates a general function of CYP20-3 in disaggregation of protein aggregates (Romano et al. 2004). A specific contribution of CYP20-3 to the formation of the plastidic CSC by its chaperone function is unnecessary, since formation of bacterial and plants CSC are known to occur spontaneously without assistance of chaperones (Wirtz et al. 2010). Furthermore CYP20-3 is reported to be one of the major regulatory hubs in the chloroplast by SAT1 independent functions (Muthuramalingam et al. 2009). These less characterized functions might cause variable responses of *cyp20-3* mutants due to slightly different high light conditions used in different labs (see also below). *Arabidopsis* has been shown to adjust its metabolic network to environmental factors (*e.g.* frequency of watering, type of soil, difference in light quality) that are unique in a laboratory (Massonnet et al. 2010). It might be that a combination of these environmental factors causes a different capacity to respond to high light stress (pre-adaptation) in *cyp20-3* grown in a single lab. In this study *cyp20-3* mutants of the control group displayed no stress symptoms or pre-induced high GSH levels, ruling out the possibility that the plants were pre-adapted to oxidative stress at the beginning of the experiment. We therefore conclude that high light stress does not cause per se hypersensitivity in *cyp20-3*, but it needs a specific combination of environmental factors (including high light) to result in the phenotype observed by Dominguez-Solis et al. (2008).

References

Dominguez-Solis JR, He Z, Lima A, Ting J, Buchanan BB, Luan S (2008) A cyclophilin links redox and light signals to cysteine biosynthesis and stress responses in chloroplasts. Proc Natl Acad Sci USA 105:16386–16391

Haas FH, Heeg C, Queiroz R, Bauer A, Wirtz M, Hell R (2008) Mitochondrial serine acetyltransferase functions as a pacemaker of cysteine synthesis in plant cells. Plant Physiol 148:1055–1067

Heeg C, Kruse C, Jost R, Gutensohn M, Ruppert T, Wirtz M, Hell R (2008) Analysis of the *Arabidopsis O*-acetylserine(thiol)lyase gene family demonstrates compartment-specific differences in the regulation of cysteine synthesis. Plant Cell 20:168–185

Massonnet C, Vile D, Fabre J, Hannah MA, Caldana C, Lisec J, Beemster GTS, Meyer RC, Messerli G, Gronlund JT, Perkovic J, Wigmore E, May S, Bevan MW, Meyer C, Rubio-Dìaz S, Weigel D, Micol JL, Buchanan-Wollaston V, Fiorani F, Walsh S, Rinn B, Gruissem W, Hilson P, Hennig L, Willmitzer L, Granier C (2010) Probing the reproducibility of leaf growth and molecular phenotypes: a comparison of three *Arabidopsis* accessions cultivated in ten laboratories. Plant Physiol 152:2142–2157

Muthuramalingam M, Seidel T, Laxa M, Nunes de Miranda SM, Gärtner F, Ströher E, Kandlbinder A, Dietz K-J (2009) Multiple redox and non-redox interactions define 2-Cys peroxiredoxin as a regulatory hub in the chloroplast. Mol Plant 2:1273–1288

Romano PGN, Horton P, Gray JE (2004) The *Arabidopsis* cyclophilin gene family. Plant Physiol 134:1268–1282

Watanabe M, Mochida K, Kato T, Tabata S, Yoshimoto N, Noji M, Saito K (2008) Comparative genomics and reverse genetics analysis reveal indispensable functions of the serine acetyltransferase gene family in *Arabidopsis*. Plant Cell 20:2484–2496

Wirtz M, Birke H, Heeg C, Mueller C, Hosp F, Throm C, Koenig S, Feldman-Salit A, Rippe K, Petersen G, Wade RC, Rybin V, Scheffzek K, Hell R (2010) Structure and function of the hetero-oligomeric cysteine synthase complex in plants. J Biol Chem 285:32810–32817

Impact of Lead and Sulfur Deprivation on Soybean Plants

Oksana Sytar, Anatolij Kosyan, Nataliya Taran, and Alexander Okanenko

Abstract Lipid peroxidation products, antioxidative enzymes (superoxide dismutase and catalase) and water-soluble non-protein and water-soluble protein thiol contents were studied in soybean leaves affected by sulfate deprivation and Pb exposure. Sulfur promoted the regulation of adaptive reactions via an enhanced formation of water-soluble non-protein thiols. The reaction of soybean plants to Pb exposure was dependent on sulfur supply. At sufficient-sulfur conditions, higher water-soluble non-protein thiol compounds might contribute to better defense against Pb.

With increasing development of industrial and mining activity heavy metal pollution of soils is becoming an important global environmental problem. Among heavy metals lead (Pb) is highly phytotoxic and is a major pollutant worldwide due to its release from mining, manufacturing and urban activities. Exposure of plants to abiotic and biotic stress may disturb metabolism and induce an increased production of reactive oxygen species, *e.g.* superoxide radicals, which may result in an unspecific oxidation of proteins and membrane lipids and/or may cause DNA injury. Superoxide radicals in plant cells are usually rapidly dismutated by superoxide dismutase (SOD) to H_2O_2, which can be detoxified by catalase (CAT) and peroxidases (Polle and Rennenberg 1993; Tausz et al. 2003). High levels of Pb in plant tissues may induce oxidative stress by generating excessive levels of free radicals and reactive oxygen species (Reddy et al. 2005; Sun et al. 2010). Plant protection mechanisms against heavy metal action include the formation of phytochelatins (PCs), a class of thiol-rich peptides with the general structure $(\gamma\text{-Glu-Cys})_n$-Gly. Phytochelatins are synthesized from glutathione as precursor (Rauser 1995). Phytochelatins may form

O. Sytar (✉) • A. Kosyan • N. Taran • A. Okanenko
Department of Plant Physiology and Ecology, Faculty of Biology,
Taras Shevchenko National University of Kyiv,
64, Volodymyrs'ka St, 01601 Kyiv, Ukraine
email: o_sytar@ukr.net

L.J. De Kok et al. (eds.), *Sulfur Metabolism in Plants*, Proceedings of the International Plant Sulfur Workshop 1, DOI 10.1007/978-94-007-4450-9_33,

complexes with heavy metals, which subsequently may be transported into the vacuole, probably by tonoplast membrane ATP-binding cassette-type transporters (Salt and Rauser 1995). High levels of Pb may affect plant sulfur metabolism, since synthesis of PCs might require increased sulfate uptake and assimilation (Ernst et al. 2008). Moreover, it has been observed that high Pb level may also affect sulfolipid metabolism in plants (Guschina and Harwood 2002). In the present communication the impact of Pb and sulfate deprivation on lipid peroxidation, antioxidant enzyme activity, thiol and sulfolipid (sulfoquinovosyl diacylglycerol, SQDG) content was investigated in soybean.

Soybean seeds (*Glycine max* (L.) Merr., cultivar "Ustya") were germinated and grown on water culture with (+S) and without sulfate (–S), in a climate-controlled greenhouse for 14 days at 23°C, 65% RH and at a 16 h photoperiod. The composition of the nutrition solution was as follows: control (+S): 1.44 g l^{-1} $Ca(NO_3)_2.4H_2O$, 0.25 g $l^{-1}$$K_2HPO_4$, 0.25 g l^{-1} $Mg(NO_3)_2$; 0.3 g l^{-1} $(NH_4)_2SO_4$; sulfate deprived (–S): 1.44 g l^{-1} $Ca(NO_3)_2.4H_2O$, 0.25 g $l^{-1}$$K_2HPO_4$, 0.25 g l^{-1} $Mg(NO_3)_2$, 0.135 g l^{-1} carbamide. After 14 days, the foliage was sprayed with 0.5 mM Pb in form $Pb(NO_3)_2$ in distilled water. Control plants were sprayed by distilled water. All experiments were set up in triplicate and each replicate contained four plants. Leaves were taken from each plant in each of the three experiments (biological replication, n = 3). Each extract from one biological replicate was measured three times (analytical replication).

For Pb analyses, leaves were separated and oven dried at 80°C for 2 days. Dried plant material was acid digested with 4 ml HNO_3 : $HClO_4$ (3:1, v/v) and the digested material was dissolved in 10 ml of 0.1 N HNO_3 and analyzed for Pb content by using an atomic absorption spectrophotometer (GBC 932 plus, Australia).

For biochemical analyses, leaves separated from treated plants 6, 24, and 48 h after the treatment were used. Lipid hydroperoxides were assayed according to Droillard et al. (1987). Lipid peroxidation in leaves was determined by measuring the formation of malondialdehyde (MDA) as described by Heath and Packer (1968) with slight modifications: Leaves (0.2 g) were homogenized in 3 ml 0.1 M Tris buffer containing 0.3 M NaCl. After that, 2 ml of 20% trichloroacetic acid (TCA) containing 0.5% 2-thiobarbituric acid (TBA) and 2 ml 20% TCA were added. The mixture was heated at 95°C for 30 min and then the homogenate was centrifuged at 10,000 g for 5 min. Absorbance of the supernatant was measured at 532 nm. MDA was calculated by using an extinction coefficient of 155 mM^{-1} cm^{-1}. Leaf tissue for enzyme assays was frozen in liquid nitrogen and stored at –20°C until extraction. The frozen samples were homogenized with 10 ml 0.1 M Na_2HPO_4/NaH_2PO_4 (pH 7.0) at 4°C. The homogenate was centrifuged at 4°C at 4,000 g for 15 min. The supernatant was used for the biochemical assay. The activity of catalase (CAT) (EC 1.11.1.6) was determined as described by Havir and McHale (1987). SOD (EC.1.15.1.1) activity in extracts of soyabean leaves was based upon the indirect spectrophotometric method of Forman and Fridovich (1973). The reaction mixture contained 50 mM $Na_2CO_3/NaHCO_3$ (pH 10.2), 0.10 mM EDTA, 0.01 mM ferricytochrome c, and 0.05 mM xanthine. The assay was initiated by the addition of sufficient xanthine oxidase to produce a basal rate of ferricytochrome *c* reduction corresponding to an increase in A_{550} of 0.035 units min $^{-1}$ (V_1). After verification of

V_1, extract was added and the resulting reaction velocity (V_2) calculated. One unit of superoxide dismutase was defined as the amount of enzyme which inhibited the rate of ferricytochrome c reduction by 50% ($V_1/V_2=2$) in a 3 ml assay volume. Units of activity were calculated as a function of the reaction velocities (units = V_1/V_2- 1.06). For sulfoquinovosyl diacylglycerol analyses, plant leaves (0.2 g) were homogenized with 0.5 g glass powder and 0.5 g $Na_2(SO_4)$. The homogenate was transferred to a glass column and 3 ml acetone was added for filtration. To the filtrate 1 ml hexane : benzene (4:1, v/v) and 2 ml H_2O was added and centrifuged at 5,000 g for 5 min at RT (room temperature). After centrifugation 1 ml of the bottom layer was added to 1 ml of 0.01% azure (prepared in acetone) and 2 ml benzene, and again centrifuged at 5,000 g for 5 min at RT. The upper layer was collected and absorbance was measured at 610 nm. The calculation of SQDG was based on the standard curve of sodium dodecylsulfate according to Kean (1968). The water-soluble protein thiol content was determined as described by Sedlak and Lindsay (1968). The following components were added to 20 µl of sample: 75 µl 0.5 M Tris–HCl, 25 µl 0.01 M DTNB reagent, 400 µl methanol. Samples were centrifuged at 3,000 g for 5 min at room temperature, 3×90 µl supernatant were transferred into a flat-bottom microplate using a multichannel pipette and extinction was measured at 412 nm. Taking into account that conventional desalting methods (desalting, gel filtration) may affect thiol content (by air oxidation) proteins were removed by TCA precipitation prior to SH quantitation. For this purpose 50 µl sample/standard were added to 2×25 µl 5% TCA and then centrifuged for 15 min at 3,000 g and room temperature. 50 µl of supernatant from the 5% TCA precipitation was mixed with 200 µl 0.5 M Tris–HCl (pH 8.9) and 20 µl 0.01 M DTNB in a flat-bottom microplate and extinction was measured at 412 nm. The water-soluble protein-thiol content was calculated by subtracting the water-soluble non-protein thiol content from that of the total water-soluble thiol content (Sedlak and Lindsay 1968. Significant differences of these data were calculated using analysis of variance (ANOVA followed by Duncan's multiple range test, SIGMASTAT 9.0).

Pb spraying only resulted in high Pb levels in the leaves, and no detectable levels were observed in stems and roots (Table 1). Sulfur deprivation induced a strong decrease in the water-soluble non-protein thiol content in the leaves of soybean plants (Fig. 1a). In addition, Pb exposure resulted in a decrease in the water-soluble non-protein thiol content in both sulfate-sufficient and sulfate-deprived leaves 6 h after the exposure. In sulfate-sufficient plants the water-soluble non-protein thiol content in leaves increased further to 34% after 48 h. The water-soluble protein-thiol content in leaves of sulfur-deprived plants was about 30% decreased in Pb-exposed plants after 24 and 48 h (Fig. 1b). In leaves of sulfur-sufficient plants there were no changes in water-soluble protein-thiol content upon Pb exposure. The Pb-induced increase in water-soluble non-protein thiol contents in sulfur-sufficient plants may be required to detoxify Pb. Time-dependent decrease in glutathione (GSH) content has been shown upon Cd exposure (Dixit et al. 2001). GSH is a free thiol, which plays an important role in the cellular redox status and also serves as substrate for phytochelatin synthesis.

Table 1 Content of Pb in leaves, stems and roots of sulfur-sufficient (+S) and sulfur-deprived soybean plants (−S) after foliar spraying with 0.5 mM of $Pb(NO_3)_2$

Time after treatment	Pb content (mg g^{-1} DW)					
	Roots		Stems		Leaves	
	+S	−S	+S	−S	+S	−S
6 h	n.d.	n.d.	n.d.	n.d.	0.16±0.03	0.18±0.04*
24 h	n.d.	n.d.	n.d.	n.d.	0.15±0.04	0.19±0.10*
48 h	n.d.	n.d.	n.d.	n.d.	0.13±0.03	0.17±0.04*

n.d. not detectable
*Denote significant differences between (−S), (Pb, −S) and (+S), (Pb, +S)

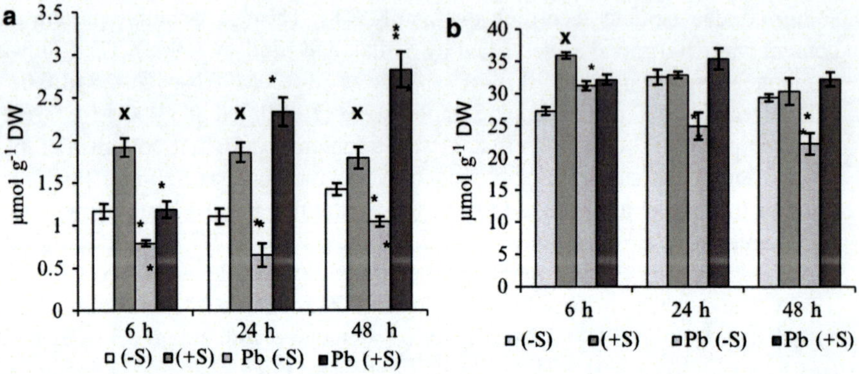

Fig. 1 The impact of Pb on water-soluble non-protein thiol (**a**) and water-soluble protein-thiol content (**b**) in leaves of soybean plants grown at sulfur-sufficient (+S) and sulfur-deprived conditions (−S). *, significant differences between (−S), (Pb, -S) and (+S), (Pb, +S) and x, significant differences between (−S) and (+S)

The lipid hydroperoxide content was lower in leaves of sulfur-deprived than sulfur-sufficient plants (Fig. 2a). A similar effect was observed in young maize leaves, where the superoxide content was lower sulfur-deprived than sulfur-sufficient plants (Bouranis et al. 2003). Pb exposure resulted in a decrease in the lipid hydroperoxide content in leaves of both sulfur-deprived and sulfur-sufficient plants after 6 h, however, it caused an increase in the lipid hydroperoxide content in leaves after 48 h. Apparently Pb negatively affected plant metabolism and resulted in an induction of oxidative stress reactions. The lipid hydroperoxide content in leaves of plants grown with different sulfur supply was differently affected by Pb, which might indicate a role of sulfur in response reactions of soybean plants at excess Pb (Fig. 2a).

MDA content increased in sulfur-sufficient plants 6 h after exposure to Pb and corresponded with a decrease in lipid hydroperoxide content. After 24 h the content of MDA was quite similar in leaves of sulfur-deprived and sulfur-sufficient plants, there was only a slight decrease in its content in leaves of sulfur-deprived Pb exposed plants. After 48 h the content of MDA decrease on 21% in leaves of sulfur-deprived

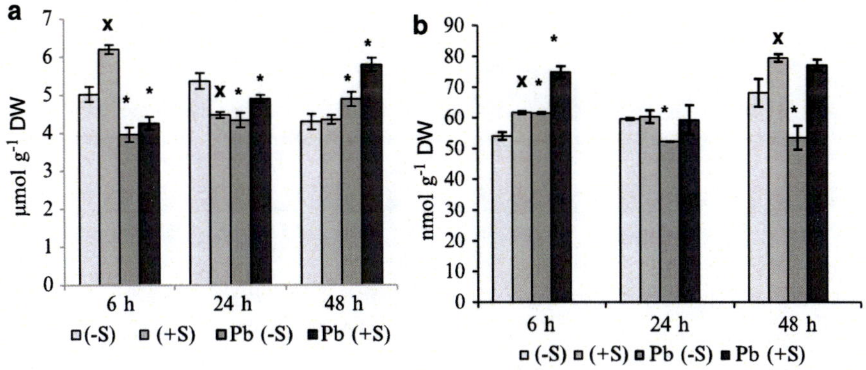

Fig. 2 The impact of Pb on lipid hydroperoxide (**a**) and MDA (**b**) content in leaves of soybean plants grown at sulfur-sufficient (+S) and sulfur-deprived conditions (–S). *, significant differences between (–S), (Pb, -S) and (+S), (Pb, +S) and x, significant differences between (–S) and (+S)

Table 2 The impact of Pb on the activity of superoxide dismutase (SOD) and catalase (CAT) in leaves of soybean plants grown at sulfur-sufficient (+S) and sulfur-deprived conditions (–S)

	SOD (units)		
	6 h	24 h	48 h
+S	18.52±0.68	18.65±0.12	18.18±0.07
–S	19.08±0.12	18.89±0.06	18.19±0.07
+S, Pb	19.91±0.16*	18.08±0.39	17.29±0.14*
–S, Pb	16.80±0.69*	18.61±0.07	18.47±0.02
	CAT (μmol O_2 min^{-1})		
	6 h	24 h	48 h
+S	2.08±0.07	4.78±0.03	5.33±0.02
–S	2.35±0.05*	4.81±0.05	4.66±0.06*
+S, Pb	2.08±0.07	2.91±0.06*	5.35±0.05
–S, Pb	1.87±0.02*	3.87±0.09*	4.65±0.15*

*Denote significant differences between (–S), (Pb, –S) and (+S), (Pb, +S)

Pb exposed plants (Fig. 2b). The latter might be the result of an activation of antioxidative enzymes, especially SOD (Table 2). It has been observed that upon Cd exposure MDA levels increased in the leaves of pea plants with concentration (Dixit et al. 2001).

Sulfur deprivation in itself had no effect on SOD activity, but here the activity of SOD was decreased in leaves of Pb-exposed soybean after 6 h (12%), but returned to control values after 24 and 48 h. In sulfur-sufficient plants Pb exposure resulted in a slight increase in SOD activity after 48 h (Table 2). CAT may eliminate the H_2O_2 produced by SOD avoiding the inhibition of SOD by H_2O_2 accumulation (Beyer and Fridovich 1987). CAT activity was slightly higher in sulfur-deprived plants 6 h after the Pb exposure with subsequent decrease below control levels after 48 h.

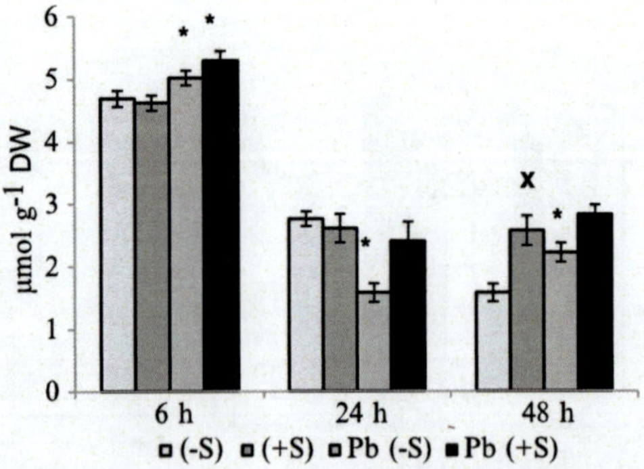

Fig. 3 Impact of Pb on the SQDG content in leaves of soybean plants grown at sulfur-sufficient (+S) and sulfur-deprived conditions (–S). *, significant differences between (–S), (Pb, -S) and (+S), (Pb,+S) and ^{x}, significant differences between (–S) and (+S)

Inhibition of CAT activity was observed in both sulfur-sufficient and sulfur-deprived plants after 24 h after the Pb exposure. After 48 h, CAT activity had returned to control (non-Pb exposed) levels in sulfur-sufficient and sulfur-deprived plants. Nevertheless, exposure of *Pisum sativum* to 0.5 or 1 mM Pb in resulted in increased levels of superoxide and increased activities of SOD and CAT in root cells (Malecka et al. 2001). Differences in experimental conditions and/or Pb tolerance may explain the differences in the obtained results.

Sulfate deprivation and Pb exposure resulted in significant increases in SQDG content of leaves after 6 h, followed by a decline (by nearly 30%) after 24 h of the exposure. In sulfur-sufficient plants the SQDG content was higher after 6 h of the Pb exposure. The observed increase in SQDG content in leaves of sulfur-deprived plants may indicate a possible stimulation of SQDG synthesis even upon sulfate deprivation. However, SQDG is considered to be a strong anion, which is able to create a complex with Pb atoms (Okanenko et al. 2003) and which may explain the decline in its content 24 h after Pb exposure (Fig. 3).

In summary, data obtained in this experiment showed that sulfate deprivation significantly affected lipid peroxidation reactions and potentially disturbed the defense against heavy metal toxicity. A balanced nutrient medium with sulfur supported more favorable changes in thiol contents, especially in free thiols, which may be useful in plant defense reactions to Pb.

References

Beyer WF, Fridovich I (1987) Assaying of SOD activity: some large consequences of minor changes in conditions. Anal Biochem 161:559–566

Bouranis DL, Chorianopoulou SN, Protonotarios VE, Siyiannis VF, Hawkesford MJ (2003) Effect of sulfate starvation on the production of reactive oxygen species and lignification of young maize leaves. Final COST action 829 workshop. Progress in Plant Sulfur Research, Braunschweig, Germany, 15–18 May 2003, p 6

Dixit V, Panday V, Sliyam R (2001) Differential antioxidative responses to cadmium in roots and leaves of pea *(Pisum sativum L.* cv. Azad). J Exp Bot 51:1101–1109

Droillard MJ, Paulin A, Massot JC (1987) Free radical production, catalase and superoxide dismutase activities and membrane integrity during senescence of petals of cut carnations (*Dianthus caryophyllus*). Physiol Plant 71:197–202

Ernst WHO, Krauss G-J, Verkleij JAC, Wesenberg D (2008) Interaction of heavy metals with the sulphur metabolism in angiosperms from an ecological point of view. Plant Cell Environ 31:123–143

Forman HJ, Fridovich I (1973) Superoxide dismutase: a comparison of rate constants. Arch Biochem Biophys 158:396–400

Guschina IA, Harwood JL (2002) Lipid metabolism in the moss *Rhytidiadelphus squarrosus* (Hedw) Warnst. from lead–contaminated and non–contaminated populations. J Exp Bot 53:455–463

Havir EA, McHale NA (1987) Biochemical and developmental characterization of multiple forms of catalase in tobacco leaves. Plant Physiol 84:450–455

Heath RL, Packer L (1968) Photoperoxidation in isolated chloroplasts. I. Kinetics and stoichiometry of fatty acid peroxidation. Arch Biochem Biophys 125:189–198

Kean E (1968) A rapid sensitive spectrophotometric method for quantitative determination of sulfatides. J Lipid Res 9:319–329

Malecka A, Jarmuszkiewicz W, Tomaszewska B (2001) Antioxidative defense to lead stress in subcellular compartments of pea root cells. Acta Biochim Pol 48:687–698

Okanenko A, Taran N, Kosyk O (2003) Sulphoquinovosyldiacylglycerol and adaptation syndrome. In: Advances in research of plant lipids: proceedings of the 15th international symposium on plant lipid (Japan), Kluwer Academic Publishers, Dordrecht, The Netherlands, pp 361–364

Polle A, Rennenberg H (1993) Significance of antioxidants in plant adaptation to environmental stress. In: Mansfield T, Fowden L, Stoddard F (eds) Plant adaptation to environmental stress. Chapman and Hall, London, pp 263–273

Rauser WE (1995) Phytochelatins and related peptides. Structure, biosynthesis, and function. Plant Physiol 109:1141–1149

Reddy AM, Kumar SG, Jyothsnakumari G, Thimmanaik S, Sudhakar C (2005) Lead induced changes in antioxidant metabolism of horsegram (*Macrotyloma uniflorum* (Lam.) Verdc.) and bengal gram (*Cicer arietinum* L.). Chemosphere 60:97–104

Salt DE, Rauser WE (1995) MgATP-dependent transport of phytochelatins across the tonoplast of oat roots. Plant Physiol 107:1293–1301

Sedlak J, Lindsay RH (1968) Estimation of total, protein-bound, and nonprotein sulfhydryl groups in tissue with Ellman's reagent. Anal Biochem 25:192–205

Sun S-Q, He M, Cao T, Yusuyin Y, Han W, Li J-L (2010) Antioxidative responses related to H_2O_2 depletion in *Hypnum plumaeforme* under the combined stress induced by Pb and Ni. Environ Monit Assess 163:303–312

Tausz M, Gullner G, Kőmives T, Grill D (2003) The role of thiols in plant adaptation to environmental stress. In: Abrol YP, Ahmad A (eds) Sulphur in plants. Kluwer Academic Publishers, Dordrecht, pp 221–244

Index

L.J. De Kok et al. (eds.), *Sulfur Metabolism in Plants*, Proceedings of the International Plant Sulfur Workshop 1, DOI 10.1007/978-94-007-4450-9,

S

T